The Physics and Chemistry of Low Dimensional Solids

NATO ADVANCED STUDY INSTITUTES SERIES

Proceedings of the Advanced Study Institute Programme, which aims at the dissemination of advanced knowledge and the formation of contacts among scientists from different countries

The series is published by an international board of publishers in conjunction with NATO Scientific Affairs Division

A	Life Sciences	Plenum Publishing Corporation
B	Physics	London and New York
C	Mathematical and Physical Sciences	D. Reidel Publishing Company Dordrecht, Boston and London
D	Behavioural and Social Sciences	Sijthoff & Noordhoff International Publishers
E	Applied Sciences	Alphen aan den Rijn and Germantown U.S.A.

Series C – Mathematical and Physical Sciences

Volume 56 – The Physics and Chemistry of Low Dimensional Solids

The Physics
and Chemistry of
Low Dimensional Solids

Proceedings of the NATO Advanced Study Institute
held at Tomar, Portugal, August 26 - September 7, 1979

edited by

LUIS ALCÁCER
Chemistry Department, Solid State Phys. Chem. Group,
Laboratório Nacional de Engenharia e Tecnologia Industrial (LNETI),
Sacavém, Portugal

D. Reidel Publishing Company

Dordrecht : Holland / Boston : U.S.A. / London : England

Published in cooperation with NATO Scientific Affairs Division

Library of Congress Cataloging in Publication Data

NATO Advanced Study Institute, Tomar, Portugal, 1979.
     The physics and chemistry of low dimensional solids.

     (NATO advanced study institute series: Series C, Mathematical and
physical sciences; v. 56).
     Includes index.
     1.   Solid state chemistry–Congresses.   2.   One-dimensional
conductors–Congresses.   3.   Organic semiconductors–Congresses.
I.   Alcácer, Luis, 1937–   II.   Title.   III.   Series.
QD478.N37   1979   541'.0421   80–18830
ISBN-13:978-94-009-9069-2   e-ISBN-13:978-94-009-9067-8
DOI: 10.1007/978-94-009-9067-8

Published by D. Reidel Publishing Company
P.O. Box 17, 3300 AA Dordrecht, Holland

Sold and distributed in the U.S.A. and Canada
by Kluwer Boston Inc., Lincoln Building,
160 Old Derby Street, Hingham, MA 02043, U.S.A.

In all other countries, sold and distributed
by Kluwer Academic Publishers Group,
P.O. Box 322, 3300 AH Dordrecht, Holland

D. Reidel Publishing Company is a member of the Kluwer Group

TABLE OF CONTENTS

LIST OF PARTICIPANTS

Alcácer, L.: *Lab. Fis. Eng. Nucl., Sacavém*
Andersen, J. R.: *Forsogsanlaeg Risφ, Roskilde*
Bechgaard, K.: *H. C. Ørsted Institutet, Kφbenhavn*
Bloch, A. N.: *Johns Hopkins University, Baltimore*
Bozio, R.: *The University, Padova*
Chaikin, P. M.: *University of California, Los Angeles*
Comès, R.: *Université Paris-Sud*
Conwell, E. M.: *Xerox Webster Research Center, Webster*
Cowan, D. O.: *Johns Hopkins University, Baltimore*
Day, P.: *Oxford University*
Delhaes, P.: *Centre de Recherche Paul Pascal, Talence-Cedex*
Engler, E. M.: *IBM T. J. Watson Research Center, Yorktown Heights*
Epstein, A. J. *Xerox Webster Research Center, Rochester*
Farges, J.-P.: *Université de Nice-Valrose*
Flandrois, S.: *C.R.P.P., C.N.R.S., Talence*
Garito, A. F.: *University of Pennsylvania, Philadelphia*
Heeger, A. J.: *University of Pennsylvania, Philadelphia*
Jérome, D.: *Université Paris-Sud*
Keller, H. J.: *Anorganisch-Chemisches Institut der Universität Heidelberg*
Kommandeur, J.: *University of Groningen*
Kuptsis, J. D.: *IBM T. J. Watson Research Center, Yorktown Heights*
MacDiarmid, A. G.: *University of Pennsylvania, Philadelphia*
Megtert, S.: *Université Paris-Sud*
Metzger, R. M.: *The University of Mississippi*
Miller, J. S.: *Occidential Research Corporation, Irvine*
Patel, V. V.: *IBM T. J. Watson Research Center, Yorktown Heights*
Pecile, C.: *The University, Padova*
Poehler, T. O.: *Johns Hopkins University, Laurel*
Potember, R. S.: *Johns Hopkins University, Laurel*
Pouget, J. P.: *Université Paris-Sud*
Pynn, R.: *I.L.L., Grenoble-Cedex*
Renard, M.: *Centre de Recherches sur les Très Basses Températures, C.N.R.S., Grenoble-Cedex*
Schad, R. G.: *IBM T. J. Watson Research Center, Yorktown Heights*
Schultz, T. D.: *IBM T. J. Watson Research Center, Yorktown Heights*
Soos, Z. G.: *Princeton University*
Thomas, G. A.: *Bell Laboratories, Murray Hill*
Tomkiewicz, Y.: *IBM T. J. Watson Research Center, Yorktown Heights*
Vettier, C.: *I.L.L., Grenoble-Cedex*
Weger, M.: *The Hebrew University, Jerusalem*
Wudl, F.: *Bell Laboratories, Murray Hill*

PREFACE

    The progress that has been made recently and the consequent increased interest in Low-Dimensional Solids has been impressive. This volume contains the latest advances concerning the chemistry and physics of, particularly, one dimensional solids, which were treated at the 1979 Advanced Study Institute, held in Tomar, Portugal.

    The subject is treated starting from basic concepts on the physics of one dimension up to the design of new materials and includes technological and scientific applications. The book therefore should be of interest to solid state physicists, synthetic chemists and scientists concerned with materials science and technology.

    Contrary to some pessimistic views, the subject continues to be of interest and very much alive indeed, and new perspectives are open particularly in the domain of new materials and conducting polymers and their technological interest.

    The Advanced Study Institute was sponsored by the NATO Scientific Affairs Division, and in Portugal, by the "Laboratório Nacional de Engenharia e Tecnologia Industrial", the Gulbenkian Foundation and the Cultural Services of the French Embassy.

LNETI, Sacavém   Luis Alcácer
April, 1980

L. Alcácer (ed.), The Physics and Chemistry of Low Dimensional Solids, ix.
Copyright © 1980 by D. Reidel Publishing Company.

PHYSICS IN ONE DIMENSION

T. D. Schultz

IBM Watson Research Center, Yorktown Heights, NY

ABSTRACT. We present two introductory lectures. In the first, the case of weak electron-electron interactions in one-dimension is discussed in both the insulating and metallic phases, with emphasis on Peierls distortions and fluctuations, the Kohn anomaly, the contrasts between one and three dimensions and between commensurate and incommensurate distortions. The case of strong electron-electron repulsions is also considered with emphasis on the $U \to \infty$ limit in the Hubbard model and the charge- and spin-density waves that can form. The two cases are contrasted. In the second lecture, the Ginzburg-Landau formalism is developed and applied to the study of fluctuations, commensurability, phasons, solitons and the interplay of one and three dimensions.

I.   INTRODUCTION

The purpose of these two lectures is to present the basic theoretical ideas that are peculiar to one-dimensional (1D) physics, ideas that will arise again and again in the subsequent lectures. I shall try to make them accessible to graduate students and to chemists who know only a little solid-state physics.

Theoretical physicists originally studied one-dimensional models [1] because they believed them to be sufficiently simpler than 3D models as to be exactly soluble, *in some sense*. The first nontrivial many-body problem that was solved exactly was the 1D antiferromagnetic Heisenberg model of spin-1/2's, solved by Bethe [2] in 1931 $(H = -J \sum_m \mathbf{S}_m \cdot \mathbf{S}_{m+1})$. Another 1D spin-1/2 problem that was solved exactly was the XY model $(H = -J \sum_m (S_m^x S_{m+1}^x + S_m^y S_{m+1}^y))$, solved by Lieb, Schultz and Mattis [3]. The one-phonon problem in a 1D *random* lattice was solved by Dyson [4], and the corresponding one-electron problem is closely related to it. The problem of many electrons in free space in one dimension has been specialized

1

L. Alcácer (ed.), The Physics and Chemistry of Low Dimensional Solids, 1–30.

in a number of ways that are related to the weak-interaction limit and then solved exactly [5], and the half-filled 1D band of electrons repelling each other when on the same lattice site (Hubbard model) was solved by Lieb and Wu [6]. The statistical mechanics of a 1D Ginzburg-Landau field was solved by Pfeuty and Lazerowicz and by Scalapino, Sears and Ferrell [7]. Of great current interest is the whole field of exact solutions to certain non-linear partial differential equations in one space (and one time) dimension, which leads to so-called kinks and solitons [8]. In all these cases, of course, the word "solved" is to be understood in a limited sense *viz. some* aspects of the problem were solved exactly.

Although exactly soluble, the 1D problems were of limited interest for several reasons: they were too different from the corresponding 3D problems to provide a suitable test for approximation methods used in 3D (e.g. mean-field approximations); they showed no phase transitions except when the interactions were of very long range; and until recently, there were few physical systems to which these models corresponded.

Now we have real 1D spin systems and 1D metals, so the study of the theory of one-dimensional systems has recently undergone an explosive growth. The 1D metals are intrinsically interesting for a number of reasons: they are many-electron systems, the electron-phonon interactions are important, they are endlessly variable, chemically, and they have many anomalous properties. In the organic metal TTF-TCNQ, for example, the anomalous properties include a high dc conductivity for which there are at least eight different explanations, unusual temperature dependences of spin susceptibility and EPR linewidth, much debated optical and infrared properties, several phase transitions that are associated with a variety of structural properties seen in diffuse X-ray and neutron scattering, and, as we shall see at this NATO Institute, interesting and anomalous pressure dependences for many of these properties.

In section II, I shall take a fundamental point of view, showing the form of the Hamiltonian of interest and discussing its various terms. In section III, I shall discuss the many-electron—phonon system under the assumption of weak interactions among the electrons, emphasizing notions like the Peierls instability, the Kohn anomaly and Peierls fluctuations. In section IV, I shall discuss the other limit, that of strong electron-electron repulsions, within the context of the Hubbard model, emphasizing the possibility of both charge-density and spin-density wave (CDW and SDW) instabilities. In section V, which will constitute the second lecture, I shall take a semi-phenomenological point of view, starting with a Ginzburg-Landau free-energy functional, discussing fluctuations, commensurability, phasons and solitons, the interplay between one and three dimensions in actual phase transitions, and other three-dimensional effects.

II.   HAMILTONIAN

The systems we shall consider have Hamiltonians of the general form

$$H = H_{el} + H_{el-el} + H_{latt} + H_{el-latt} . \qquad (2.1)$$

We discuss each of the terms, and in this way we discuss the physics underlying the models studied.

A.   One-Electron Hamiltonian

The one-electron Hamilton H_{el} derives its particular form from the fact that partially filled orbitals are associated with well-separated chains or stacks of atoms or molecules. In the organic metals, for example, one has stacks of planar molecules in which the interplanar distance may be $\gtrsim 3$ Å while the spacing between the neighboring stacks may be two or three times that much. The molecular orbital of interest, because it is partially filled in these solids, is a π orbital (Fig. 1). Because of the differences in separations and the directionality of the π orbitals, there is appreciable overlap of the orbitals from successive molecules within the stack but very much smaller overlap between orbitals on neighboring stacks. The matrix element of the one-electron Hamiltonion between successive orbitals parallel to the stacks (the "transfer integral"), which we call $t_{\parallel}$ or t_b (in TTF-TCNQ the stacking axis is called the b axis by crystallographers) is thus assumed to be much greater than those transverse to the stacks ($t_{\perp}$ or t_a and t_c). It is here, and often only here that the one-dimensionality of the problem enters; i.e. the other terms of H are three-dimensional. In modeling the electronic behavior of these systems, we neglect the lower-lying molecular orbitals, which are assumed to be doubly-filled, and the higher-lying molecular orbitals, which are assumed to be empty.

For a single stack, the one-electron wave functions can then be well-approximated by Bloch sums

$$\psi_{k_b}(\mathbf{r}) = M^{-1/2} \sum_m e^{ik_b mb} \phi_\pi(\mathbf{r} - m\mathbf{b}) \qquad (2.2)$$

having energies

$$\epsilon(k_b) = 2t_b \cos k_b b \qquad (2.3)$$

where $\phi_\pi(\mathbf{r} - m\mathbf{b})$ is the π molecular orbital at the mth site. For a chain of M sites and periodic boundary conditions, the wave number k_b takes on the M values $0, \pm 2\pi/b, \pm 4\pi/b, \ldots$. If $t_b < 0$, we have a "normal" one-dimensional energy band and if $t_b > 0$ we have an "inverted" band, as shown in Fig. 2. Because the density of states is constant with k_b (one for each $2\pi/Mb$), the density as a function of energy has divergences (integrable) at the band edges.

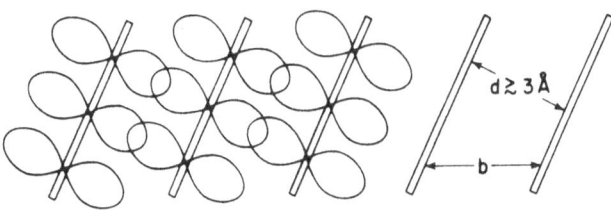

Fig. 1. Molecular stack showing π-orbital overlaps (schematic).

This is quite different from higher dimensions where, because the band edges occur at isolated points whose neighborhoods are a smaller fraction of **k** space, the singularities are of lower order and $N(\epsilon)$ is *not* divergent.

In many-body problems, it is convenient to write the Hamiltonian in the language of second quantization. If C_{ms} and $C_{ms}^\dagger$ are operators that annihilate and create an electron respectively at site m with spin s, then $C_{ms}^\dagger C_{ms} = n_{ms}$ is the number operator for such an electron and

$$H = \epsilon \sum_{m,s} C_{ms}^\dagger C_{ms} + t_b \sum_{m,s} (C_{ms}^\dagger C_{m+1,s} + h.c.) \qquad (2.4)$$

Now let us consider a two-dimensional crystal consisting of a sheet of parallel identical stacks separated one from the next along the c direction. Just as it was convenient to work with states of a definite k_b instead of the site number m, it is now convenient to introduce k_c instead of the stack number n. The wave functions will be Bloch sums characterized by (k_b, k_c) and the energy will be a function of (k_b, k_c). In general, if t_c is small but non-vanishing, then *in principle* $\epsilon(\mathbf{k})$ is cosinusoidal in k_c as well as k_b, the surfaces of constant energy are "warped" (Fig. 3), and the singularities in $N(\epsilon)$ at the band edges are smeared out. Also, the one-electron matrix elements between states of different (k_b, k_c) depend on both the k_c's as well as the k_b's. *In fact*, the uncertainty in energy due to scattering within the stacks (by phonons, impurities, electron-electron interactions, etc.) may be greater than t_c, a measure of the transverse bandwidth, so that the concept of banding in the transverse direction may not be useful. While there may be some coherent mixing between adjacent stacks, which will affect the density of states, the notion of extended states and associated energy bands in that direction would not be meaningful in determining the appropriate state density.

It is interesting to consider the special case of independent stacks, *i. e.* $t_c = 0$, in this language. This limit is, of course, simplest to understand in (k_b, n)

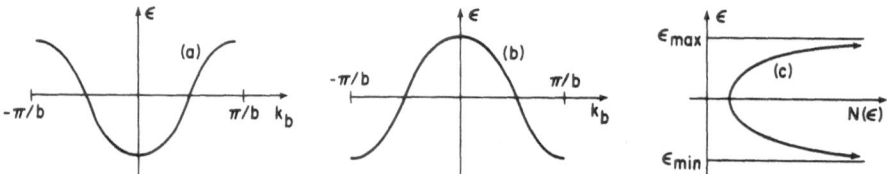

Fig. 2. One-dimensional tight-binding band: (a) normal, (b) inverted, and (c) density of states, showing $(\epsilon - \epsilon_0)^{-1/2}$ singularities at $\epsilon_{min}, \epsilon_{max}$.

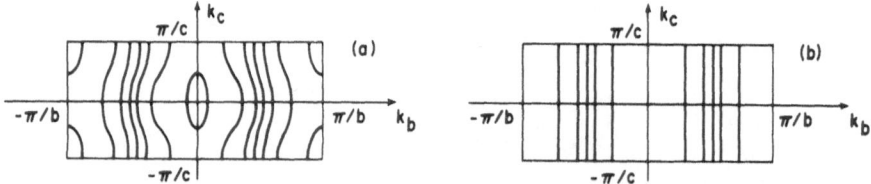

Fig. 3. Brillouin zone of 2D tight-binding band: surfaces of constant, equally-spaced energies for (a) $t_c \ll t_b$ and (b) $t_c = 0$.

space, but to make contact with out intuition developed for 3D bands, we simply note that in (k_b, k_c) space, the energy is independent of k_c so the surfaces of constant energy are flat (Fig. 3b), the density of states $N(\epsilon)$ remains singular, and one-electron matrix elements are independent of the k_c's involved.

In TTF-TCNQ and other charge-transfer compounds where both kinds of stacks are to be modeled with nearly free electrons (or holes), it is believed that the two kinds of stacks are dissimilar, *i. e.* that one is normal and the other is inverted. For two non-interacting stacks, one of each kind, the situation is shown in Fig. 4. Because the electrons in the TCNQ band have been transferred from the TTF band, the number of holes in the latter (and therefore the range they span in k_b) equals the number of electrons in the former. For this reason, the Fermi energy is where the bands cross. If $t_a \neq 0$, then, in principle, there is a mixing between states of the same k_b, its effect is strongest when the states are degenerate or nearly so, and a "hybridization gap" develops at the Fermi surface (the hybridized bands are shown as dashed curves). For an infinite, alternating set of such stacks, the gap goes to zero at the zone boundary $k_a = \pm\pi/a$, yielding a "zero-gap semiconductor." If interactions in the third direction are similarly included, there are overlapping bands, leading to a semi-metal. However, because of disorder and other sources of scattering, such coherence is probably not extended in the transverse directions, except possibly in systems where either t_a or t_c is appreciable, and then only at low temperatures where the thermal disorder is small.

B.   Electron-Electron Hamiltonian

The two-body interaction between electrons is, of course, the Colomb interaction between the electronic charge distributions of the π molecular orbitals as screened by electrons in the other filled orbitals. These interactions occur both within and between stacks and have the form

$$H_{el-el} = \sum_{\mathbf{R},\mathbf{R}',s,s'} V_{\alpha\alpha'}(\mathbf{R}-\mathbf{R}') \; n_{Rs} \, n_{R's'} \qquad (2.5a)$$

where α, α' identify the molecular types at $\mathbf{R}$, $\mathbf{R}'$. In the study of a single stack, the effects of all other stacks are either neglected or implicitly assumed simply to modify some of the interactions within each stack. In this case the

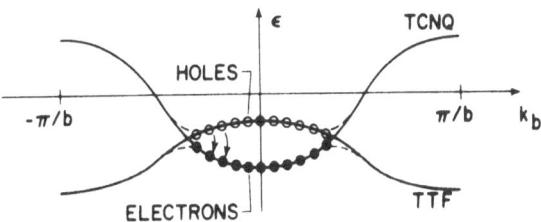

Fig. 4.   Normal (TCNQ) and inverted (TTF) 1D tight-binding bands showing charge transfer, ϵ_F at crossing, hybridization effects (---) and gap.

interaction term will have the form

$$H_{el-el} = \sum_{m,\mu,s,s'} V_\mu \, n_{ms} \, n_{m+\mu,s'} \tag{2.5b}$$

which, with H_{el}, is the "extended Hubbard model." Often this interaction is replaced by one involving only an on-site interaction:

$$H_{el-el} = U \sum_m n_{m\uparrow} \, n_{m\downarrow} \; . \tag{2.5c}$$

Here $U \neq V_0$ but rather $U \simeq V_0 - V_1$, to reflect the energy required to bring two electrons that are on adjacent sites onto the same site. We note that the terms $n_{ms} n_{ms} = n_{ms}$ do not appear explicitly because they are included in ϵn_{ms}.

C.   Lattice Hamiltonian

The Hamiltonian for the lattice vibrations has the same general form as for any harmonic lattice:

$$H_{latt} = \sum_{\mathbf{q},\beta} \omega_{q,\beta} \, (b_{q\beta}^\dagger \, b_{q\beta} + \frac{1}{2}) \tag{2.6}$$

where β enumerates the branches, the $\mathbf{q}$'s are three-dimensional and $b_{q\beta}$ and $b_{q\beta}^\dagger$ destroy and create vibrational or librational quanta. Although these systems are electronically one-dimensional or nearly so, there is no such simplification in H_{latt}; the normal modes are three-dimensional, i. e. there is appreciable dependence of the frequencies, polarizations, etc. on the transverse components of $\mathbf{q}$. What complicates the analysis so much, and therefore what demands simplification, is the large number of branches. There are, of course, as many branches as there are degrees of freedom within a unit cell. In TTF-TCNQ, with two TCNQ molecules of 20 atoms each and two TTF molecules of 14 atoms each, the 68 atoms lead to 204 branches. In certain symmetry directions, it may be possible to divide these into acoustic branches, in which the unit cells move more-or-less rigidly, librational branches, in which the molecules rotate, and optical branches, in which there is either intracellular or intramolecular motion. But for general directions of $\mathbf{q}$, even these simplifications are not possible.

D.   Electron-Lattice Hamiltonian

The electron-lattice interactions that are usually considered arise because of the modulation of one or another of the parameters in H_{el} when molecules are displaced or deformed. For example, the on-site energy ϵ at site m can be modulated by an intramolecular distortion of amplitude u_m, giving rise to an interaction term of the form

$$H_{el-latt} = \lambda \sum_{m,s} u_m \, C_{ms}^\dagger C_{ms} \, , \tag{2.7a}$$

or it can be modulated by displacements $u_{m\pm1}$ of the molecules on neighboring sites, giving rise to a term of the form

$$H_{el-latt} = \lambda \sum_{m,s} (u_{m+1} - u_{m-1}) \, C_{ms}{}^{\dagger} C_{ms} \, . \qquad (2.7b)$$

The transfer integral between sites m and $m+1$ can also be modulated by changes in the positions and hence separation of the corresponding molecules, giving rise to a term of the form

$$H_{el-latt} = \lambda \sum_{m,s} (u_{m+1} - u_m) \, C_{ms}{}^{\dagger} C_{ms} \, . \qquad (2.7c)$$

We show these schematically in Fig. 5. The different kinds of interactions lead to minor differences which we can neglect for the purposes of these lectures.

III. WEAKLY INTERACTING ELECTRONS

A. Low-Temperature Phase – Peierls Instability

In this section we shall neglect the Coulomb interaction. Furthermore, to start with the simplest case, we consider the ground state of only one stack. To see an important effect of $H_{el-latt}$, imagine a periodic distortion of the lattice characterized by some wave number q:

$$u_m = M^{-1/2} \left(u_q e^{iqmb} + \text{c.c.} \right) \qquad (3.1)$$

and, for simplicity, imagine that u_m describes some rigid molecular displacement. Then $H_{el-latt}$ couples the electronic state at k to those at $k\pm q$. The effect will be small unless the coupled states are nearly degenerate, $i.\,e.$ $\epsilon_{k\pm q} \simeq \epsilon_k$ The effect is to open a gap in the one-electron spectrum at $\pm q/2$, as

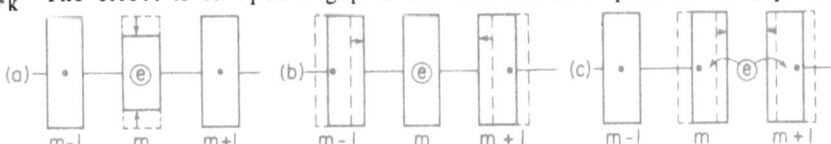

Fig. 5. Three mechanisms for electron-lattice coupling.

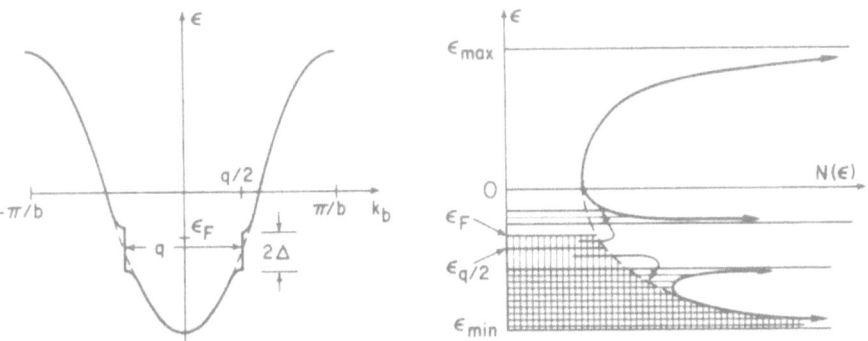

Fig. 6. Effect of distortion having wave number q on spectrum and density of states (assuming $\epsilon_F > \epsilon_q/2$).

a band originally filled to a Fermi energy ϵ_F, the effect in general will be to depress most of the filled states but to raise others, as shown, or *vice versa*. Only if $\epsilon_{q/2} = \epsilon_F$, or equivalently if $q = 2k_F$, are all filled states lowered and no filled states raised. Thus there is a gain in total electronic energy that is optimum (neglecting the variation of the coupling g_q with q) for $q = 2k_F$. When one calculates the total change in energy of the Fermi sea due to a distortion of amplitude u_q, one finds that it goes like $(g_q^2 u_q^2/\epsilon_F) \ln (g_q u_q/\epsilon_F)$ due to the singular density of states at the edge of the gap. The lattice elastic energy of course varies like $\omega_q u_q^2/2$ where ω_q is some frequency (it is not necessarily a normal-mode frequency because u_q does not necessarily refer to a normal mode). The change in total energy is thus of the form

$$\delta<H>_{u_q} = \frac{1}{2}u_q^2 \left[\omega_q + const. \times (g_q^2/\epsilon_F) \ln (g_q u_q/\epsilon_F)\right] . \tag{3.2}$$

We see that this change will be negative for sufficiently small u_q, regardless of how small the electron-lattice coupling g_q is. As a function of u_q, the ground-state energy has the form of the $T=0$ curve in Fig. 7. There is an optimum u_q which yields a gap

$$\Delta = g_q u_q \propto \epsilon_F\, e^{-\omega_q/g_q^2 N(\epsilon_F)} \tag{3.3}$$

and an energy lowering ("stabilization energy") $\propto N(\epsilon_F)\Delta^2$. What we have established is that the system is unstable against *any* deformation of wave number $2k_F$ that couples to the electrons. This is the famous Peierls instability [9]. It is analogous to a Jahn-Teller distortion in a single molecule, except that use is made of the continuum of many-electron energies. Just which deformation within each unit cell gives the maximum stabilization energy is a quantitative question. It clearly need not correspond to a normal vibrational or librational mode.

Where do normal 3D arguments break down in 1D? In 3D, we might argue that a distortion wave of amplitude u_q induces a charge-density wave of amplitude $\rho_q = \chi_q u_q$, where χ_q is an appropriate susceptibility of the electronic system. Furthermore, the resulting interaction between the distortion wave and the CDW yields an interaction energy $\propto u_q \rho_q \propto \chi_q u_q^2$ which is quadratic in u_q. In 1D, at T = 0 °K, one finds that χ_q is divergent, because of the singular density of states at the gap edges, so the 3D arguments break down.

The Fermi sea in the presence of the periodic distortion has a charge density that is modulated with wave number $q_p = 2k_F$. This is due to the

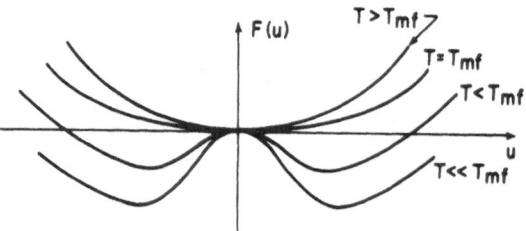

Fig. 7. Electron free-energy *vs. u* for four temperatures (schematic).

modulation of all the occupied wave functions, but especially of those near $\pm k_F$. The amplitude of the charge-density-wave modulation is proportional to u_q or to Δ. There will be a Coulomb interaction between such CDW's, but the potential will fall off exponentially with the distance $R_\perp$ between the stacks when $R_\perp \gg \lambda_p$, λ_p being the wavelength of the CDW. This is because as one goes further and further from an infinitely long CDW, the CDW behaves like a higher and higher multipole, and since there is no limit to how high the multipole, the potential falls off faster than any inverse power of $R_\perp$.

The electronic properties in a Peierls-distorted state should be largely those of a narrow-gap semiconductor with a T-dependent gap. The conductivity and spin susceptibility should decrease exponentially with decreasing temperature. The thermopower (see Chaikin's lectures) should also reflect the semiconducting behavior. If the entire superlattice could slide, it would however lead to a collective type of conductivity, Fröhlich or "sliding" conductivity [10]. Both imperfections and forces between CDW's on different kinds of chains will oppose this, so there has been some debate as to whether this is observed in fields large enough to "de-pin" the distortion.

Now consider how this discussion is modified at finite temperatures. First, there is an obvious feed-back loop: As the temperature is raised, more and more electrons are thermally excited across the gap. This lowers the stabilization energy for a distortion of a given amplitude, which in turn decreases the optimum amplitude. This in turn reduces the gap and thereby increases the number of electrons thermally excited. At each temperature there is, of course, an optimum amplitude, but this decreases with increasing temperature. There comes a temperature at which the optimum amplitude goes to zero, the mean-field transition temperature T_{mf}. We say "mean-field" because we have assumed that all relevant distortions can be characterized by a single distortion amplitude i. e. all fluctuations from perfect periodicity have been ignored. For a one-dimensional system, this is a drastic assumption leading to a phase transition at this finite temperature when in fact there is no long-range order at any $T > 0$. Nevertheless, the concept of a free energy as a function of a single distortion amplitude turns out to be surprisingly useful as we shall see in section V.

At each temperature above 0 °K the free energy has approximately the form

$$F(u_q) = a(T) u_q^2 + b(T) u_q^4 \tag{3.5}$$

which we represent in Fig. 7. We see that $a(T_{mf}) = 0$ and that the sign of $a(T)$ is the same as that of $T - T_{mf}$. Calculations analogous to those leading to (3.2) yield

$$T_{mf} \propto e^{-\omega_q / 2N(\epsilon_F)g_q^2} \tag{3.6}$$

It is important to realize that the actual transition temperature T_c in a 3D system of 1D stacks could be much lower than T_{mf} because of the fluctuations we have neglected. The actual phase transition is due to the coupling of the

1D fluctuations occurring in a 3D crystal; it cannot be determined from purely 1D considerations, which would give $T_c = 0$. We return to this later in section V-E.

Let us now turn to the question of commensurability. Until now, we have neglected the effect of terms that are of higher order in the interaction on the splitting of the states at $\pm k_F$. But if $q_p = nb^*/m$, where $b^* = 2\pi/b$ is a vector of the reciprocal lattice and n/m is a rational fraction, then there is an additional mixing of the states at $\pm k_F$ due to such a higher-order interaction. This is most clearly seen if $n/m = 1/2$, because then the state k_F mixes with $-k_F$ not only through $-q_p$ but also through $+q_p$ (since, in this case, $k_F + q_p = 3b^*/4 \equiv -b^*/4 = -k_F$). The case n/m = 1/3 is shown in Fig. 8. Thus, whenever $2k_F$ is commensurate with b^*, there is an enhancement of the stabilization energy, an enhancement which incidentally decreases exponentially with the order m of the commensurability [11]. Furthermore, when the CDW is commensurate with the lattice, different relative positions of the wave and the lattice are really different, i.e they can not be made to look the same simply by selecting appropriate points from which to look. The stabilization energy is consequently not constant but periodic as the wave slides, with periodicity λ_p/m. There are optimum positions for the wave and a barrier to the sliding, which also decreases exponentially with m. As $m \to \infty$ we reach the incommensurate case for which the gap is smallest and for which there is no barrier to sliding.

We might imagine an experiment in which the amount of charge transfer can be varied continuously, perhaps by applying pressure as Jérome will discuss in his lectures. As $2k_F$ varies continuously, how does the wave vector q_p of the optimum distortion vary? Since among all the perfectly periodic distortions the commensurate ones are enhanced, we might imagine a q_p that has an infinite number of discontinuities as it jumps from one commensurate value to another, but with some of the low-order commensurate values stabilized over a significant range of $2k_F$'s — a so-called Devil's Staircase. The actual behavior will be different because the optimum distortions will turn out not to be perfectly periodic as we shall see in section V-D [12].

When a superlattice (the periodic distortion) forms, there are implications for both the electronic band structure and for the phonon bands. This is simplest to see graphically if the distortion is commensurate and of low order, say $m = 3$. Then the Brillouin zone for the electrons breaks into three mini-zones which we show in a reduced-zone scheme in Fig. 9a. Perturbations with

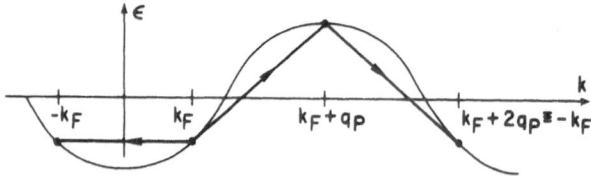

Fig. 8. First- and second-order matrix elements connecting k_F and $-k_F$ if $q_p = 2\pi/3$.

$q = 0$ can now induce transitions that are vertical within the reduced-zone scheme, although the matrix elements will be appreciable only near the new gaps, where the mixing is largest. An analogous behavior occurs in the phonon spectrum (Fig. 9B), where the "gap" opening at $b^*/3$ can be quite large. To understand this, consider the lowest branch, where the phonons in the limit $q \rightarrow 0$ correspond to the translation of the crystal. Since this requires no energy, $\omega_{phon}(q) \rightarrow 0$ as $q \rightarrow 0$. In the next branch, the "phason" branch, the phasons in the limit $q \rightarrow 0$ correspond to the translation of the periodic distortion, i. e. of the superlattice. In the commensurate case, where there is an energetically optimum phase for the periodic distortion, there is a frequency associated with small translations of the distortion which we call ω_0. Then $\omega_{phas}(q) \rightarrow \omega_0$ as $q \rightarrow 0$. In the incommensurate case, it requires no energy to displace the distortion, so that $\omega_{phas}(q) \rightarrow 0$ as $q \rightarrow 0$. We discuss this more mathematically in section V-D.

Until now, we have assumed that the electrons move in a perfectly ordered lattice. Often this is not the case, either because of impurities or alloying, or because of the fields produced by random fluctuations on other stacks. Such imperfections can have two effects. First [13], by destroying k as a good quantum number, they prevent mixing of pairs of states k and $k \pm q$ and thus decrease the energy splitting that results from a periodic distortion. Equivalently, for states of a definite k, the imperfections smear out the energy and make the mixing of, say $-k_F$ and $k_F \pm q$ less effective than when they were strictly degenerate. This smears out the singularities in $N(\epsilon)$ on both sides of the gap which increases $a(T)$ at any T, decreases T_{mf} and $\Delta(T)$, and thereby decreases T_c. Imperfections can also tend to pin the local phase of Peierls fluctuations in the neighborhood of each imperfection. Since the imperfections are randomly spaced, the random phase-pinning inhibits the development of long-range order within a single stack and can also inhibit the correlation of phases on different stacks that is essential for a sharp 3D phase transition [14]. This effect is most clearly seen in the quasi-1D metal KCP where no sharp transition is seen and the correlation length transverse to the stacks never becomes infinite as $T \rightarrow 0$ [14a, 15].

In concluding this discussion of the low-temperature phase, I should like to make some comments on what happens if the electronic motion is genuinely three-dimensional, albeit very anisotropic. First, suppose all the stacks are identical. For the purely 1D case, as we have remarked, the energy surfaces

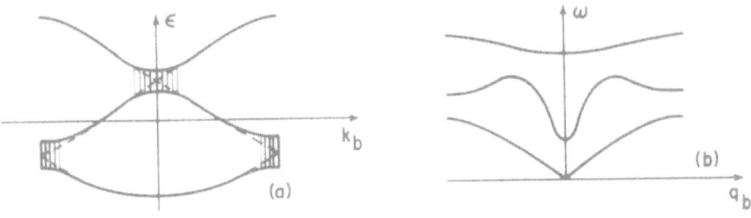

Fig. 9. (a) Electron mini-zones and Peierls gaps and (b) phonon, phason and amplitude-modulated branches, all in reduced zone of superlattice.

including the Fermi surface are all flat and the matrix element of the perturbation caused by a periodic distortion of wave vector $\mathbf{q}$ is independent of $q_\perp$.
Thus a gap is created over the whole Fermi surface by any $\mathbf{q}$ for which $q_b = 2k_F$. The best $q_\perp$ is determined by other considerations. If there are weak
transverse transfer integrals $t_\perp$, the Fermi surface is not flat but periodic. Now
although a single $\mathbf{q}$ (say with $q_\perp = 0$, to fix ideas) will still produce a gap for
all $k_\perp$, the energy of this gap will vary with $k_\perp$ and the Fermi energy will lie in
the gap for only a small range of $k_\perp$, a range that shrinks to zero as u_q (and
therefore Δ) tends to zero (Fig. 10a). Consequently, the stabilization energy
goes more rapidly to zero as $u_q \to 0$ or $g_q \to 0$ than in 1D. However, in case
the Fermi surface is, say, perfectly sinusoidal, then the two Fermi surfaces
"nest", i. e. one can be displaced into the other with a single $\mathbf{q}$ (Fig. 10b) and
the behavior is similar to that of flat Fermi surfaces (except that $q_\perp$ is now
determined, too) [16]. Second, suppose the stacks are of two kinds that
alternate. If both are normal or both are inverted, there is no mixing between
the bands of the two kinds of stacks at the Fermi energy because they have
different k_F's. If one is normal and the other inverted, then, as we have
discussed, a hybridization gap can form at the Fermi energy in an undistorted
lattice. A Peierls distortion would have to mix states on both sides of this gap
and so is no longer so energetically favorable. Of course, if there is a Peierls
distortion creating a Peierls gap, then the states that would hybridize would no
longer be degenerate and hybridization would no longer be so energetically
favorable. In other words, the hybridization and Peierls tendencies are compe-
titive. This is believed to explain some interesting effects in TSeF-TCNQ at
low temperatures [14, 17].

B.   High-Temperature Phase – Kohn Anomaly and Peierls Fluctuations

Above the Peierls transition temperature, we are dealing with a metallic
phase. Some notions carry over from the theory of the weakly interacting 3D
electron gas such as those of quasi-particles replacing the bare electrons (but
having nearly real self-energies near the Fermi energy), of screened interactions
between the quasi-particles, and of an enhanced spin susceptibility due to
exchange interactions. But the singularity in the response of the electron gas to
perturbations having $q = 2k_F$ is also reflected in the frequency spectrum of the
phonons. If one considers a phonon-mode of wave number q and bare frequency ω_q and asks how the presence of the electron gas modifies the vibrations of

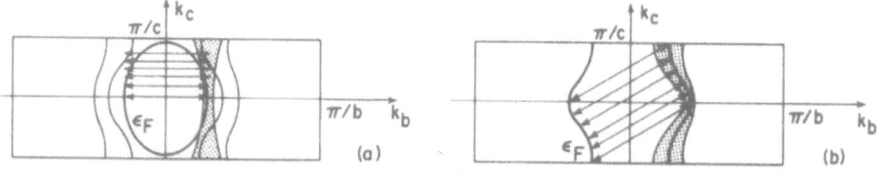

Fig. 10.   (a) Some energy surfaces (incl. Fermi surface) for anisotropic 3D
electrons.   Arrows $(q_b, 0)$ show mixing of degenerate states.   Gap region
(shown on one side) shows states whose original energies are in gap.   (b) when
Fermi surface for quasi-1D electrons are sinusoidal.   Arrows $(q_b, \pi/c)$ show
nesting of Fermi surfaces.   Gap region is defined as in (a).

such a mode, one can show [9b] (neglecting all electron-electron interactions) that the phonon has a renormalized frequency Ω_q given by

$$\Omega_q^2 = \omega_q^2 - \frac{\omega_q g_q^2}{M} \sum_k \frac{f(\epsilon_k) - f(\epsilon_{k-q})}{\epsilon_{k-q} - \epsilon_k - \Omega_q} \tag{3.7}$$

where $f(\epsilon)$ is the Fermi distribution. The last term simply reflects the response, calculated with time-dependent perturbation theory, of the electron gas to a perturbation of wave number q and frequency Ω_q. If we set $\Omega_q = 0$, this term is precisely what occurs in calculating the stabilization energy of a Peierls distortion at temperature T. The solution of (3.7) leads to a spectrum for each T which we show schematically in Fig. 11a. The lowering of Ω_q for $q \simeq 2k_F$ is the so-called Kohn anomaly. It was first discussed by Kohn for three dimensions where the effect is much less pronounced because the last term of (3.7) though non-analytic is not divergent. At sufficiently low temperatures $\Omega_q^2 \propto \omega_q^2$ $\ln(T/T_{Kohn})$ where, not surprisingly, $T_{Kohn} = T_{mf}$.

In this discussion we have neglected phonon-phonon interactions due both to lattice anharmonicities and to interactions *via* the interacting electron gas. If such interactions are included, then it is likely that the $2k_F$ phonon never becomes completely soft. Instead, at lower temperatures there begins to appear another kind of mode near $2k_F$ that is quasi-static, a Peierls fluctuation. If inelastic and elastic neutron scattering were done (elastic scattering has not been done due to too small samples and resulting low intensities), one would expect the intensity of neutron scattering at $q_b = 2k_F$ to behave as in Fig. 11b, in analogy to the central peak seen in ferroelectrics. We believe that it is the Peierls fluctuations, and not the softening of $2k_F$ phonons, that is the counterpart of the Peierls distortion. From the point of view of phonons, such quasi-static fluctuations can be viewed as a collective mode in the interacting phonon gas. With decreasing temperature, the amplitude and correlation length of these fluctuations can be expected to grow. When the correlations are of sufficiently long range, then the interactions between fluctuations on nearby stacks are strong enough to begin to lock the relative phase of the fluctuations and a "cross-over" to 3D behavior occurs. At slightly lower temperatures the actual 3D phase transition occurs.

In the presence of long- (but not infinite-) range fluctuations, the one-electron spectrum will begin to develop a "pseudo-gap" [18], the precursor of a Peierls gap (Fig. 12). If the pseudo-gap is pronounced enough compared

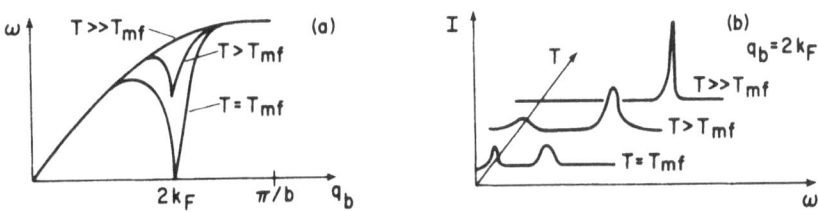

Fig. 11. (a) Kohn anomaly at three temperatures from simple theory. (b) Intensity *vs.* ω expected for neutron scattering $2k_F$ phonon.

with the temperature, it could be expected [*18, 19*] to lower the spin susceptibility and to have other effects that depend on $N(\epsilon_F)$.

IV.   STRONG ELECTRON-ELECTRON REPULSION

We now consider the case of strong electron-electron repulsions [*20*]. We
confine ourself to a discussion of the 1D Hubbard model having the Hamiltonian

$$H_{Hubb} = t\sum_{m,s}(C_{m+1,s}{}^{\dagger}C_{ms} + \text{h.c.}) + U\sum_{m}n_{m\uparrow}n_{m\downarrow} \qquad (4.1)$$

although possible electron-phonon effects through terms like (2.7) will be
mentioned briefly as will effects of the extended Hubbard model. The statement that electron-electron repulsions are strong means simply that t/U is very
small. The limit $t/U = 0$ can be reached, however, in at least two different
ways, by $U \rightarrow \infty$ and by $t \rightarrow 0$. We shall consider the first of these in detail
and then say a few words about the second.

A.   Strong Repulsion Because U → ∞

When $U \rightarrow \infty$, no two electrons can occupy the same site regardless of
their spin. The spins are completely decoupled both from the spatial motion
and from one another because two electrons never know each other's spin. The
wave functions are therefore of the form

$$\Psi(m_1s_1, m_2s_2,..., m_Ns_N) = \mathscr{A}\psi(m_1,...,m_N)\,\chi(s_1,...,s_N) \qquad (4.2)$$

where $\mathscr{A}$ is the operator that antisymmetrizes the product function $\psi\chi$ under all
permutations of particles. The total energies depend on ψ, not χ Since the ψ's
must vanish if $m_i = m_j$ and must be products of plane waves otherwise, they
are just the wave functions of a spinless free Fermi gas, *i.e.* Slater determinants
of plane waves. The choice of single-particle k's is slightly complicated because it is the many-body antisymmetrized wave functions that must satisfy
appropriate periodic boundary conditions, but for most purposes one can
neglect these complications. The system energies will be those of such a Fermi
gas, *i.e.* $E(k_1,...,k_N) = \Sigma_i \epsilon(k_i)$ if there are N particles. In the ground
state, since *each* particle must go into a different spatial state, *the Fermi wave
number k_F is twice as large as for non-interacting electrons.*

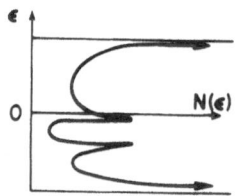

Fig. 12.   Pseudo-gap in electron density-of-states for $T < T_{mf}$, due to Peierls
fluctuations.

Many of the properties of this system are very similar to those of a non-interacting electron gas. The system has infinite conductivity unless a dissipative mechanism is introduced. The charge density, while uniform *on the average*, shows short-range correlations that oscillate with a periodicity b/ν, where $\nu = N/M$ is the number of electrons per site. When coupled to the lattice, this Fermi gas also has a tendency to suffer a Peierls distortion [21]. The Peierls wave number is now $q_p = \nu b^*$ corresponding to a Peierls wavelength $\lambda_p = b/\nu$. Since q_p is just twice that for free electrons, it is said that the system shows a "$4k_F$ instability", although it must be borne in mind that in this phrase k_F refers to the Fermi wave number *if the same density of electrons were nearly free*, even though they may be strongly interacting.

The behavior of the spins is, of course, entirely different from that of the weakly-interacting electron gas. If $U = \infty$, then the spins should exhibit a Curie susceptibility ($\sim 1/T$) for all T. But of course $U \neq \infty$, although it may be large. In this case, either of two electrons on neighboring sites with anti-parallel spin can hop virtually to the site of the other and back, resulting in a lowering of their combined energy. If the electrons have parallel spins, no such virtual hopping is allowed. Thus the energy when the spins of adjacent electrons are antiparallel is slightly lower than when parallel, implying an effective antiferromagnetic interaction $J_1 S_n \cdot S_{n+1}$ between spins when on adjacent sites (n numbers the particles, not the sites). A detailed calculation shows that $J_1 = 4t^2/U$. When the electrons are i sites apart, there is still such an antiferromagnetic interaction but J_i falls off exponentially with distance i. If the spin energies are small compared with the electron kinetic energies, as they should be if $t/U << 1$, then the rapid electron motion should replace the spatially-dependent spin interactions with an average interaction $\overline{J}$, leading then to a uniform Heisenberg antiferromagnetic chain of spin-1/2's for the spin degrees of freedom. As far as I know, these conclusions are plausible but have never been shown to follow rigorously for the Hubbard model with arbitrary filling [20a, 22]. Although the ground state and, in principle, all the other states of this spin problem were found by Bethe [2], our knowledge of the spin susceptibility χ_{spin} comes from the extrapolations made by Bonner and Fisher [23] from their calculations on relatively short chains. They found a $\chi_{\text{spin}}(T)$ that is Curie-like at high temperatures, has a broad peak at about 0.64J, and then decreases to 0.405 μ_B^2/J as $T \to 0$, corresponding to a singlet ground state and a gap for elementary excitations (μ_B is, here, the Bohr magneton).

Because there can be only one spin excitation on each site, the gas of spin excitations is *also* equivalent to a gas of spinless fermions whose average number is half the number of spins. For the XY model, these pseudo-fermions do not interact, but for the Heisenberg model, there is a two-body interaction between pseudo-fermions on neighboring sites of a strength that is comparable to the bandwidth. Thus we have a third kind of fermion to be added to the usual spinning electron and the spinless fermion representing the spatial motion of infinitely repulsive electrons. This fermion gas, like the other two, should also have a tendency to a Peierls instability. This is the so-called spin-Peierls instability [24]. If such a distortion occurs, one has a spin-density wave. In some ways this is analogous to a chain of dimers, each in a singlet state, although none of the spins is localized so the actual many-spin wave function is much more complicated than for a dimer chain. The wavelength (wave vector) of such a deformed state is just twice (half) that of the charge-density wave for the $U \to 0$ case. The spin-Peierls instability is therefore at $2k_F$, the same as the Peierls CDW (and the Overhauser SDW [22]) instability for $U \to 0$. When, in TTF-TCNQ, only the $2k_F$ distortion was known, Torrance, who was convinced that U was large, believed the $2k_F$ distortion to be the CDW and so initiated the search for a SDW distortion at k_F [26]. The finding of an instability at $4k_F$ was, nevertheless, considered by some to be a vindication of Torrance's views, only the nature of the $2k_F$ wave having been misidentified. Several other explanations have since been offered for the $4k_F$ instability, and the subject remains controversial.

B.   Strong Repulsion Because t → 0

If $t \to 0$, the electrons behave like classical particles, being perfectly localized with no cost in energy. It is not the exclusion principle now that will keep them more-or-less uniformly spaced; rather such uniform spacing must come from off-site repulsions (V_1, V_2, ... > 0) in the extended Hubbard model. Now the question of whether or not the lattice will distort to stabilize this Wigner lattice will depend on the detailed nature of the coupling. The case of small but non-vanishing t and realistic choices of V_1 has recently been studied in detail both numerically and analytically by P. Maldague [27], who finds charge-density waves with an essentially sinusoidal envelope even when the density of electrons is not a low-order rational fraction of the site density.

C.   Contrasting Weak and Strong Electron-Electron Interactions

To summarize this lecture we contrast the behavior for the limiting cases of weak and strong electron repulsions in Table I.

TABLE I.   Contrasting Weak and Strong Coupling Behavior

	Weak ($U \simeq 0$)	Strong ($t/T \simeq 0$)
Conductivity	No e-e scattering, need dissipative mech.	Need dissipative mech. despite e-e scattering.
Spin susceptibility	Small, less T-dependent	Large, more T-dependent
Optical properties	Will be discussed in other lectures	
CDW's	$2k_F$	$4k_F$
	not simultaneous	could be simultaneous
SDW's	$2k_F$	$2k_F$
Phase transitions	The phase transition behaviors would be different but they could be obscured by complications going beyond one-chain models.	

V.   GINZBURG-LANDAU FREE-ENERGY FUNCTIONAL

A.   Introduction

In the first lecture we have treated the lattice in a very crude way, at least in the low-temperature phase. We have assumed a lattice distortion that was static and characterized by a single wave number q. Such a treatment misses all kinds of static effects that are not simply periodic and misses a variety of dynamic effects such as the existence of various kinds of collective motions and the role of fluctuations. In this section we generalize the mean-field approach by means of the Ginzburg-Landau formalism. One considers not only one perfectly periodic distortion but *arbitrary* distortions that are in some sense *near* some ideal distortion. Such distortions will have not just a few degrees of freedom but in an infinite solid, infinitely many degrees of freedom. Either the free energy as a function*al* of the distortion can be computed using some many-body theory, or the form of the free-energy functional can be postulated consistent with the symmetries of the problem, leaving a small number of coefficients to be determined by fitting predictions to experiment. In either case, one can then do both the statistical mechanics (including fluctuations) and the classical dynamics of this distortion field. This approach has several advantages: the fluctuations can be included, the 1D-3D crossover and the phase transitions can be described, the interplay of the crystal lattice and the superlattice together with the related question of commensurability *vs.* incommensurability can be studied, and collective behavior such as sliding conductivity, phasons and solitons can be described and investigated. The disadvantages are that we are restricted to temperatures below T_{mf} and the resulting theory is still not really a fundamental theory. Being semi-phenomenological, the theory can not be used for a detailed verification of a microscopic theory or a detailed determination of microscopic parameters. Nevertheless, the approach has so many successes that we wish to devote this lecture to formulating the approach for 1D systems (subsection B), calculating the free energy and certain correlation functions of such a system with proper inclusion of fluctuations (subsection C), investigating phasons and solitons in 1D (subsection D), and finally considering a 3D lattice of 1D stacks (subsection E).

B.   Order Parameter and Free-Energy Functional

The fundamental quantity in a Ginzburg-Landau theory is the "order parameter" defined at the sites $\mathbf{R}$ in the solid. For the order parameter one can take some distortion amplitude $u(\mathbf{R})$ at site $\mathbf{R}$ or the electronic gap $\Delta(\mathbf{R})$ near $\mathbf{R}$ or the CDW amplitude $\rho(\mathbf{R})$ near $\mathbf{R}$. Since, in some rough sense they are all proportional, we can choose any one as the order parameter and call it $\phi(\mathbf{R})$. Just which choice we make will affect only the coefficients in the resulting free-energy functional. If we consider $\mathbf{R}$ to have a component $R_{\parallel}$ along the stacks and a vector $\mathbf{R}_{\perp}$ transverse to the stacks, then it is customary (though not essential) to replace the discrete variable $R_{\parallel}$ by a continuous variable y. Also, until subsection E we consider only a single stack and so neglect $\mathbf{R}_{\perp}$.

It is also useful to define a complex rather than a real order parameter, because the latter has simpler properties under translation. To do this, imagine a uniform periodic distortion with $q \simeq q_p$:

$$\phi_q(y) = A_q \cos (qy + \varphi_q) \tag{5.1}$$

where A_q and φ_q are the amplitude and phase of the distortion. This suggests that we define a complex order parameter for this distortion

$$\psi_q(y) = A_q e^{+i\varphi_q} e^{iqy} \equiv \psi_q e^{iqy} \tag{5.2}$$

so that

$$\phi_q(y) = \frac{1}{2} (\psi_q(y) + \psi_q^*(y)) = Re \, \psi_q(y) \tag{5.3}$$

The electronic free-energy in the presence of such a distortion will be assumed to have the form

$$F_q = \left[a(T) + c(T) (q-q_p)^2 \right] |\psi_q|^2 + b(T) |\psi_q|^4 \tag{5.4}$$

i. e. it depends only on the amplitude, not the phase of the real order parameter. We shall discuss the modifications arising from commensurability and the resulting phase dependence later. We have included the terms already written down in (3.5) when $q = q_p$ and have added the $c(T)$ term to take into account the cost in free energy when q differs from its optimum value. The coefficients $a(T)$, $b(T)$ and $c(T)$ are either to be calculated using some theory of the electron gas in the presence of the distortion $\phi_q(y)$ or are to be treated as phenomenological parameters chosen to fit experiment.

If we now have a general, rather than a perfectly periodic, distortion, but one involving only q's near q_p, then $\phi(y)$ can be similarly decomposed:

$$\phi(y) = \frac{1}{2(2\pi)^{1/2}} \int_{q\simeq q_p} dq(\psi_q e^{iqy} + \psi_q^* e^{-iqy}) \equiv \frac{1}{2}[\psi(y) + \psi^*(y)] \tag{5.5}$$

and the free-energy generalizing F_q will now be a functional of ψ_q:

$$F[\psi_q] = \int dq[a(T) + c(T)(q-q_p)^2]|\psi_q|^2 \tag{5.6}$$

+ higher-order terms.

In terms of $\psi(y)$, this is

$$F[\psi(y)] = \int dy \, f[\psi(y)] \tag{5.7a}$$

where

$$f[\psi(y)] = a(T)|\psi(y)|^2 + c(T) \, | \, (\frac{1}{i}\frac{d}{dy} - q_p)\psi(y) \, |^2 \tag{5.7b}$$

+ higher-order terms.

This can be simplified if we let $\chi(y) = e^{iq_p y}\psi(y)$, giving

$$f[\chi(y)] = a(T)\,|\,\chi(y)\,|^2 + c(T)\,|\,\frac{1}{i}\frac{d\chi}{dy}\,|^2 \qquad (5.7b')$$

+ higher-order terms.

In these equations we have been deliberately vague about the terms of fourth-order because, in addition to the straightforward generalization to include a term like $b(T)\int dq\,|\psi_q|$ in $F[\psi_q]$, there are other terms coupling different q's that are essential. These will be discussed later.

For continuous y and for a discrete lattice in almost all cases, it is possible to define a complex order parameter, because a change of phase leads to a genuinely different pattern (Fig. 13). However, for the special case that $q_p = b^*/2$, corresponding to dimerization, a shift of phase is only a change of amplitude (Fig. 13b); the order parameter for the "half-filled-band" case is therefore real.

Clearly the free-energy functionals $F[\psi_q]$ and $F[\psi(y)]$ we have defined in terms of ψ_q and $\psi(y)$ could be more general and must be more specific. We have neglected, for example, higher derivatives than the second in the second-order terms and all derivatives in the fourth-order terms. These neglects are probably unimportant, but what of the terms involving no derivatives? An important guide to the allowable terms in $F[\psi_q]$ comes from symmetry. For example, the free energies for two deformations that differ only by a crystal lattice translation must be the same. Now if we slide a deformation $\psi(y)$ a distance $y_0 = mb$, the new deformation is $\overline{\psi}(y) = \psi(y-y_\rho)$. The same translation changes ψ_q into $\overline{\psi}_q = \psi_q\exp(iqy_0)$ and ψ_q^* into $\overline{\psi}_q^* = \psi_q^*\exp(-iqy_0)$. A general term in the free-energy functional suffers the change

$$\psi_{q_1'}^*\psi_{q_2'}^*...\psi_{q_1}\psi_{q_2}...$$

$$\qquad (5.8)$$

$$\rightarrow \psi_{q_1'}^*\psi_{q_2'}^*...\psi_{q_1}\psi_{q_2}...e^{i(q_1'+q_2'+...-q_1-q_2-...)mb}$$

If the free energy is to be invariant, we require that

$$q'_1 + q'_2 + ... - q_1 - q_2 -... = 0, \pm b^*, \pm 2b^*, ... \qquad (5.9)$$

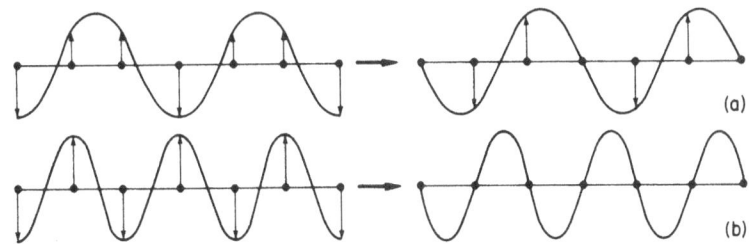

Fig. 13. Effect of shifting phase of periodic distortion by $\pi/2$ in discrete lattice if (a) $q_p = b^*/3$ and (b) $q_p = b^*/2$.

Taking 0 on the right gives the "normal" condition, one that makes no reference to the presence of a crystal lattice and the only one that would be allowed if each stack were treated as a translationally invariant continuum. Taking $\pm b^*$, $\pm 2b^*$, etc. on the right gives conditions that are analogous to the wavevector conservation in an *Umklapp* process (the analogy is even stronger if one imagines ψ_q as destroying wave number q and $\psi^*_{q'}$ as creating wave number q').

Let us first consider the kinds of terms allowed in F that have only one q. Inclusion of only these terms would give the mean-field results. The *normal* condition gives the terms like $|\psi_q|^2$, $|\psi_q|^4$, $|\psi_q|^6$, etc. The *Umklapp* condition with $\pm b^*$ and the requirement that F be real gives, among others, terms of the form

$$\frac{1}{2} \int dq \delta(q - \frac{b^*}{n})(D_q(T)\psi_q^n + \text{c.c}) \propto | D_{b^*/n}\psi_{b^*/n}^n | \cos(\delta + n\varphi) \quad (5.10)$$

where $n = 2, 3, 4, \ldots$ and δ is the phase of the complex coefficient $D_{b^*/n}$. Thus there are additional terms that can be added to the free energy for commensurate q's. These terms enhance the stabilization energy at the optimum phases $\varphi = [-\delta + (2m+1)\pi]/n$ and produce a barrier of height $2|D\psi^n|$ for sliding the deformation. Similar additional commensurability terms will occur in F from $\pm 2b^*$, $\pm 3b^*$, etc.

The important advance of Ginzburg-Landau theory over mean-field theory is that we are not restricted to a single q, although the $c(T)$ term will insure that $q_{i} \simeq q_p$ for all i. In second-order, there are no solutions to (5.9) other than $q_1 = \bar{q}_1$ (recall that $|q_i| \leq b^*/2$). However, in all higher orders there are sets of different q's that are near the single-q solutions we have already discussed. These terms represent fluctuations from the mean-field solution. For example a *normal* fourth-order term that includes fluctuations is

$$\frac{1}{2\pi} \int D(q_1, q_2, q_1', q_2'; T)\psi^*_{q'_1}\, \psi^*_{q'_2}\psi_{q_1}\psi_{q_2}\delta(q_1' + q_2' - q_1 - q_2) \quad (5.11)$$

If we assume that $q_1 \simeq q_2 \simeq q_1' \simeq q_2' \simeq q_p$ in (5.11) and neglect the q dependences of $D(q_1, q_2, q_1', q_2' ; T)$, then this has the particularly simple form in y space

$$D \int dy |\psi(y)|^4 \quad (5.12)$$

which is perhaps the simplest way to include fluctuations in the normal fourth-order terms. Similarly, in third-order, while there is no normal term, there is an *Umklapp* term of the form

$$\frac{1}{2(2\pi)^{1/2}} \int \cdots \int dq_1 dq_2 dq_3$$

$$\quad (5.13)$$

$$[D(q_1, q_2, q_3; T)\psi_{q_1}\psi_{q_2}\psi_{q_3} + \text{c.c}]\delta(q_1 + q_2 + q_3 - b^*)$$

which will be important if, in the $c(T)$ term, $q_p \simeq b*/3$. This will describe fluctuations around the commensurate distortion with $n = 3$. If we again neglect the q dependence of $D(q_1,q_2,q_3;T)$ and ψ_q, we obtain

$$\frac{1}{2} \int dy \left[D(\psi(y))^3 e^{ib*y} + \text{c.c.} \right] \tag{5.14}$$

Similar generalizations of (5.10) exist for all n. In subsection D, we shall use this form to discuss the case of near-commensurability.

C. Exact Calculations of Free Energy and Correlation Functions

$F[\psi(y)]$ can be considered to be the system free-energy subject to the constraint that the order parameter be a specified function $\psi(y)$. Then $\exp(-\beta F[\psi])$ is the *relative* probability of finding this order-parameter function. To evaluate the system free-energy, we must evaluate the partition sum over all order-parameter functions $\psi(y)$. When the sites are discrete, this is just

$$Z = \int ... \int d\psi_1 ... d\psi_M e^{-\beta F(\psi_1,...,\psi_M)} \tag{5.15}$$

When the sites form a continuum, we write this as a so-called functional integral or "path integral"

$$Z = \int \mathcal{D}\psi(y) e^{-\beta \int_0^{Mb} f[\psi(y),d\psi/dy]dy} \tag{5.16}$$

which is defined from (5.15) and (5.7) with an appropriate limiting process. Similarly, if we wish to calculate the correlation between the order parameters at y_1 and y_2, we must calculate the ratio of two functional integrals:

$$<\psi(y_1)\psi(y_2)> = \int \mathcal{D}\psi(y) e^{-\beta F} \psi(y_1)\psi(y_2)/Z \tag{5.17}$$

These functional integrals may look very strange, especially if one has not had years developing intuitive feelings about them. However, there is a connection with an entirely different problem, *viz.* that of a single quantum-mechanical (QM) particle, which can provide that intuition quickly. The connection depends on Feynman's formulation [28] of the problem of the partition function $Z(\beta)$ for such a particle, a problem about which our intuition should be strongly developed. Feynman showed that if the particle has the classical Lagrangian $L(x, dx/dt)$, then

$$Z(\beta) = \int \mathcal{D}x(\tau) e^{-\int_0^\beta L(x,idx/d\tau)d\tau} \tag{5.18}$$

where $\beta = 1/kT$, $\tau = it$ is imaginary time, and the functional integration or path integration extends over all paths $x(\tau)$ that start and end at the same point. A comparison of (5.18) with (5.16) and (5.7) shows the correspondences in Table II. The last relationship of Table II, which is very important, is not so obvious. It depends on the fact that in the ground state the correlation between the positions of the QM particle at two times a large *imaginary* time apart decays exponentially with the imaginary-time difference, and the decay rate is proportional to the excitation energy to the lowest state that is connected to the ground state by a matrix element of x (assuming $<x> \neq 0$ in the ground state).

TABLE II.   Correspondence of 1D Ginzburg-Landau and Feynman's
Single Quantum-Mechanical Particle.

1D Ginzburg-Landau	Feynman QM Particle
y	τ
$M \to \infty$	$\beta \to \infty$
$\psi(y)$	$x(\tau)$
$f(y) = a|\psi|^2 + b|\psi|^4$	$V(x) = a|x|^2 + b|x|^4$
c	particle mass
(correlation length ξ)$^{-1}$	First excitation energy

We make two remarks: First, if ψ is complex, then the analogy requires a complex x, i. e. x is a 2D vector or the QM particle moves in two dimensions. Second, the temperature in the Ginzburg-Landau problem should not be confused with that in the Feynman problem; the former temperature enters into the latter problem through the parameters $a(T)$, $b(T)$, etc. that characterize $V(x)$ and the parameter $c(T)$ that characterizes the QM particle's mass.

Now let us apply our intuitive ideas about a single QM particle to get insight into a Ginzburg-Landau chain. Consider first the case that $q_p = b^*/2$, i. e. a chain that dimerizes in the mean-field approximation. This is like a QM particle in a potential well having the shape of one of the curves in Fig. 7. For $T > T_{mf}$, there is only one well and a clear gap between the ground and first excited states, implying an exponential decay in imaginary time with a finite decay constant for the QM particle and therefore a similar exponential decay along the chain of the correlation function $<\psi(y_1)\,\psi(y_2)>$ for the Ginzburg-Landau problem. For $T < T_{mf}$, the QM potential has two wells. The splitting of the ground and first excited states goes to zero exponentially with increasing well-depth so that the correlation length ξ grows exponentially with decreasing T.

Consider now the case that q_p is incommensurate so that (5.7b) with (5.12) again applies but ψ is complex. Now our quantum mechanical particle moves in a two-dimensional potential. For $T > T_{mf}$, the potential has a single minimum at $x = 0$, but at lower temperatures it is a wine-bottle potential (Fig. 14). At very low temperatures, the particle moves near the very bottom of the wine bottle, which strongly localizes it radially, at a radial distance $|a/2b|^{1/2}$, but not angularly. Thus the correlation of the *amplitudes* $<|\psi(y_1)|\,|\psi(y_2)|>$ is very long-range but the phases are correlated over only a short range. The ground QM state has $\ell_z = 0$ while the first excited states have $\ell_z = \pm 1$. The excitation energy goes to zero only because the effective moment of inertia goes like $|a/2b|$, which increases linearly as T decreases if $a(T)$ decreases linearly. This leads to a correlation length increasing like $1/T$ as $T \to 0$.

It is worth remarking that when the trough of the wine bottle is so deep that only the phase is really variable, a case we shall discuss in much more detail in the next subsection, the problem is similar to that of a chain of classical spins with XY-model coupling. This analogy is often useful in thinking about the coupling between chains, particularly if one has developed some intuition about spin systems.

Finally let us consider that q_b is commensurate of order n (i. e. $q_p = b*/n$). Then, as we have seen, an additional term like (5.14) can occur in the free energy that will add n ripples along the trough in the wine-bottle potential at low temperatures. If the ripples are shallow, they will have no significant effect on the excitation energies of the QM particle. If the ripples are deep, the excited states may become nearly degenerate with the ground state because of the small tunneling matrix element between the trough minima. The situation is clearly complicated, but one can see the possibility of a crossover from $1/T$ behavior to exponential behavior for $\xi(T)$.

D. Phasons and Solitons

Let us consider the wine-bottle regime of a single chain more carefully, i. e. we assume that the amplitude of the order parameter is fixed but that the phase may vary. We shall investigate phase functions that make F stationary, thus neglecting fluctuations. But we shall also study the dynamical behavior of the phase field. We consider the incommensurate case first because the results will be familiar. We then generalize to the commensurate case where we shall find some very interesting, new-looking results [29].

If, generalizing (5.2), we substitute $\psi(y) = |\psi(y)| \exp[i(q_p y + \varphi(y))]$ into (5.7) with the fourth-order term (5.12), neglect variations in $|\psi(y)|$ and

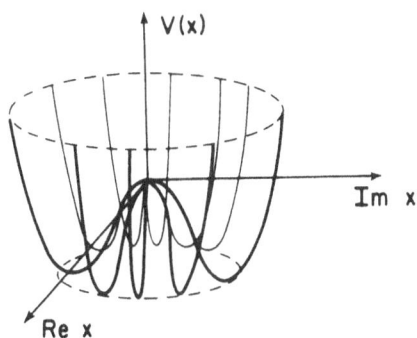

Fig. 14. Wine-bottle potential of particle corresponding to Ginzburg-Landau potential for complex order parameter.

integrate by parts, we obtain a free-energy functional of $\varphi(y)$:

$$F[\varphi] = c|\psi|^2 \int (d\varphi/dy)^2 dy + \text{const.} \tag{5.19}$$

The mean-field approximation corresponds to replacing the functional integral (5.13) by $\exp(-\beta F[\varphi_0(y)])$ where $\varphi_0(y)$ is the "best" $\varphi(y)$, *i. e.* the one that minimizes $F[\varphi]$. The Euler-Lagrange equation for those $\varphi(y)$'s that make F stationary is

$$2c|\psi|^2 \frac{d^2\varphi}{dy^2} = 0 , \tag{5.20}$$

which has the solutions

$$\varphi(y) = \varphi_0(y) + ky , \tag{5.21}$$

where k is an arbitrary constant. The solution that minimizes F has $k = 0$, yielding $\psi = |\psi| \exp[i(q_p y + \varphi)]$, the familiar mean-field result.

So far this is a static deformation, but we can generalize the Ginzburg-Landau approach to include dynamics if we define a Lagrangian for the order-parameter field by introducing a kinetic-energy term:

$$\mathcal{L} = \frac{1}{2}\mathcal{M} \int \left(\frac{d\phi}{dt}\right)^2 dy - F[\phi(y)] \tag{5.22}$$

where $\mathcal{M}$ is an effective mass. The equation of motion implied by this Langrangian is the wave equation

$$\mathcal{M} \frac{\partial^2\varphi}{\partial t^2} - 2c|\psi|^2 \frac{\partial^2\varphi}{\partial y^2} = 0 \tag{5.23}$$

having the solutions

$$\varphi(y,t) = Re\, e^{i(ky-\omega_k t-const.)} \tag{5.24}$$

with the dispersion relation

$$\omega_k = \left(2c|\psi|^2/\mathcal{M}\right)^{1/2} k \tag{5.25}$$

We sketch the wave:

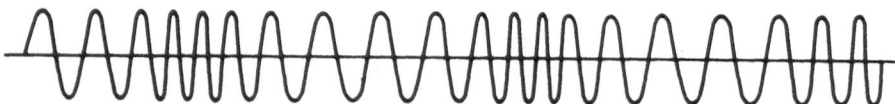

In regions where the points of zero phase are denser (sparser), we can think of q as being locally greater (smaller). With time, these regions can move. These periodic, small-amplitude variations in local q are just the phason modes we have discussed qualitatively earlier. We see that $\omega_k \to 0$ as $k \to 0$, as we previously argued from general principles. It is important to realize that though this discussion is simple in y space, it would have been much more difficult in q space.

Let us now turn to the case where q_p is nearly commensurate of order n so that, generalizing (5.14), we may add to $F[\varphi]$ the term

$$-D|\psi|^n \int \cos\left[n((q_p-q_n) + \varphi(y))\right] dy \qquad (5.26)$$

where $q_n = b^*/n$ is the nearby commensurate wave number. We have assumed for simplicity that arg $D = 0$. The Euler-Lagrange equation for $\varphi(y)$ that make $F[\phi]$ stationary is now

$$2c|\psi|^2 \frac{d^2\varphi}{dy^2} - 2nD|\psi|^2 \sin\left[n(\varphi(y) + (q_p-q_n)y)\right] = 0 \qquad (5.27)$$

If we make the substitution $\overline{\varphi} = \varphi + (q_p-q_n)y$, so that $\overline{\varphi}$ measures the deviation from a perfectly periodic commensurate wave with wave number q_n, then the $\overline{\varphi}$'s making $F[\varphi]$ stationary satisfy

$$2c|\psi|^2 \frac{d^2\overline{\varphi}}{dy^2} - nD|\psi|^n \sin n\overline{\varphi} = 0 . \qquad (5.28)$$

This equation recalls the equation for the classical pendulum

$$m\ell^2 \frac{d^2\theta}{dt^2} - mg \sin \theta = 0 \qquad (5.29)$$

where θ is the pendulum angle measured from the upward direction (*i. e.* equilibrium at $\theta = (2m + 1)\pi$). Our intuition (as well as exact solutions) for this physical problem also helps to understand the spatial variation of the order parameter. In Fig. 15 we have plotted two solutions of the pendulum angle θ vs. t. In the periodic curve, we see small-amplitude vibrations around the equilibrium position $\theta = \pi$. In the other curve the pendulum starts at the top ($\theta = 0$) with extremely small velocity, staying there until roughly t_1. Then it falls rapidly, swinging back up to the top ($n\theta = 2\pi$) at roughly t'. This motion is continued periodically and is of course entirely different from the small amplitude solution. Reinterpreting the abscissa as the position y along the chain and the ordinate as the phase $\overline{\varphi}$, we see that until the position y_1 the phase $\overline{\varphi} \simeq 0$, *i. e.* the wave is essentially locked to the underlying lattice by its commensurability energy. Then between y_1 and y_2 there is a phase slip of 2π after which it is essentially locked again. This pattern is repeated periodically

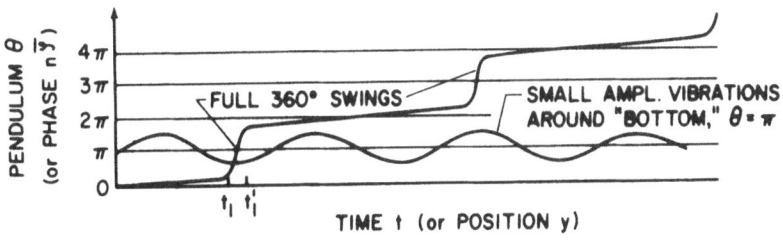

Fig. 15. Small-amplitude vibrations (phasons) and full-circle swings (solitons) of pendulum (of φ field near commensurate wave number).

along the chain. The analytic solutions having just one phase slip of 2π (-2π) are called solitons (antisolitons). This behavior can perhaps be understood as what happens if two combs whose densities of teeth are nearly commensurate are pressed against one another, as in Fig. 16.

When such static solutions $\overline{\phi}(y)$ are substituted into F, there is a term of the form $2c|\psi|^2 (q_p-q_n)\int \overline{\phi}'(y)\,dy$, the only term in which q_p appears explicitly. Since $\int \overline{\phi}'\,dy/2\pi = [\overline{\phi}(Mb)-\overline{\phi}(0)]/2\pi$ is the number of solitons in the chain, the quantity $2c|\psi|^2 (q_p-q_n)$ acts like a chemical potential for solitons and determines the optimum number (i. e. the number that minimizes the free energy). If the local q is proportional to a charge density which, on the average, must equal that corresponding to q_p, then this constraint would take the form $0 = \int (d\overline{\phi}/dy)dy$ or $\overline{\phi}(Mb)-\overline{\phi}(0) = (q_p-q_n)Mb$.

In analogy with (5.23), we can also obtain an equation for the dynamical behavior in the commensurate case:

$$\mathcal{M} \frac{\partial^2\overline{\phi}}{\partial t^2} - 2c|\psi|^2 \frac{\partial^2\overline{\phi}}{\partial y^2} + nD|\psi|^n \sin n\overline{\phi} = 0 \tag{5.30}$$

This is the sine-Gordon equation, a non-linear partial differential equation that reduces to the well-known Klein-Gordon equation when the sine term is linearized. It has exact analytical solutions that correspond to solitons and antisolitons, moving both to the left and right, with a number of interesting properties. We simply summarize a few:- 1) Isolated solitons and antisolitons move without changing their shape or dissipating. 2) They pass through one another without changing their shape, the only evidence of the "collision" being a phase change. 3) In a Peierls-distorted system, there is associated with a soliton or antisoliton a charge of one sign or the other, corresponding to the fact that locally q is increased or decreased. 4) They can be pinned by defects. 5) The solitons and antisolitons can each form a lattice, as we saw in studying the static solutions. Due to the fluctuations that we have neglected, such a lattice would form only at T = 0 °K. Of course, if the solitons are on a chain that is part of a 3D system with interactions in the transverse directions, a soliton lattice (actually a super-super lattice) can form at finite T. 6) The lattice of solitons can vibrate, each vibrational mode being a kind of phason mode which has also been called a "muron" mode (vibration of the domain wall) [30, 12].

We should mention that serious attempts have been made, initiated by the work of Krumhansl and Schrieffer [31], to see how much of the exact free-

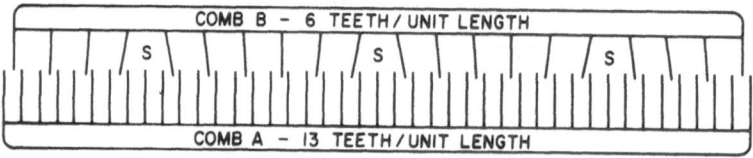

Fig. 16. Two nearly commensurate combs. Solitons at S.

energy obtained using the Feynman analogy can also be obtained by treating the system as a gas of phonons, solitons and their interactions.

All that we have said about solitons has been about *phase* solitons. However in a dimerized system, changes of phase are equivalent to changes of amplitude (Fig. 13b), so one must consider amplitude solitons. The occurrence of such solitons has been proposed for the quasi-one-dimensional semiconductor polyacetylene to explain some of its transport and magnetic properties [32].

E.   A 3D Lattice of 1D Chains.

We conclude these lectures with a few remarks based on Ginzburg-Landau considerations, about the behavior of a 3D lattice of 1D chains.

1. Within the Ginzburg-Landau free energy there will now be terms of the form

$$J(\mathbf{R}, \mathbf{R}') \int \psi_R^*(y) \, \psi_{R'}(y') \, dy \tag{5.31}$$

where $\mathbf{R}$ and $\mathbf{R}'$ are now transverse vectors locating two different chains. If we are in a regime where the $|\psi|$'s saturate, then this is equivalent to an interaction proportional to $\cos\left[\theta_R(y) - \theta_{R'}(y')\right]$ *i. e.* it is like a lattice of continuously distributed two-dimensional (*i. e.* XY-model) spins where $\theta_R(y)$ is the orientation in the xy plane of spin space of the spin at $(y, \mathbf{R})$. Furthermore, these interactions frequently compete. If, for example, charge-density waves on nearby stacks want to be out of phase with one another, as expected for simple Coulomb interactions, it is like having antiferromagnetic interactions between nearby pairs of XY-model spins. The competition of such antiferromagnetic interactions with one another can lead to very exotic arrangements of spins, as in the rare-earth metals, arrangements that can also change continuously with temperature or pressure. It is precisely such competing interactions that are believed to explain the interesting temperature-dependence of the superlattice in TTF-TCNQ [15, 33]. In addition, in TTF-TCNQ one of the transverse superlattice cell parameters approaches commensurability with the crystal lattice as the temperature is lowered ($n = 4$). This leads to the possibility of a commensurate-incommensurate transition which was analyzed for this material using solitons by Bak and Emery [33].

2. In addition to occurring because a superlattice cell parameter becomes commensurate with a crystal lattice parameter, solitons can also occur in quite incommensurate 3D systems. This is because each chain's distortion is commensurate with the superlattice occurring on all the other chains. Thus a term like $\psi_F^* \psi_Q + c.c.$ in TTF-TCNQ leads to a term in the Ginzburg-Landau free energy like $\cos[\theta_F(y) - \theta_Q(y)]$ which, for fixed θ_Q, gives a cosinusoidal potential having the superlattice periodicity for a particular $\theta_F(y)$. Such solitons are like dislocations in the superlattice. Thus because of 3D effects, solitons are another possible form of excitation on individual chains, in addition to phonons, phasons, etc., even in the absence of external commensurability. Such solitons might be present because of thermal excitation or they might be present

to provide the needed phase slip between randomly placed imperfections that randomly pin the local phase.

3. Three-dimensional interactions such as (5.31) play a crucial role in causing phase transitions. Purely 1D systems, as we have remarked, show no phase transitions; they simply develop domains of correlated phase that are of average length $\xi(T)$, and well below T_{mf} (but above T_c) the amplitude $|\psi|$ can be considered saturated. In what we shall call a quasi-1D system, we imagine a coupling between the phases of these domains on adjacent stacks of strength $J_\perp$/unit length. Then we would expect a phase transition at such a temperature that

$$J_\perp \, \xi(T) \simeq kT \tag{5.32}$$

This is exactly what is found in the 2D Ising model, which was solved by Onsager [34] for arbitrary $J_\perp$ and $J_\parallel$. In particular, it was found that for $J_\perp \ll J_\parallel$, the specific heat $C(T)$ shows the behavior sketched in Fig. 17. Although $C(T)$ has a broad peak at about T_{mf} that is characteristic of a purely 1D chain at such a temperature, there is a crossover to 2D behavior and a 2D phase transition marked by a weak logarithmic singularity in $C(T)$ at a lower temperature, T_c. To test (5.32), we note that the correlation length of a one-dimensional Ising model at low temperature is $1/2 \exp(2\beta J_\parallel)$. The criterion (5.32) would give $kT_c \simeq 2J_\parallel \ln(kT_c/J_\perp)$ while the exact solution of the 2D Ising model gives $\sinh(2J_\parallel/kT_c) \sinh(2J_\perp/kT_c) = 1$, which reduces for $J_\perp \ll kT_c \ll J_\parallel$, to precisely this result with an equality. We note that T_c is very insensitive to $J_\perp$, a consequence of the rapidly changing nature of $\xi(T)$. The Ising model corresponds, of course, to a real order parameter. For incommensurate cases, where we have seen that $\xi(T) \sim 1/T$, T_c is much more sensitive to the value of $J_\perp$.

4. In contrast to the quasi-1D case, where we make little error by assuming that $|\psi|$ saturates and that $\xi(T)$ is unaffected by the 3D interactions until (5.32) is approximately satisfied, in what we shall call the anisotropic 3D case, the growth of $|\psi|$ is itself affected by the 3D interactions. Then T_{mf} and T_c will be much closer together.

F.   Concluding Remarks

In concluding this second lecture, we may summarize by saying that sufficiently far below T_{mf} many interesting properties relating to the formation of the superlattice (e. g. the phase transition, the superlattice structure, the elementary distortional excitations, the role of commensurability, etc.) can be treated using the Ginzburg-Landau free-energy functional. However, the

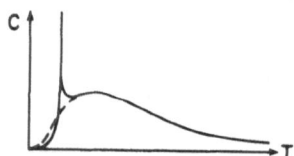

Fig. 17. Specific heat of quasi-1D (———) and 1D (---) Ising models.

choice of order parameters and the actual free-energy functional remain pheno-menological in the absence of a satisfactory many-body theory. Also, the dynamic interplay of electrons and higher-frequency phonons (*e. g.* the Kohn effect, isotope effect on T_c, etc.) are not dealt with within this formalism.

ACKNOWLEDGMENTS

It is a pleasure to acknowledge many valuable conversations over the years with numerous IBM colleagues, and, especially recent conversations with Per Bak. I am also indebted to Paul Chaikin and Hanoch Gutfreund for very pertinent comments after the lectures.

REFERENCES

1. For a collection of reprints and critical discussion of exactly soluble problems in one dimension, see Lieb, E. H. and Mattis, D. C.: 1966, *Mathematical Physics in One Dimension*, Academic Press, New York and London.
2. Bethe, H. A.: 1931, Z. Physik **71**, pp. 205-226.
3. Lieb, E. H., Schultz, T. D. and Mattis, D. C.: 1961, Annals of Physics **16**, pp. 407-466.
4. Dyson, F. J.: 1953, Phys. Rev. **92**, pp. 1331-1338.
5. For a review of the whole body of work on the Luttinger model and its generalizations, see J. Sólyom: 1979, Adv. in Phys. **28**, pp. 201-303.
6. Leib, E. H. and Wu, F. Y.: 1968, Phys. Rev. Lett. **20**, pp. 1445-1448.
7. (a) Lajzerowicz, J. and Pfeuty, P.: 1971, J. de Phys. **32**, C5a, pp. 193-194, and (b) Scalapino, D. J., Sears, M. and Ferrell, R. A.: 1972, Phys. Rev. B **6**, pp. 3409-3416
8. For review articles and contributions of current interest, see *Solitons and Condensed Matter Physics*, Springer Solid-State Sciences Vol. 8 (A. R. Bishop and T. Schneider, Eds.), Springer-Verlag, Berlin, Heidelberg and New York, 1978.
9. (a) Peierls, R. E.: 1953, *Quantum Theory of Solids*, Clarendon Press, Oxford, p. 108ff. and (b) Rice, M. J. and Strässler, S.: 1973, Solid State Commun. **13**, pp. 125-128.
10. (a) Fröhlich, H.: 1954, Proc. Roy. Soc. **A223**, pp. 296-305; (b) Allender, D., Bray, J. W. and Bardeen, J.: 1974, Phys. Rev. B **9**, pp. 119-129; and (c) For a review within a perturbation-theoretic framework, see Sham, L. J.: 1979, in *Highly Conducting One-Dimensional Solids* (J. T. Devreese, R. P. Evrard and V. E. van Doren, Eds.), Plenum Press, New York and London.
11. Lee, P. A., Rice, T. M. and Anderson, P. W.: 1974, Solid State Commun. **14**, pp. 703-709.
12. von Boehm, J. and Bak, P.: 1979, Phys. Rev. Lett. **42**, pp. 122-125, and to be published.
13. Schuster, H. G.: 1974, Solid State Commun. **14**, pp. 127-129.
14. Sham, L. J. and Patton, B. R.: 1976, (a) Phys. Rev. Lett. **36**, pp. 733-736 and (b) Phys. Rev. B **13**, pp. 3151-3153; and (c) Imry, Y. and Ma, S.: 1975, Phys. Rev. Lett. **35**, pp. 1399-1401.

15. For X-ray and neutron studies of ordering in several quasi-1D solids, see Comès, R. and Shirane, G.: 1979, in *Highly Conducting One-Dimensional Solids, op. cit*, pp. 17-67.

16. Horovitz, B., Gutfreund, H. and Weger, M.: 1975, Phys. Rev. B **12**, pp. 3174-3185.

17. Brom, H. B., Schultz, T. D., Tomkiewicz, Y. and Gill, W. D., submitted for publication.

18. (a) Lee, P. A., Rice, T. M. and Anderson, P. W.: 1973, Phys. Rev. Lett. **31**, pp. 462-465 and (b) Rice, M. J. and Strässler, S.: 1973, Solid State Comm. **13**, pp. 1389-1392.

19. Rybaczewski, E. F., Smith, L. S., Garito, A. F., Heeger, A. J. and Silbernagel, B. G.: 1976, Phys. Rev. B **14**, pp. 2746-2756.

20. For reviews of various aspects of this limit, see (a) Torrance, J. B.: 1977, in *Chemistry and Physics of One-Dimensional Metals,.*, (H. J. Keller, Ed.) Plenum Press, New York and London, 137-166, and (b) Emery, V. J.; 1979, in *Highly Conducting One-Dimensional Metals, op. cit.*, pp. 247-303.

21. Bernasconi, J., Rice, M. J., Schneider, W. R. and Strässler, S.: 1975, Phys. Rev. B **12**, pp. 1090-1092.

22. Klein, D. J. and Seitz, W. A.: 1974, Phys. Rev. B **10**, pp. 3217-3227.

23. Bonner, J. C. and Fisher, M. E.: 1964, Phys. Rev. **135**, pp. A640-A658.

24. See, for example, Pytte, E.: 1974, Phys. Rev. B **10**, pp. 4637-4642.

25. (a) Overhauser, A.W.: 1959, Phys. Rev. Lett. **4**, pp. 462-465; (b) Madhukar, A.: 1974, Solid State Commun. **15**, pp. 921-924, and (c) Weger, M. And Gutfreund, H.: 1979, Solid State Commun. (in press).

26. Private communications to many workers in field.

27. Maldague, P.: 1979, Bull. Am. Phys. Soc. **23**, pp. 354-355 and to be published.

28. Feynman, R. P. and Hibbs, A. R.: 1965, *Quantum Mechanics and Path Integrals*, McGraw Hill, New York. For a detailed discussion of this application, see Ref. [7b].

29. The Ginzburg-Landau approach to superlattice dynamics, commensurability, etc. was treated in a series of papers by W. L. McMillan in The Physical Review: (a) 1975, **12**, pp. 1187-1196; (b) 1975, **12**, pp. 1197-1199; (c) 1976, **14**, pp. 1496-1502; (d) 1977, **16**, pp. 4655-4658; and Bhatt, R. N. and McMillan, W. L.: 1975, Phys. Rev. B **12**, pp. 2042-2048.

30. Turkevitch, L. A. and Doniach, S.: 1979, Bull. Am. Phys. Soc. **24**, pp. 263-264.

31. Krumhansl, J. A. and Schrieffer, J. R.: 1975, Phys. Rev. B **11**, pp. 3535-3545.

32. (a) Goldberg, I. B., Crowe, H. R., Newman, P. R., Heeger, A. J. and MacDiarmid, A. G.: 1979, J. Chem. Phys. **70**, pp. 1132-1136; (b) Su, W. P., Schrieffer, J. R. and Heeger, A. J.: 1979, Phys. Rev. Lett. **42**, pp. 1698-1701; (c) Rice, M. J.: 1979, Phys. Lett. **71A**, pp. 152-154; (d) Weinberger, B. R., Kaufer, J., Heeger, A. J., Pron, A. and MacDiarmid, A. G.: 1979, Phys. Rev. B **20**, pp. 223-230; and (e) Tomkiewicz, Y., Schultz, T. D., Brom, H. B., Clarke, T. C. and Street, J. B.: 1979, Phys. Rev. Lett. **43**, pp. 1532-1536.

33: (a) Bak, P. and Emery, V. J.: 1976, Phys. Rev. Lett. **36**, pp. 978-982 and (b) Bak, P.: 1976, Phys. Rev. Lett. **37**, pp. 1071-1074.

34. Onsager, L.: 1944, Phys. Rev. **65**, pp. 117-149.

BASIC PHYSICAL CONCEPTS OF LOW DIMENSIONAL SOLIDS

Gordon A. Thomas

Bell Laboratories, Murray Hill, NJ 07974 USA

A review will be presented utilizing a few properties of linear chain compounds to illustrate some elementary physical ideas related to charge density waves.

I. INTRODUCTION AND EXAMPLES OF LOW DIMENSIONAL MATERIALS

Let us begin by summarizing ten points to be covered in this review. First, to give a specific idea of the sort of material to be discussed, we shall look at a model of the compound TTF(SeCN)$_{.58}$. This charge-transfer salt is an example of a collection of molecular chains that is metallic, then undergoes a distortion and becomes insulating. Second, we shall see how electron spin paramagnetic resonance measurements can determine the location of the mobile charges. Third, we shall present the basic relationship between the wavelength of the chain distortion and the amount of charge that is transferred in the formation of the material. Fourth, we shall go through some of the principles of X-ray scattering measurements and see the way in which they can be used to determine the distortion wavelength. Fifth, we shall enumerate the types of forces between the molecular chains and then, sixth, discuss some results of these couplings, including the observation of a change in the shape of the unit cell. Seventh, we shall discuss basic concepts of fluctuations above a phase transition as they show up in X-ray scattering and, eighth, as they affect the electrical properties and the magnetic susceptibility. The structural studies show, generally, that the fluctuations above the metal-insulator transition are toward the ground state, which is composed of distorted chains that are ordered three dimensionally.

31

L. Alcácer (ed.), The Physics and Chemistry of Low Dimensional Solids, 31–51.
Copyright © 1980 by D. Reidel Publishing Company.

This information suggests that the precursive fluctuations decrease the electrical conductivity and the magnetic suscepti- bility. Finally, we shall discuss the way in which an increase in the energy of the most energetic electrons (making a "better" metal) tends to increase the ordering temperature (making a "better" insulator) or, in other words, "catch 22".

Let us begin by looking at a specific example of a low dimensional material before getting into questions of general principles. Figure 1(a) shows a model of an ensemble of stacks of TTF molecules as they occur in the tetragonal crystal struc- ture[1-3] of TTF(SeCN)$_{.58}$. The point I want to make here is that it is clear from looking at the three-dimensional model and the schematic illustrations in Fig. 1(b) that the material is quasi-one-dimensional as a result of the chain-like arrangement of the molecules. A large number of analogous crystal struc- tures have been determined and discussed.[4-20] Since the material exists in a three-dimensional solid, it is also not surprising that the coupling between chains plays an important role.

MOLECULAR STACKS: SIDE VIEW

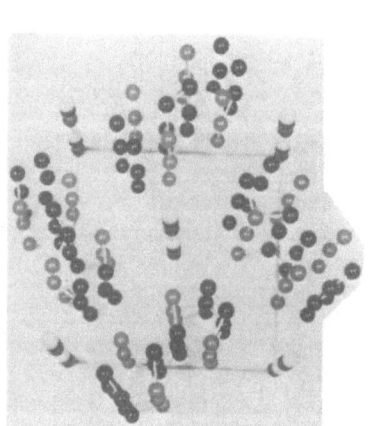
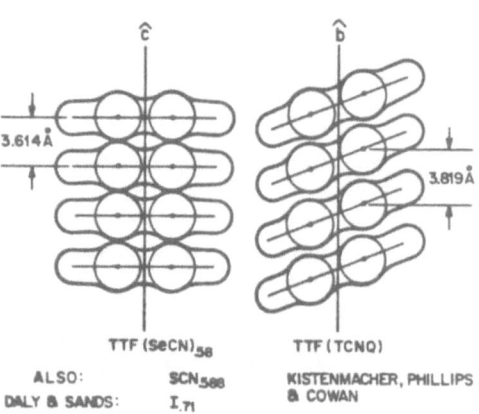

Fig. 1. Left: example of a low dimensional solid: tetragonal TTF(SeCN)$_{.58}$. Stacks of TTF molecules, as in schematic side view, right, exhibit one-dimensional properties. The crystal structures of various TTF compounds have been studied by the authors listed.

As this compound grows from solution into a crystal, there is a charge transfer from the TTF molecules to the SeCN. On chemical grounds, one might guess that this charge transfer was complete in the sense of filling each SeCN molecule. This would leave behind a fraction of a charge for each TTF molecule exactly equal to the fraction of SeCN's compared to TTF's.

II. LOCATION OF CHARGE CARRIERS

The question of where the charges lie between these two chains can be studied further using EPR[21-27] with a result that supports the hypothesis that we would make on chemical grounds. Figure 2 illustrates how we arrive at this conclusion. It is a complicated figure perhaps, but the point that I want to make is very simple: the free electrons are entirely on the TTF chains. We see no EPR signal of the sort that we expect from SeCN. Our expectations are based on similar studies of the spacial asymmetry of the EPR absorption by an unpaired spin on TTF molecules in other compounds. The graph in Fig. 2 shows the EPR absorption signal, actually its derivative with respect to magnetic field, as a function of the applied magnetic field. There are four traces, and I would like to concentrate in particular on the top two. In the first one, curve a, the magnetic field is perpendicular to the chain axis; that is, the direction of the field lies in the plane of the TTF molecule. The small, narrow line that is in this top trace to the right is the absorption due to our free-electron reference, and the large signal that is shaded is shifted in magnetic field substantially. The shift is the same as that in a variety of compounds containing TTF when the magnetic field is oriented in this way. When we tip the magnetic field around so it is parallel to the chain axis, that is, perpendicular to the TTF molecular plane, we see a shift in the TTF resonance to the free-electron position, again as in a variety of TTF compounds we have studied. Curve c, the result for a powder, is just an average over all directions. The reference signal is shown, without a sample, in curve d. The important point is that we do not see a signal characteristic of the SeCN molecules. We would expect a shifted signal in curve b if there were free electrons that could sample the Se environment. We have, then, evidence beyond our chemical intuition that the pseudohalide molecular orbitals are filled and that there is a partially filled band formed by the overlapping orbitals of the TTF molecule.

III. CHARGE TRANSFER AND THE DISTORTION WAVELENGTH

This array of TTF molecules is an excellent example of a

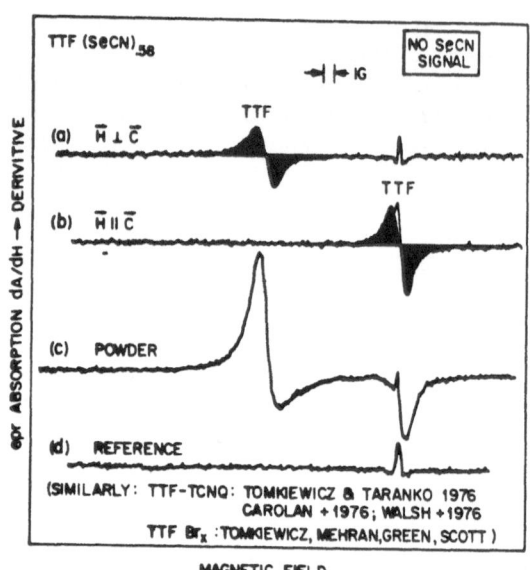

Fig. 2. Example of EPR absorption spectra as a function of magnetic field H for different configurations of H relative to a model one-dimensional system.

coupled set of chains containing free electrons. Such a chain has a tendency to distort,[28-37] that is to have a variation in the spacing between molecules as you go along the chain. Figure 3 illustrates why we expect such a distortion to occur and why the wavelength λ of the distortion is directly related to the amount of charge transfer x, according to

$$\frac{c}{\lambda} = \frac{x}{2} ,$$ (1)

where c is the spacing between TTF molecules. The origin of this relation is the wavelength of the free electrons on the chain. The electron momentum $\hbar k'$ is related to its wavelength by h/λ h being Planck's constant; we often speak of a normalized k vector or electron momentum, that is c/λ, which is dimensionless. The smooth curve in Fig. 3 is the energy as a function of k that is presumably well known[38] for the conduction electrons in a band.

Let us consider the potential existence of a chain distortion, often referred to as a charge density wave or Peierls

WAVE LENGTH OF CHAIN DISTORTION λ

 AMOUNT OF CHARGE TRANSFER X

ELECTRON MOMENTUM : $\hbar k' = h/\lambda$

NORMALIZED : $k = c/\lambda$

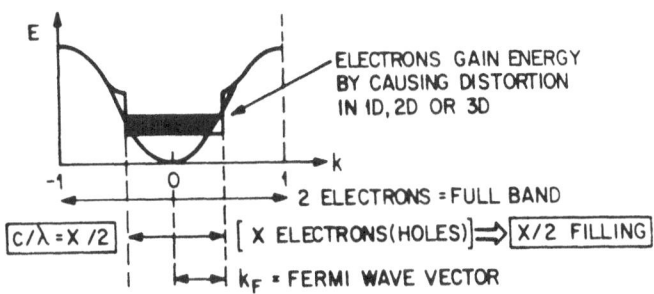

ORGANICS:	c/λ	X/2		
TTF-TCNQ	.295, .590		POUGET, COMES, SHIRANE, SHAPIRO	1976,76
			KHANNA, DENOYER, GARITO, HEEGER	1977,77
			KAGOSHIMA, ISHIGURO, ANZAI	1976
TSeF - TCNQ	.315		WEYL, ENGLER BECHGARD,	1976
HMTSeF - TCNQ	.37		JEHANNO, ETEMAD	
TTF(SCN).588	.294	.588/2	THOMAS, MONCTON, WUDL,	1978
			KAPLAN, LEE	
HMTTF - TCNQ	.36		MEGTERT, POUGET, COMES,	1978
			GARITO, BECHGARD, FABRE,	
			GIRAL	
TTF(SeCN).580	.290 (λ~3.5C)	.580/2	TMWK & L	1979

Fig. 3. Relationship between chain distortion wavelength and amount of charge transfer x.

distortion,[28] causing a gap in this electron spectrum. The electrons always gain energy by causing such a distortion at the right wavelength. This is something that occurs in one dimension, two dimensions, or in three dimensions. In this case we are talking about one particular direction in the material in which the electrons' energy is reduced by an amount equal to the region shaded in Fig. 3. This is the difference between the band of energy that the electrons fill when there is no gap compared to the amount that they fill in the presence of the gap. Recall that in order to fill an entire energy band we usually need to transfer two electrons. If we transfer x electrons, we have a fraction x/2 filling of the band. In the case of TTF, the outermost electron orbitals are filled before the compound is formed and a fraction of an electron per molecule is trans-ferred.[39-48] In this case it is convenient to deal with the small unfilled fraction of the band and describe this part as filled with holes. These holes, in an amount x per TTF molecule, are the carriers of which we shall speak in describing all of the properties to be discussed below. Holes behave precisely as positively charged electrons would.

In order to create a gap as shown in the figure at the energy of the most energetic holes, called the Fermi energy, the lattice distortion must have the same wavelength as these holes. Thus, the wave vector of the distortion is $c/\lambda = x/2$ = the frac-tion of the band filled by holes. This wave vector is also twice the Fermi wave vector, $2k_F$, as shown in Fig. 3. We should note carefully at this point that these relationships apply to the simple case in which two electrons or holes (with opposite spins) can occupy each state in the Fermi distribution. In the case of an extremely strong Coulomb repulsion among the electrons[48-56] the distortion wave vector would be $4k_F$. The compound TTF(SeCN)$_{.58}$ is one example of the simple case.

Distortions of the lattice in chain compounds have been observed in a number of organic materials as listed in the bottom part of Fig. 3. The values for c/λ are shown, along with names of those who have done the work. The second point I want to make, related to Fig. 3, is that we assume in most cases (where I have left blanks in the column under x/2) that we determine the amount of filling from the wave vector c/λ that is observed for the distortion. We do not need to make that assumption in the case of the SeCN and SCN compounds because we can measure independently the fraction of filling through the EPR experiments and chemical considerations, as I have discussed. Let me go on now to the determination of the value of x and the distortion wavelength using X-ray scattering.

IV. SCATTERING MEASUREMENTS

According to Bragg's law, an X-ray (or neutron) beam of wavelength λ_X will be scattered by a periodic array of atoms at an angle θ determined by the spacing between them. The scatter-ing occurs constructively when an integral number n of wavelength makes up the path difference $2d \sin\theta$ of X-rays scattered from different planes of atoms, i.e. $n\lambda_X = 2d \sin\theta$. Thus spots of scattered intensity appear at angles which can be converted, using the known λ_X, to values of the periodicity d of the mole-cules or distortions in the crystal. The common variable used

to display the results is the momentum or wave-vector change in
the scattering event $2\pi/d$.

Figure 4 shows the scattered X-ray intensity along the
chain direction as a function of $2\pi/d$ for TTF(SeCN)$_{.58}$. The
scattering is obtained by rotating a single crystal to measured
angles θ and a detector simultaneously to 2θ with respect to an
incident X-ray beam. Scattering from three periodicities within
the crystal are seen. First, the strong peaks from the TTF
chains at $n = 1,2,\ldots$ are used to define the unit 1 for the scale.
We also see strong peaks at .580 n due to the SeCN chains. These
chains are incommensurate with the TTF ones, that is, .580 is not
a simple fraction. This value is also a measure of the ratio of
TTF to SeCN, i.e. $x = .580$.

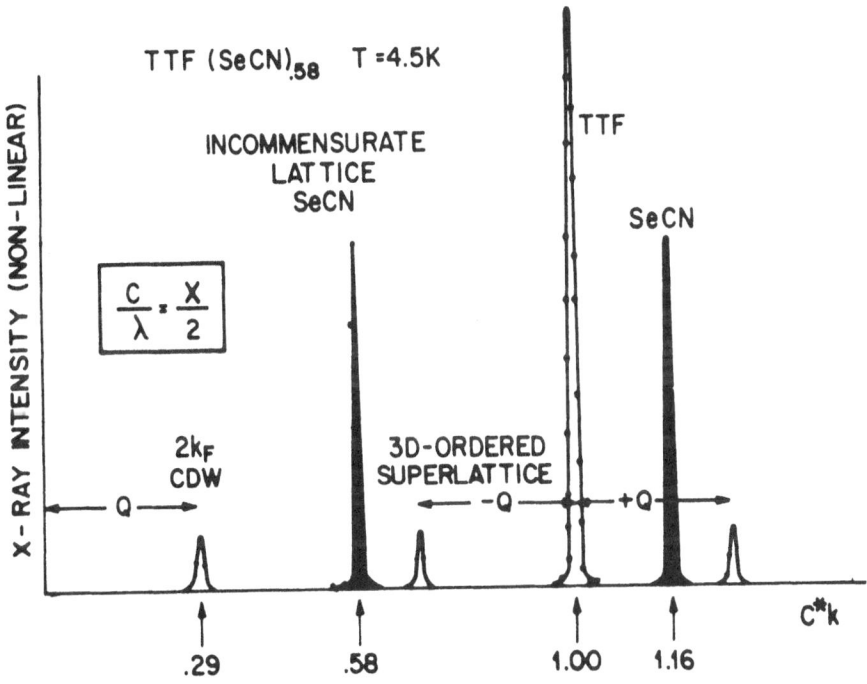

Fig. 4. Schematic presentation of X-ray intensity as a function
of scattering wave vector k.

The third periodicity in Fig. 4 is the most interesting one,
as it arises from the distortion of the TTF chains. The data
shown are taken at low temperatures in the ordered state where
the sharp spots indicate an ordered 3-dimensional distortion
superimposed on the TTF chains. The normalized wave vector of

this superlattice along the chains is .290. This value completes
our illustration of the basic relationship between the amount of
band filling and the distortion wavelength λ. For emphasis, I
have indicated the key equation in the box in Fig. 4. These
results show directly that both $x/2 = .580/2 = .290$ and $c/\lambda =$
.290, within the experimental error.

V.   FORCES BETWEEN CHAINS

     As we mentioned above, there is a three-dimensionally
ordered superlattice at low temperatures. That result, in itself,
is evidence of strong coupling of the chain distortions. However,
studies of the crystal structure have shown us more about the
particular types of forces involved. The example of TTF(SeCN)$_{.58}$
allows us to illustrate one of the types of coupling which in
principle is quite general, but in practice has been observed
only rarely. We shall emphasize this strain coupling. The
central equation is shown in the box in Fig. 5 and relates the
size of the strain $\delta\beta$ to the magnitude of the charge density wave.

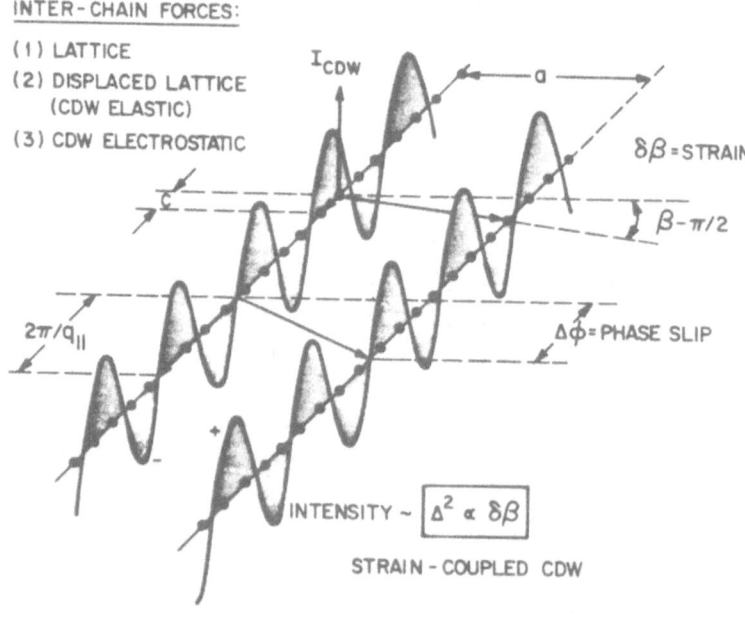

Fig. 5.   Illustration of general forces between chains.

     In general, there are three types of interchain forces,[57-60]
as listed at the top of Fig. 5. One is the simple lattice force
between molecules as they sit in the chains. The second type is

the lattice coupling between the molecules in their displaced positions with the periodicity of the charge density wave. The final type is the electrostatic force between regions of positive or negative charge which builds up in the charge density wave. In the case of most chain-like materials, such as for example TTF-TCNQ, it is sufficient to consider only the electrostatic interactions[57-59] in order to describe the interchain interactions.

However, in general one must consider all three of these forces and in the case of the TTF pseudohalide compounds, SeCN and SCN, we have observed the effects of the strain types of interchain forces. These forces show up because the position of one chain with respect to its neighbor in the absence of the charge density wave distortion is not the same as the ideal interchain orientation in the presence of this distortion. That is, the chains would like to have different slips with respect to each other when the charge density wave forms, and, as you might guess, the extent to which one chain would slip with respect to another is determined by how big the charge density wave is. Qualitatively it is clear that as the intensity of the charge density wave grows, one would expect the interchain slip (measured quantitatively by the strain or the change in the angle β) would grow proportionally. We call this a strain coupled charge density wave.

VI.   THREE-DIMENSIONAL ORDERING AND STRAIN COUPLING

As is well known, an isolated chain will always distort at zero temperature, but only the coupled chains will have a phase transition.[28-37] As a result of the coupling, we can observe static distortions at finite temperatures. Let me go on now to a specific observation of this ordering and the associated coupling of the charge density waves, as illustrated in Fig. 6. The main point that I want to make is the new observation of the strain coupling and the verification of the relationship[60] between the intensity of the charge density wave and the magnitude of the strain. Again, the equation is given, for emphasis in a box in the figure. What is shown in Fig. 6 is a plot of the charge density wave intensity as a function of temperature on the left axis and, on the right axis with the same temperature scale, the splitting of the (002) X-ray peak, which is a direct measure of the angle β and thus of the slip between the chains. First note, again, that at low temperatures there is long-range order in the chain distortion, that is, sharp X-ray spots arising from the formation of three-dimensional charge density waves. The crystal structure under these conditions is monoclinic.

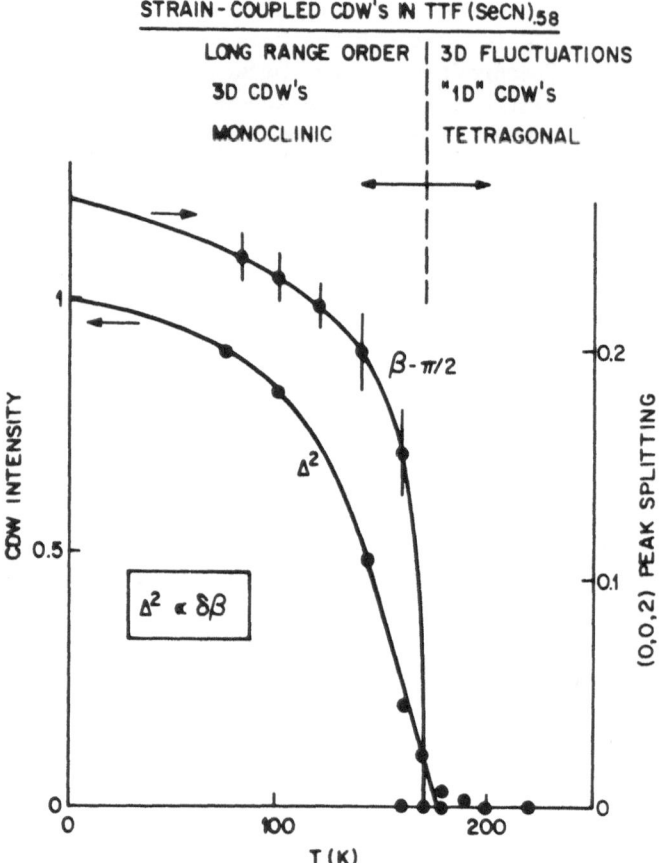

STRAIN - COUPLED CDW's IN TTF (SeCN)$_{.58}$

LONG RANGE ORDER	3D FLUCTUATIONS
3D CDW's	"1D" CDW's
MONOCLINIC	TETRAGONAL

Fig. 6. Example of effects of forces between chains: growth of CDW intensity and intrinsic strain as a function of temperature.

There is a well-defined phase transition temperature and above that temperature there are fluctuations in the charge density toward the static low temperature state. The crystal structure of TTF(SeCN)$_{.58}$ in this temperature region is tetragonal. At lower temperatures, as you can see from the graph, the magnitude of the intensity of the charge density wave signal grows, that is Δ^2 increases. There is also a change in the angle β; the chains slip with respect to each other by an increasingly large amount. It is clear from the graph that there is a direct relationship between the magnitude of δ^2 and the changes in the angle β. The details of the transition point, that is the temperature region where the static three-dimensional order of the charge density waves goes to zero in this system, has not been fully investigated as yet. I would like to shift, instead, to a discussion of the region in which there are fluctuations.

VII. FLUCTUATIONS

The point that I want to emphasize here is that above the
ordering temperature there are fluctuations toward the ordered
charge density wave state. As in the case of two-dimensional
materials,[61] it is useful to keep in mind the distinction that
a charge density wave is a static lattice distortion and that the
fluctuations are low energy phonons which have a special character in that they have the same principal wavelength as the static
distortion into which they condense. The fluctuations become
increasingly one dimensional as the temperature is raised away
from T_c: the coherence between chains decreases as the amplitude
of the fluctuations decreases. Generally the fluctuations retain
some three-dimensional character which we measure quantitatively
as a coherence length. This length is inversely proportional
to the width of the X-ray scattering peak (with the resolution
function of the experimental apparatus accounted for separately).

To illustrate these fluctuations I have shown in Fig. 7 the
X-ray scattering intensity as a function of the scattering wave
vector at a temperature above the ordering point. The shaded
region is the peak that occurs at the same wave vector as the
three-dimensionally ordered scattering that we see at lower
temperatures: i.e., at a reduced wave vector of .290 relative
to the scattering from the TTF. Experimentally, as we scan with
the X-rays perpendicular to the chain axis direction, we find the
pattern shown in the inset. The plot is scattering intensity
versus the perpendicular wave vector, $k_\perp/a*$ (again normalized).
Clearly there is a peak in this direction, indicating that the
fluctuations involve some coupling between chains, although the
peak is many times broader in this transverse direction, indicating a one-dimensional nature. This pattern, with a long
coherence length in the chain direction and a short one between
chains, is common to the chain structure compounds.

In this fluctuating state, the time averaged position of the
TTF molecules is such that all are in their equilibrium position,
i.e. with no distortion. However, if we take a snapshot at any
instant of time, the molecules will be displaced in a way that
exhibits some remnant of the order that will become static at
lower temperatures.

VIII. ELECTRICAL CONDUCTIVITY AND MAGNETIC SUSCEPTIBILITY

These fluctuations are related to other properties that we
observe, and two of these properties are shown in Fig. 8. Here,
I have drawn squares again around the principal points that I
wish to emphasize in the figure. First, in the ordered state
there is an electronic energy gap which provides a good description

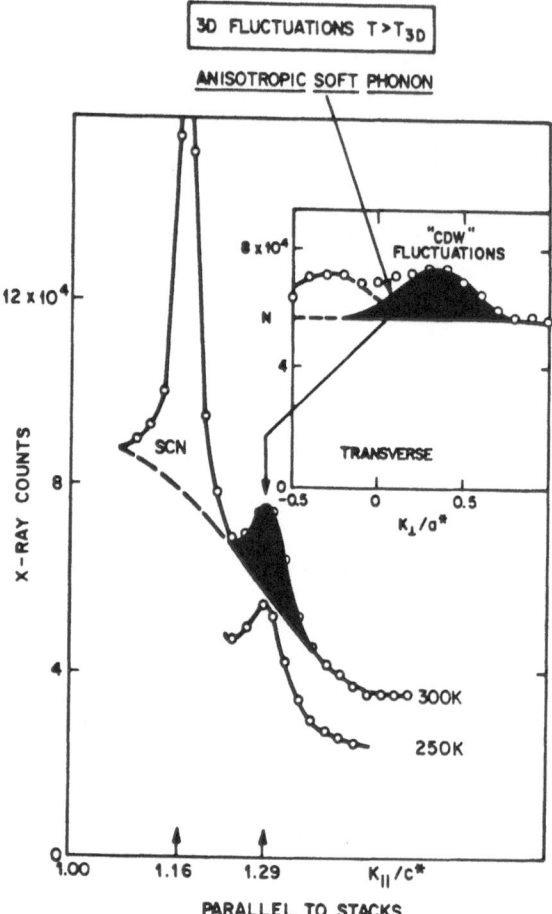

Fig. 7. Example of fluctuations above the ordering temperature in a linear chain compound. X-ray intensity as a function of wave vector is shown along the chain direction and, in the inset, perpendicular to it.

qualitatively, and quantitatively to some extent, of the electrical conductivity and the magnetic susceptibility. These two variables are shown as a function of temperature in the figure. The other point is that above the ordering temperature. The fluctuations change the properties in the direction of the ordered state. We have seen clearly that these fluctuations act in this way in the X-ray scattering. In the case of the conductivity, the same modification occurs, that is fluctuations which increase the resistivity.[62-68] In general, above the ordering temperature, the system fluctuates toward the ordered state; in this case the ordered state is an insulator.

We define the ground state as an insulator because there is

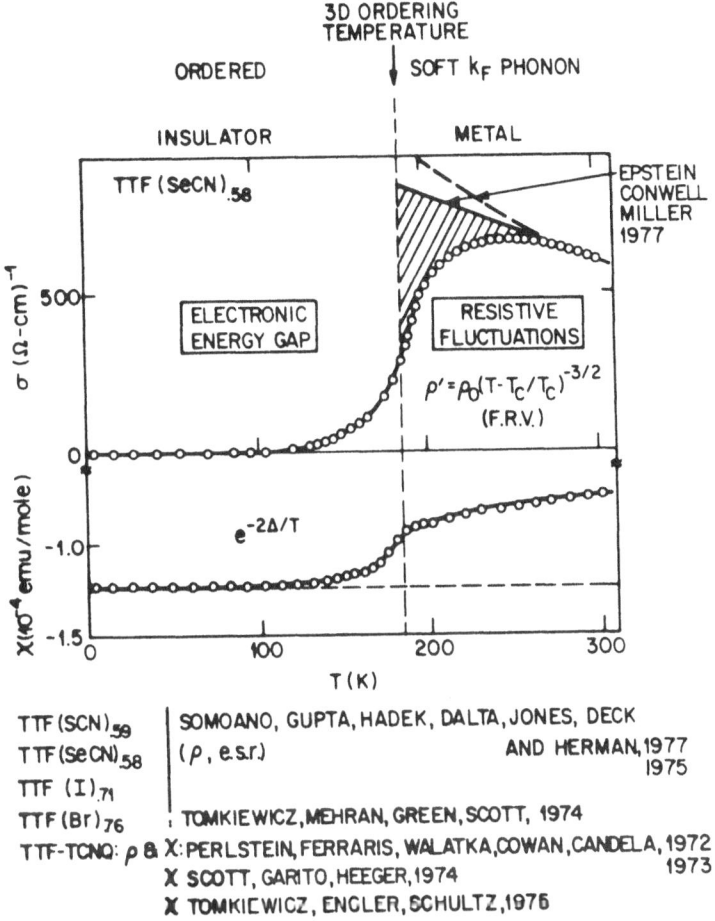

Fig. 8. Example of electrical conductivity σ and magnetic susceptibility χ as a function of temperature in a linear chain compound. Other work is listed below the graph.

an energy gap at the Fermi energy. Finite energy is required to excite an electron into a state where it can move freely (or in which it can become unpaired and contribute to the magnetic susceptibility). In the sense that the energy gap is small, a material like TTF(SeCN)$_{.58}$ is a semiconductor. Based on these data, we don't have a quantitative measure of the functional form of the conductivity due to the resistive fluctuations because we can't determine the conductivity in their absence. However, it is clear that the conductivity is dropping above the ordering temperature and that there is a large reduction from some "normal" conductivity behavior shown by the region shaded in the figure. The "normal" behavior has been discussed by a number of authors[69-72] as has the form of the resistive fluctuations.[78-79]

Theoretically, the added resistivity, ρ', varies as a constant times the temperature difference from the transition temperature to the minus three halves.

The magnetic susceptibility, I think, is less well understood in terms of the behavior of the fluctuations above the transition temperature. In both the magnetic susceptibility and the conductivity, the behavior below the transition is characterized by excitations over the energy gap which produce the signal that we observe. I have listed, at the bottom of Fig. 8, some systems where measurements[81-92] similar to those shown in the figure have been made. I would particularly like to emphasize the very nice work[2] on the same pseudohalide systems that I have used as examples. There have also been studies of a very similar nature on the susceptibility of the bromine compound of TTF, as listed, and by a large number of workers on TTF-TCNQ. In these other materials the behavior we have stressed here occurs generally: that is, resistive fluctuations and a drop in the susceptibility as we approach the ordering temperature with an electronic energy gap below T_c.

IX.   THE ORDERING TEMPERATURE AND THE FERMI ENERGY

The final point that I would like to make is what we know about the factors that determine the magnitude of the three-dimensional ordering temperature.[3] One might naturally expect that in a system with ordering that is driven by the most energetic conduction electrons (those at the Fermi energy E_F), the ordering temperature T_c would scale with the Fermi energy. The key relationship is shown inside a box in Fig. 9, and again we shall not go into its derivation;[93-98] E_F is simply the characteristic scaling energy that we expect in these systems. We write the equation using E_F^* to incorporate the proportionality constant in the relationship into E_F. The quantity λ_p in the exponent is related to the strength of the coupling between the electrons and the lattice (it is dimensionless, not a wavelength). This equation applies, in principle, to one, two or three dimensions. It is an equation that is called the McMillan equation[99] in the case in which it relates the superconducting transition temperature to the Debye frequency and an analogous exponential factor. In the case of the spin-Peierls system,[100-102] a similar equation also describes the transition temperature.

Let us consider the limited evidence we have supporting this fundamental relationship between the ordering temperature and the Fermi energy for the system of TTF chains. Figure 9 shows two quantities as a function of the spacing between sulfur atoms along the TTF chains. Data points are given for four compounds containing TTF chains, the iodine, the SeCN, the SCN, and the

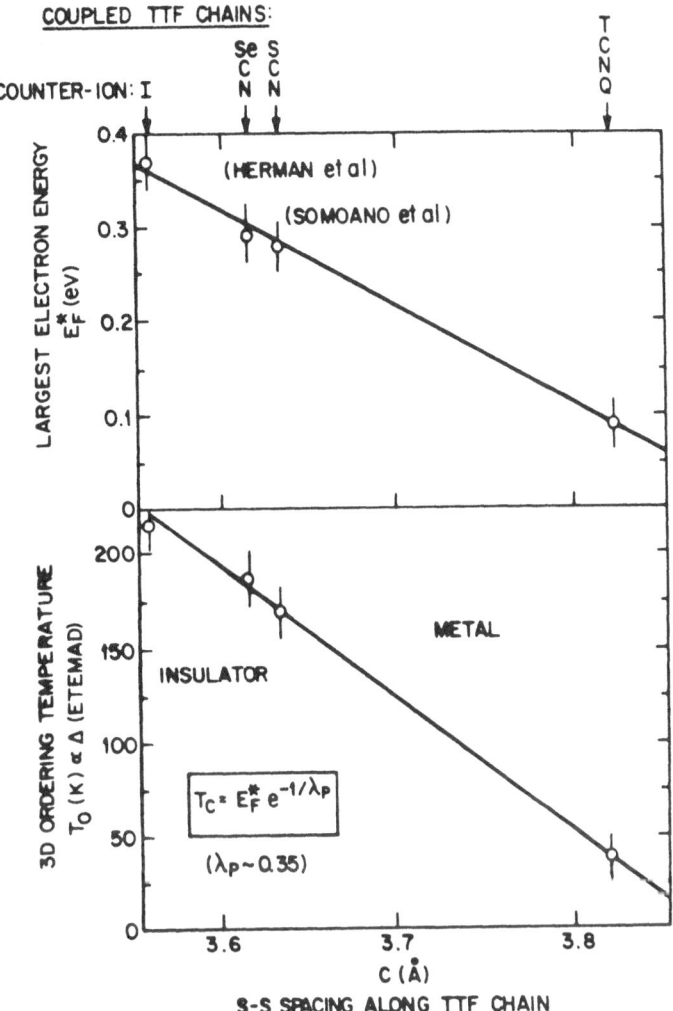

Fig. 9. Examples of the variation in characteristic energies of a series of chain compounds with different spacings along the chains.

TCNQ salts. In the top half is the Fermi energy which can be estimated in a variety of ways.[32-37] The error bars are mostly systematic ones related to what method one uses to determine the Fermi energy of the carriers on the TTF chain. The trend is the same regardless of the experimental approach. Theoretical calculations of E_F have also been carried out[45-47] and the measurements are in reasonable agreement. The results in Fig. 9 show exactly what you would expect intuitively. That is, as you move the TTF molecules closer together the band width (and the Fermi energy in this case) becomes larger.

In the bottom half of Fig. 9, the corresponding variation of the temperature is plotted. The ordering temperatures can be determined from the knee in the magnetic susceptibility or sometimes from the peak in the derivative of the conductivity or from the temperature at which the ordered X-ray scattering intensity goes to zero. The value of T_c is a little more complicated in the case of the TCNQ compound. What I have plotted is the transition temperature of the TTF chain at 38 K as we deduce it from ESR studies. In all cases we wish to consider the TTF chains and we see that for a small sulfur-sulfur spacing along the chain we find a large band width and a high transition temperature. In fact, if you plot E_F^* as a function of T_c you get a straight line, as we would expect from the relationship between T_c and E_F^*. If we assume $E_F = E_F^*$, we obtain a value of $\lambda_p = .35$ for all of these compounds. It may be, to some extent, an accident that λ_p is approximately constant, but the constant λ_p may occur because it is primarily determined by the interaction between the conduction electrons and the TTF molecules which are common to all these compounds. In addition, the amount of band filling is comparable in these salts.

Now, I would like to mention also that similar behavior occurs for the energy gap Δ in the insulating state. Generally we expect that the ordering temperature will scale with the energy gap and therefore also with the Fermi energy. The scaling between T_c and Δ is something that has been discussed by a number of workers, particularly by Etemad.[103]

The bottom section of Fig. 9 is a phase diagram in the temperature density plane. In principle, if we put together a TTF chain in a compound with similar coupling, at a given chain spacing, we get an insulator, provided we are in the region marked insulator or a metal in the region marked metal. We have referred to this result as "catch 22": a better quasi-one-dimensional metal tends to become insulating at a higher temperature.

With the results of Fig. 9 in mind, I would like to summarize. There is a coupling between chains which can produce a number of effects. One of these is a change in shape of the crystal's unit cell, another is a three-dimensional ordering. The characteristic wavelength of the distortion is directly related to the most energetic conduction electrons, and this Fermi energy is related, in turn, to the temperature where the distortion occurs.

I would like to acknowledge with thanks my collaborators. First, and perhaps foremost, is Fred Wudl, who with Marty Kaplan has made the samples, without which we could not, of course, have done our experiments. Next is Dave Moncton with whom I have done the X-ray scattering measurements. Some of the theoretical work I have discussed has been done with Patrick Lee. Some of the early experiments on electrical conductivity were done with Dave Schafer. The EPR results have come from a collaboration with Mickie Walsh and Lou Rupp, and a number of the measurements of conductivity, susceptibility, and scattering have been aided substantially by Jan Mock. The magnetic susceptibility measurements have been made in collaboration with Frank DiSalvo.

REFERENCES

1. F. Wudl, D. E. Schafer, W. M. Walsh, Jr., L. W. Rupp, Jr., F. J. DiSalvo, J. V. Waszczak, M. L. Kaplan, and G. A. Thomas, J. Chem. Phys. *66*, 377 (1977).

2. R. B. Somoano, A. Gupta, V. Hadek, M. Novotny, M. Jones, T. Datta, R. Deck, and A. M. Hermann, Phys. Rev. B *15*, 595 (1977).

3. G. A. Thomas, F. Wudl, F. J. DiSalvo, W. M. Walsh, Jr., L. W. Rupp, Jr., and D. E. Schafer, Solid State Commun. *20*, 1009 (1976).

4. R. Comès, M. Lambert, H. Launois, and H. R. Zeller, Phys. Rev. B *8*, 571 (1973).

5. K. Krogmann, Angew. Chem. Int. Ed. Engl. *8*, 35 (1969).

6. For reviews see G. Shirane, Rev. Mod. Phys. *46*, 437 (1974); J. D. Axe, Trans. ACA, Feb. 1 (1971); Proc. Conf. Neutron Scattering, Gatlinburg (1976) (ORNL Conf. 760601 P1); B. Dorner and R. Comès, in *Dynamics of Solids and Liquids by Neutron Scattering*, T. Springer (ed.) [Springer, Berlin (1977)].

7. J. Harada and G. Honjo, J. Phys. Soc. Jpn. *22*, 45 (1967).

8. R. Comès, M. Lambert, and A. Guinier, C. R. Acad. Sci. Paris *266*, 959 (1968).

9. F. Denoyer, R. Comès, and M. Lambert, Solid State Commun. *8*, 1979 (1970).

10. R. Currat, R. Comès, B. Dorner, and E. Wiesendanger, J. Phys. C 7, 252 (1974).

11. T. J. Kistenmacher, T. E. Phillips, and D. O. Cowan, Acta Crystallogr. Sect. B *30*, 763 (1974).

12. R. H. Blessing and P. Coppens, Solid State Commun. *15*, 215 (1974).

13. A. J. Schultz, G. D. Stucky, R. Craven, M. J. Schaffman, and M. B. Salamon, J. Am. Chem. Soc. *98*, 5191 (1976).

14. A. J. Schultz and G. D. Stucky, J. Am. Chem. Soc. *98*, 3194 (1976).

15. F. Denoyer, R. Comès, A. F. Garito, and A. J. Heeger, Phys. Rev. Lett. *35*, 445 (1975).

16. S. Kagoshima, H. Anzai, K. Kajimura, and T. Ishiguro, J. Phys. Soc. Jpn. *39*, 1143 (1975).

17. G. Shirane, S. M. Shapiro, R. Comès, A. F. Garito, and

A. J. Heeger, Phys. Rev. B *14*, 2325 (1976); S. M. Shapiro,
G. Shirane, A. F. Garito, and A. J. Heeger, Phys. Rev. B
15, 2413 (1977).

18. J. P. Pouget, S. K. Khanna, F. Denoyer, R. Comès, A. F.
    Garito, and A. J. Heeger, Phys. Rev. Lett. *37*, 437 (1976);
    S. K. Khanna, J. P. Pouget, R. Comès, A. F. Garito, and
    A. J. Heeger, Phys. Rev. B *16*, 1468 (1977).

19. W. D. Ellenson, R. Comès, S. M. Shapiro, G. Shirane, A. F.
    Garito, and A. J. Heeger, Solid State Commun. *20*, 53 (1976);
    W. D. Ellenson, S. M. Shapiro, F. Shirane, and A. F.
    Garito, Phys. Rev. B *16*, 3244 (1977).

20. S. Kagoshima, T. Ishiguro, and H. Anzai, J. Phys. Soc. Jpn.
    41, 2061 (1976).

21. W. M. Walsh, Jr., L. W. Rupp, Jr., D. E. Schafer, and
    G. A. Thomas, Bull. Am. Phys. Soc. *19*, 296 (1974).

22. Y. Tomkiewicz, E. M. Engler, and T. D. Schulz, Phys. Rev.
    Lett. *35*, 456-459 (1975).

23. Y. Tomkiewicz, A. R. Taranko, and J. B. Torrance, Phys. Rev.
    Lett. *36*, 751-754 (1976).

24. Y. Tomkiewicz, A. R. Taranko, and E. M. Engler, Phys. Rev.
    Lett. *37*, 1705-1708 (1976).

25. Y. Tomkiewicz, B. Welber, P. E. Seiden, and R. Schumaker,
    Solid State Commun. *23*, 471-475 (1977).

26. Y. Tomkiewicz, A. R. Taranko, and J. B. Torrance, Phys. Rev.
    B *15*, 1017-1023 (1977).

27. D. Jérome, G. Soda, J. R. Cooper, J. M. Fabre, and L.
    Giral, Solid State Commun. *22*, 319-325 (1977).

28. R. E. Peierls, *Quantum Theory of Solids,* Clarendon, Oxford
    (1964).

29. H. Frölich, Proc. R. Soc. London A *223*, 296 (1954).

30. P. A. Lee, T. M. Rice, and P. W. Anderson, Phys. Rev. Lett.
    31, 462 (1973).

31. P. A. Lee, T. M. Rice, and P. W. Anderson, Solid State
    Commun. *14*, 703 (1974).

32. *Lecture Notes in Physics,* Vol. 34, *One-Dimensional Con-
    ductors,* Springer, New York (1975).

33. *Low Dimensional Cooperative Phenomena,* H. J. Keller (ed.),
    Plenum Press, New York (1975).

34. *Chemistry and Physics of One-Dimensional Metals,* H. J.
    Keller (ed.), Plenum Press, New York (1977).

35. *Proc. Conf. Organic Conductors and Semiconductors,* Siofok,
    Hungary (1976), in *Lecture Notes in Physics,* Vol. 65,
    Organic Conductors and Semiconductors, Springer, New York
    (1977).

36. L. N. Bulaevskii, Usp. Fiz. Nauk. *115*, 263 (1975).

37. A. J. Berlinsky, Contemp. Phys. *17*, 331 (1976).

38. C. Kittel, *Introduction to Solid State Physics* (Wiley,
    New York, 1976).

39. F. Wudl, G. M. Smith, and E. J. Hufnagel, J. Chem. Soc.
    Chem. Commun. (1970), 1425-1436.

40. D. L. Coffen, J. Q. Chambers, D. R. Williams, P. E. Garrett, and N. D. Canfield, J. Am. Chem. Soc. *93*, 2258-2268 (1971).
41. D. L. Coffen, Tetrahedron. Lett. (1970), 2633-2635.
42. S. Hünig, G. Küsslich, H. Quast, and D. Scheutzow, Justus Liebigs Ann. Chem. (1973), 310-323.
43. M. G. Miles, J. D. Wilson, and M. H. Cohen, U. S. Patent No. 3,779,814, December 18 (1973).
44. J. Ferraris, D. O. Cowan, V. Walatka, Jr., and J. H. Perlstein, J. Am. Chem. Soc. *95*, 948-949 (1973).
45. F. Herman, D. R. Salahub, and R. P. Messmer, Phys. Rev. B *16*, 2453-2465 (1977).
46. F. Herman, Physica Scripta *16*,303-306 (1977).
47. A. J. Berlinsky, J. F. Carolan, and L. Weiler, Solid State Commun. *15*, 795 (1974).
48. J. B. Torrance and B. D. Silverman, Phys. Rev. B *15*, 788 (1977).
49. V. J. Emery, Phys. Rev. Lett. *37*, 1227 (1976).
50. J. B. Torrance, in *Chemistry and Physics of One-Dimensional Metals*, H. J. Keller (ed.), Plenum, New York (1977).
51. P. A. Lee, T. M. Rice, and R. A. Klemm, Phys. Rev. B *15*, 2984 (1977).
52. J. Kondo and K. Yamaji, J. Phys. Soc. Jpn. *43*, 424 (1977).
53. J. Hubbard, Phys. Rev. B *17*, 494 (1978).
54. A. A. Ovchinnikov, Soc. Phys. JETP *37*, 176 (1973).
55. J. Bernasconi, M. J. Rice, W. R. Schneider, and S. Strässler, Phys. Rev. B *12*, 1090 (1975).
56. J. B. Torrance, Phys. Rev. B *17*, 3099 (1978).
57. P. Bak and V. J. Emery, Phys. Rev. Lett. *36*, 978 (1976).
58. P. A. Lee and H. Fukuyama, Phys. Rev. B *17*, 542 (1978).
59. P. Bak, Phys. Rev. Lett. *37*, 1071 (1976).
60. G. A. Thomas, D. E. Moncton, F. Wudl, M. Kaplan, and P. A. Lee, Phys. Rev. Lett. *41*, 486 (1978).
61. F. J. DiSalvo, Jr. and T. M. Rice, Physics Today *32*, No. 4, 32 (1979).
62. For a review, see R. D. Parks, AIP Conf. Proc. *5*, 630-638 (1972).
63. See, for example, F. C. Zumsteg, and R. D. Parks, Phys. Rev. Lett. *24*, 520-524 (1970); J. W. Shacklette, Phys. Rev. B *9*, 3789-3792 (1974).
64. R. A. Craven and R. D. Parks, Phys. Rev. Lett. *31*, 383-386 (1973).
65. Y. Suezaki and H. Mori, Prog. Theor. Phys. *41*, 1177-1189 (1969); and Phys. Lett. *28A*, 70-71 (1968).
66. P. M. Horn and D. Rimai, Phys. Rev. Lett. *36*, 809-813 (1976).
67. P. M. Horn and D. Guidotti, Phys. Rev. B *16*, 491-501 (1977).
68. D. Guidotti, P. M. Horn, and E. M. Engler, in *Organic Conductors and Semiconductors, Lecture Notes in Physics*, No. 65, L. Pál, G. Gruner, A. Jánossy, and J. Sólyom (eds.) Akadémiai Kiadó, Budapest, and Springer-Verlag, Berlin (1977).

69. E. M. Conwell, Phys. Rev. Lett. *39,* 777-780 (1977).

70. A. J. Epstein, E. M. Conwell, D. J. Sandman, and J. S.
     Miller, Solid State Commun. *23,* 355-358 (1977).

71. A. J. Epstein and E. M. Conwell, Solid State Commun. *24,*
     627-631 (1977).

72. A. J. Epstein, E. M. Conwell, and J. S. Miller, in
     *Conference on Synthesis and Properties of Low-Dimensional
     Materials, 1977,* Ann. N.Y. Acad. Sci., New York (1978).

73. G. A. Thomas, D. E. Schafer, F. Wudl, P. M. Horn, D. Rimai,
     J. W. Cook, D. A. Glocker, M. J. Skove, C. W. Chu, R. P.
     Groff, J. L. Gillson, R. C. Wheland, L. R. Melby, M. B.
     Salamon, R. A. Craven, G. DePasquali, A. N. Bloch, D. O.
     Cowan, V. V. Walatka, R. E. Pyle, R. Gemmer, T. O. Poehler,
     G. R. Johnson, M. G. Miles, J. D. Wilson, J. P. Ferraris,
     T. F. Finnegan, R. J. Warmack, V. F. Raaen, and D. Jérome,
     Phys. Rev. B *13,* 5105-5110 (1976).

74. R. P. Groff, A. Suna, and R. E. Merrifield, Phys. Rev.
     Lett. *33,* 418-421 (1974).

75. J. P. Ferraris and T. F. Finnegan, Solid State Commun. *18,*
     1169-1172 (1976).

76. P. E. Seiden and D. Cabib, Phys. Rev. B *13,* 1846-1849 (1976).

77. H. Gutfreund and M. Weger, Phys. Rev. B *16,* 1753-1755 (1977).

78. H. Fukuyama, T. M. Rice, C. M. Varma, and B. I. Halperin,
     Phys. Rev. B *10,* 3775 (1974).

79. H. Fukuyama, T. M. Rice, and C. M. Varma, Phys. Rev. Lett.
     33, 305 (1974).

80. Marshall J. Cohen, L. B. Coleman, A. F. Garito, and A. J.
     Heeger, Phys. Rev. B *10,* 1298-1307 (1974).

81. S. Etemad, T. Penney, E. M. Engler, B. A. Scott, and P. E.
     Seiden, Phys. Rev. Lett. *34,* 741-744 (1975); S. Etemad,
     E. M. Engler, T. D. Schultz, T. Penney, and B. A. Scott,
     Phys. Rev. B *17,* 513-528 (1978).

82. W. N. Hardy, A. J. Berlinsky, and L. Weiler, Phys. Rev. B
     14, 3356-3370 (1976).

83. R. A. Craven, Y. Tomkiewicz, E. M. Engler, and A. R.
     Taranko, Solid State Commun. *23,* 429-433 (1977).

84. Marshall J. Cohen, L. B. Coleman, A. F. Garito, and A. J.
     Heeger, Phys. Rev. B *13,* 5111-5116 (1976).

85. J. C. Scott, A. F. Garito, and A. J. Heeger, Phys. Rev. B
     10, 3131-3139 (1974).

86. J. E. Gulley and J. F. Weiher, Phys. Rev. Lett. *34,* 1061-
     1064 (1975).

87. S. Etemad, T. Penney, and E. M. Engler, Bull. Am. Phys.
     Soc. *20,* 496 (1975).

88. J. C. Scott, S. Etemad, and E. M. Engler, Phys. Rev. B
     17, 2269-2275 (1978).

89. R. A. Craven, M. B. Salamon, G. DéPasquali, R. M. Herman,
     G. Stucky, and A. Schultz, Phys. Rev. Lett. *32,* 769-772
     (1974).

90. M. B. Salamon, J. W. Bray, G. DePasquali, R. A. Craven,

G. Stucky, and A. Schultz, Phys. Rev. B *11*, 619-622 (1975).

91. P. M. Horn, R. M. Herman, and M. B. Salamon, Phys. Rev. B *16*, 5012-5015 (1977).

92. A. N. Bloch, T. F. Carruthers, T. O. Poehler, and D. O. Cowan, in *Chemistry and Physics of One-Dimensional Metals NATO Advanced Study Institutes Series (Physics)*, Vol. 25, H. J. Keller (ed.), Plenum Press, New York (1977), and private communication.

93. D. Allender, J. W. Bray, and J. Bardeen, Phys. Rev. B *9*, 119-129 (1974).

94. B. R. Patton and L. J. Sham. Phys. Rev. Lett. *33*, 638-641 (1974).

95. L. J. Sham and B. R. Patton, *Proc. German Physical Society Summer School on One-Dimensional Conductors, Saarbrucken, July 1974*, H. G. Schuster (ed.), Springer-Verlag, Berlin (1975).

96. P. A. Lee, T. M. Rice, and P. W. Anderson, Solid State Commun. *14*, 703-709 (1974).

97. A. Luther and V. J. Emery, Phys. Rev. Lett. *33*, 589 (1974).

98. P. A. Lee, Phys. Rev. Lett. *34*, 1247 (1975); S.-T. Chui and P. A. Lee, Phys. Rev. Lett. *34*, 315 (1975).

99. W. L. McMillan, Phys. Rev. *167*, 331 (1968).

100. J. S. Jacobs, J. W. Bray, H. R. Hart, Jr., L. V. Interrante, J. S. Kasper, G. D. Watkins, D. E. Prober, and J. C. Bonner, Phys. Rev. B *14*, 3036 (1976).

101. J. S. Jacobs, H. R. Hart, Jr., L. V. Interrante, J. W. Bray, J. S. Kasper, G. D. Watkins, D. E. Prober, W. P. Wolf, and J. C. Bonner, Physica B *86*, 655 (1977).

102. L. V. Interrante, J. W. Bray, H. R. Hart, Jr., J. S. Kasper, and P. A. Piacente, J. Am. Chem. Soc. *99*, 3523 (1977).

103. S. Etemad, Phys. Rev. B *13*, 2254-2261 (1976).

TRANSPORT IN QUASI-ONE-DIMENSIONAL SOLIDS

Paul M. Chaikin

Dept. of Physics, Univ. of Cal., Los Angeles, CA 90024

The aim of this tutorial lecture is to present an intuitive picture of the basic transport properties of solids with particular application to quasi-one-dimensional solids. After a series of elementary derivations, useful equations are given for increasingly complex situations. The experimental data for several compounds investigated in this field are compared with these equations in order to illustrate characteristic behaviors and show what transport experiments can tell us about these materials.

I. INTRODUCTION

The subject of transport properties of solids is too vast to do it justice in a short lecture or article.[1] The basic transport coefficients which we will deal with are: conductivity (σ), thermoelectric power (S), thermal conductivity (κ), Hall coefficient (R_H) and magneto-resistance $\Delta\rho(H)/\rho$. Although all of these measurements are important only two, the conductivity and thermopower have been extensively investigated in the field of quasi-one-dimensional conductors. This gives some indication of the relative difficulty of these measurements on fragile, small, anisotropic samples.

II. DRUDE THEORY

Most of our present understanding of transport stems from the work of Drude,[2] who developed his theory from the kinetic theory of gases. Although in detail this theory is hard pressed to account for observed properties, due to its simplifying assumptions, it is essentially the way that scientists visualize and describe the transport processes. The model assumes that we can

L. Alcácer (ed.), The Physics and Chemistry of Low Dimensional Solids, 53–75.
Copyright © 1980 by D. Reidel Publishing Company.

treat the solid as if it were a box filled with a certain density of particles which are in a gaslike state and have charge, mass, and specific heat.

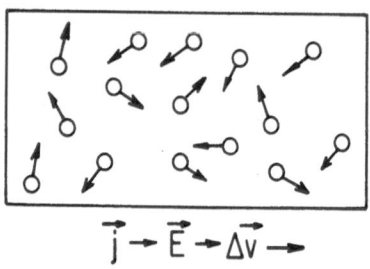

$$\vec{j} \rightarrow \vec{E} \rightarrow \vec{\Delta v} \rightarrow$$

Fig. 1. Particles with mass m and charge q in a box. Arrows indicate random velocities due to thermal motion with $\vec{E} = 0$. When $\vec{E}$ is applied particles gain additional velocity $\vec{\Delta v}$.

The electric conductivity is defined by the ratio of current density to electric field or $\vec{j} \equiv \sigma \vec{E}$. If we have a box of charged particles as shown in Fig. 1 there is a random thermal motion of the particles with zero average velocity and hence no current, before the electric field is applied. When the electric field is applied each particle accelerates and the average velocity is not zero. We now come to a basic point of the theory. We assume that there is an average collision time τ such that a particle accelerates freely up to this time, then it is scattered and has (on the average) no velocity or momentum immediately after the scattering. The equations are:

$$m \frac{\partial \vec{v}}{\partial t} = q\vec{E} , \qquad <\vec{\Delta v}> = qE\tau/m$$

$$\vec{j} \equiv nq<\vec{v}> = \frac{nq^2\tau}{m} \vec{E} \qquad\qquad\qquad\qquad \text{I}$$

$$\sigma = nq^2\tau/m$$

where we have used the definition of current as particle current times charge per particle (with particle current = density x<v> and <v> = average initial velocity + <Δv>). Since the scattering processes must take momentum out of the system (to slow the particles down) particle-particle scattering is not effective. The collisions must involve scattering off "something else" which will transfer momentum out of the box.

The thermal conductivity is defined as the ratio between the heat current and the temperature gradient $U \equiv -\kappa \vec{\nabla} T$. In Fig. 2 we imagine that we have the same box of particles as before, but now in the presence of a temperature gradient, such that the right

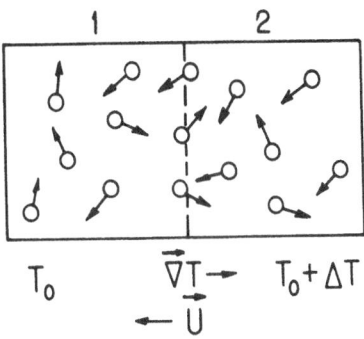

Fig. 2. Same box of particles in a temperature gradient. Heat flows from hot side (2) to cool side (1) although there is no net particle flow.

hand side (2) is warmer than the left hand side (1), to calculate the heat current (which is the particle current times the heat per particle) we construct an imaginary boundary between these two sides. In fact, in this situation there is no particle current since an equal number of particles per second pass from right to left as return. However, a particle going from 2 to 1 takes heat out of 2 as does a particle going from 1 to 2 (it must absorb heat in 2 in order to come to equilibrium with it). Therefore, the calculation proceeds by looking at the particle and heat flow from 2 to 1 and doubling it.

The particle current crossing the imaginary boundary from 2 to 1 is the average (rms) velocity along the gradient, times 1/2 n, since at any point half the particles are going left and half right. The amount of heat transported across the barrier by each particle is its specific heat (c) times the temperature difference (ΔT). The particles are assumed to thermalize at each collision giving as ΔT a mean free path (ℓ) times ($\vec{\nabla}T$). Putting these factors together we have

$$U_{2-1} = \frac{n}{2} <-v_x> \times c\Delta T = - \frac{n}{2} <v_x> c\ell_x \partial T/\partial x$$

$$U = 2 U_{2-1} = nc <v_x^2> \tau(-\partial T/\partial x) \qquad\qquad \text{II}$$

$$\kappa = \frac{1}{3} C v\ell$$

where we have taken ℓ_x as $v_x \tau$ and set $v_x^2 = v^2/3$. This general result for the thermal conductivity in terms of the total specific heat (C), the velocity and the mean free path is what we will use for several different situations.

It is interesting to note that if we take the classical approximations $C \sim 3nk_B$, $mv^2 \sim 3k_BT$, the value for κ is: $3k_B^2T(n\tau/m)$. All of the parameters which characterize a particular material (n, τ, and m) appear here in exactly the same ratio as in the electrical conductivity. If we divide by σ we obtain only fundamental constants times the temperature

$$\frac{\kappa}{\sigma} = 3 \frac{k_B^2}{e^2} \equiv L_c T \qquad\qquad\qquad III$$

This is known as the Wiedemann-Franz law (L_c is the classical Lorenz number) and played a vital role in showing that the basic idea of the Drude theory was correct. It essentially tells us the number of degrees of freedom per charge and indicates that the transport is proceeding by single carriers. As we shall see later, if the transport is the result of collective motion, the degrees of freedom are reduced and the Wiedemann-Franz law will break down.

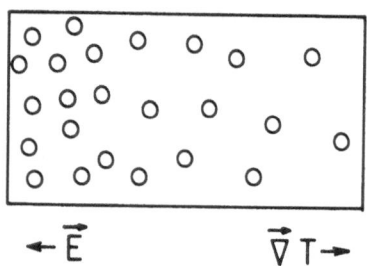

Fig. 3. In a temperature gradient the particle density changes creating a thermoelectric voltage which depends on the sign of the charge q, here assumed negative. The density variation is greatly exaggerated.

The thermopower is the coefficient relating the electric field produced by application of a temperature gradient ($\vec{E} \equiv S\vec{\nabla}T$). In Fig. 3 the configuration of the gas of particles in this situation is illustrated. The density will be reduced on the hot side and this will produce an electric field which will depend on the charge of the particles. It is difficult to proceed with calculations from this model and much more convenient instead to introduce the concept of the Peltier heat.

The Peltier configuration is shown in Fig. 4. In a sample with no temperature gradient an electrical current is set flowing. If the particles carry heat, this will produce a heat current. The ratio of heat to electrical current is the Peltier coefficient ($\vec{U} \equiv \pi\vec{j}$). Thompson in the last century and Onsager[3] more recently have shown quite generally that the Peltier and thermopower S (Seebeck) coefficients are related by a factor of temperature

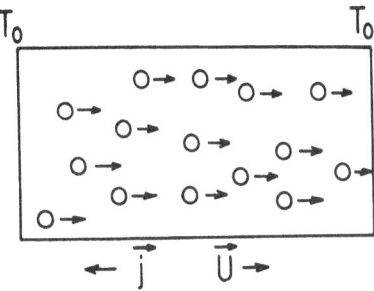

Fig. 4. With no temperature gradient an electrical current will transport heat across the box. In this schematic the particles are negatively charged so that the electric and heat currents flow oppositely.

($\Pi = ST$).[1] The heat current can be written here as particle current times heat per particle or specific heat per particle times temperature. The rest is straightforward:

$$\vec{j} = nq\vec{v}, \quad \vec{U} = nvcT = \frac{cT}{q}\vec{j}$$

$$\pi = cT/q \ , \ S = c/q \ . \qquad\qquad\qquad \text{IV}$$

More generally we may write: $S = $ (heat per particle)$/qT$.

Classically the specific heat per particle is $c \sim k_B$ which gives a thermopower of $S \sim k_B/e$ for electrons.

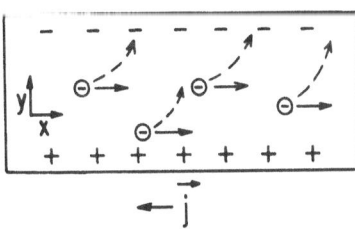

Fig. 5. In the presence of a magnetic field H coming out of the plane of the paper, negative charges moving with positive v_x are deflected in the y direction. When the charge builds up on top and bottom, an electric field develops to cancel the magnetic force. The charges again move along the axis when the force along y is zero.

The configuration for the Hall effect is shown in Fig. 5. A current is initially flowing in the x direction. A magnetic field is then applied along the z direction and due to the Lorentz force, $\vec{F} = q(\vec{E}+\vec{v}x\vec{B}/c)$, the charges are deflected along the y direction until the build-up of charge on one surface and the lessening of charge on the other surface sets up an electric field which exactly compensates the Lorentz force. At this point there is no longer a current in the y direction and steady state has been established. The relevant equations for the electric field which is established are:

$$F_y = 0 = q(E_y - v_x H/c)$$

$$E_y = v_x H/c , \qquad j_x = nqv_x$$

$$E_y = j_x H/nqc$$

The definition of the Hall coefficient is $E_y \equiv R_H j_y H_z$. For this simple case we therefore have:

$$R_H = 1/nqc \qquad\qquad\qquad\qquad\qquad V$$

If there is only one charge carrier in the system (as in the model used above), after the field E_y has been established the carriers again flow directly across the sample in the x direction. Since this was the situation before the application of the magnetic field, the conductivity must be the same and hence there is no magneto-resistance. This is a particularity of a one carrier system.

The results of the Drude theory for the transport coefficients are presented in Table I in the general sense in which we will use them for more realistic circumstances and for the classical limit in which they were originally derived. It is interesting to note that only two of these S and R_H, depend on the sign of the charge carriers.

In order to apply these formulas to a particular material we must specify the density and charge of the carriers as well as their specific heat, velocity, and scattering time. If we were to specify a metal, for example, we imply a degenerate Fermi gas consisting of a temperature independent density of electrons with a specific heat $C_{e\ell} = \Pi \frac{2}{3} k^2 TN(\varepsilon_F)$, where $N(\varepsilon_F)$ is the density of states at the Fermi energy. In the simplest approximation (the free electron model) the density of states in three dimensions is $N(\varepsilon_F) = (3/2)(n/\varepsilon_F)$ and the velocity is the Fermi velocity v_F with $\varepsilon_F = 1/2 \, mv_F^2$. For one dimension $N(\varepsilon_F) = \frac{1}{2}(n/\varepsilon_F)$. The transport coefficients expected for a metal are given in Table II.

Table I. Defining relations and Drude results for the transport coefficients.

Transport Coefficient	Defining Relation	Single Carrier	Classical Statistics
σ	$\vec{j} \equiv \sigma \vec{E}$	$nq^2\tau/m$	$nq^2\tau/m$
κ	$\vec{U} \equiv \kappa(-\vec{\nabla}T)$	$\frac{1}{3}Cv\ell$	$3k_B^2 T \frac{n\tau}{m}$
			or $\kappa/\sigma = L_c T$
S	$\vec{E} \equiv S\vec{\nabla}T$	$\frac{c}{q}$ or $\frac{heat}{qT}$	k_B/q
R_H	$E_y \equiv R_H j_x H_z$	$1/nqc$	$1/nqc$
$\Delta\rho(H)/\rho$	$\frac{\rho(H)-\rho(0)}{\rho(0)}$	0	0

Table II. Transport coefficients for the metallic state in a free electron Drude model.

Transport Coefficient	Metals	Comments
σ	$ne^2\tau/m$	(increases as T decreases)
κ	$\frac{\pi}{3}k^2 T \frac{n\tau}{m} = \sigma L T$	$L = (\pi^2/3)(k^2/e^2)$
S	$\frac{\pi^2}{2}\frac{k}{e}\left(\frac{kT}{\epsilon_F}\right)$	(for 1-d) $\frac{\pi^2}{6}\frac{k}{e}\left(\frac{kT}{\epsilon_F}\right)$
R_H	$1/nec$	
$\rho(H)$	0	

The properties which characterize a simple metal are thus: a conductivity whose temperature dependence is essentially determined by τ and would be generally expected to increase as temperature is decreased (the material becomes more ordered and scattering lifetime becomes longer at low temperature), a thermal conductivity which obeys the Wiedemann-Franz law but differs from the classical case by the small difference of $(\pi^2/3)(k^2/e^2)$ compared to $3k^2/e^2$, a small linear thermopower, classical values for Hall coefficient, and no magnetoresistance.

We can now see to what extent the quasi-one-dimensional compounds agree with the typical metallic properties we have derived. The one compound on which all of these transport measurements have been made is TTF-TCNQ. The conductivity has been measured in a large number of laboratories.[4],[5] From 300K to 60K the conductivity increases as we would expect for a metal. If we assign the temperature dependence entirely to the scattering time τ we would find that it varies faster than T^{-2}.

In this same temperature region the thermopower is seen to be quite linear in T (with some deviation below $\sim$ 150K) and negative.[6] This result confirms the metallic behavior with electron-like carriers and the slope allows the calculation of the Fermi energy which corresponds to $\varepsilon_F \sim 0.14$ eV.

The Hall coefficient at 300 K is approximately -4.5×10^{-11} v-cm/A-gauss.[7] Again the carriers appear to be electronic (negatively charged) and the carrier density calculated from this value corresponds to 0.58 electrons per TCNQ, precisely what is found from x-ray scattering results.[8] There is a slight variation of ($\sim$ 25%) of R_H between 300K and 60K, but one can probably assume that the carrier concentration is largely independent of temperature, another characteristic of a metal. Above 60K the magnetoresistance is less than 0.05%.[9]

The true test of our transport model lies in the thermal conductivity of the Wiedemann-Franz law. Unfortunately the thermal conductivity is dominated in TTF-TCNQ by the lattice contribution and we are only interested in the electronic contribution. There is a well known metal-insulator transition at 54K and κ has been measured through this temperature.[10] By extrapolating from below (where κ is almost entirely due to the lattice) the contribution from the metallic electrons can be obtained. The experimental result is that the electronic part of κ corresponds to a conductivity of $\sim 9000(\Omega\text{-cm})^{-1}$ which is the value that the authors suggest for the metallic state and gives a conductivity ratio of $\sigma_{max}/\sigma_{R.T.} \sim 30$. The implication is that the intrinsic conductivity of TTF-TCNQ at its maximum has this value and that it is entirely due to single particle (Drude-like) transport.

However, the thermal conductivity experiment also shows a significant change at $\sim$ 39K where later x-ray and neutron[11] work have indicated another structural transition. At the lower temperature the electronic conductivity is too small to account for this change and it is therefore the result of a change in the lattice contribution. Since a structural change also takes place at 54K, it is no longer possible to directly obtain the electronic κ at this temperature. The question whether the transport is completely single particle or not still remains open.[12]

To summarize the transport results above the Peierls transition temperature, assuming the most direct interpretation of the experiments, the Hall coefficient, thermopower, and thermal conductivity, which are the clearest tests of metallic behavior, are in incredibly good agreement with the simplest theory discussed above. TTF-TCNQ appears from these tests to be a regular metal (in fact a lot more regular than copper). However, the mean free path is extremely short[5] ($\sim$ 1 lattice spacing at 300K), the scattering time has an unusual dependence ($T^{-\alpha}$, $\alpha \stackrel{>}{\sim} 2$), the Wiedemann-Franz law can be argued, we know that the system has two carriers and is quasi-one-dimensional and the large thermal expansion of the crystals makes the thermopower and conductivity temperature dependences less straightforward.[13]

The scattering time for use in our model can be calculated quantum mechanically from Fermi's Golden Rule:

$$\frac{1}{\tau} = \frac{2\pi}{\hbar} \; |{<}V{>}|^2 \; N(\varepsilon F) \qquad\qquad\qquad VI$$

where the matrix element $<V>$ is a perturbation which couples different particle states. If the potential V is uniform or periodic the momentum or k states are eigenstates of the system and are not mixed by V. If a general coordinate x_i of the "background" particle (ion or molecule) which produces the potential changes, it is possible to have a non-zero matrix element. We can expand the potential as:

$$V = V_o \frac{\partial V}{\partial x_i} \; (\Delta x_i) + \frac{1}{2} \frac{\partial V}{\partial x_i^2} \; (\Delta x_i)^2 + \cdots (\Delta x_i)^b \qquad VII$$

Choosing the lowest non-vanishing term we have $1/\tau \propto {<}(\Delta x_i)^b{>}^2$. Each displacement x_i has a characteristic frequency ω_i. If $kT > \omega_i$ then the average amplitude of the displacement squared is proportional to the temperature from the equipartition theorem $[(\Delta x_i)^2 \propto kT]$. The scattering time therefore varies as:

$$\tau \propto (1/T)^b \qquad\qquad\qquad\qquad VIII$$

in the high temperature limit and varies more quickly when kT is less than ω_i. For many metals above 100K the first term in Δx_i above is non-vanishing (b = 1) and the conductivity therefore goes as 1/T. Several authors have proposed that the b = 2 term,[14] or a combination of terms[15] can explain the conductivity of TTF-TCNQ in the metallic state.

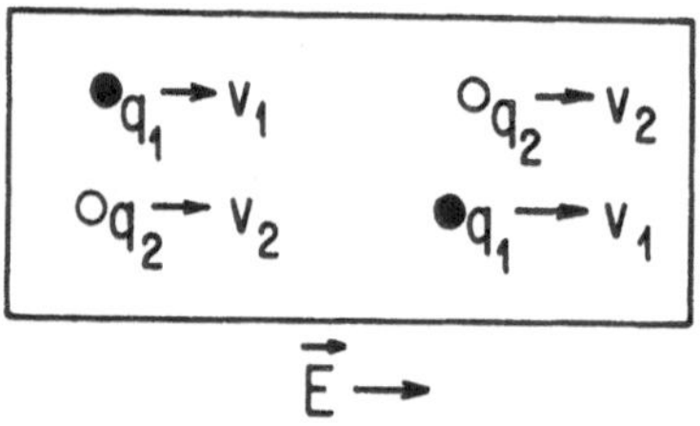

Fig. 6. In the two carrier system particles with different charges, masses, and scattering times respond to the electric field with different velocities.

We will now discuss what modifications arise from having a two carrier system as shown in Fig. 6. The velocity of each particle is again given by its appropriate $q\tau E/m$. The electric current is then the density of each type of carrier times its charge times its velocity, or:

$$\vec{j} = n_1 q_1 v_1 + n_2 q_2 v_2 = \left(\frac{n_1 q_1^2 \tau_1}{m_1} + \frac{n_2 q_2^2 \tau_2}{m_2} \right) E = (\sigma_1 + \sigma_2) E \qquad \text{IX}$$

The conductivities from each carrier simply add. Similar arguments show that the thermal conductivities add as well.

For the heat flow we need the particle flow times heat per particle.

$$U = n_1 v_1 c_1 T + n_2 v_2 c_2 T = \left(\frac{n_1 q_1 \tau_1 c_1 T}{m_1} + \frac{n_2 q_2 \tau_2 c_2 T}{m_2} \right) E$$

$$= (\sigma_1 S_1 T + \sigma_2 S_2 T) \frac{1}{\sigma_1 + \sigma_2} \qquad \text{X}$$

In the last line we have used our previous results from Table I, $\sigma_i S_i = (n_i q_i^2 \tau_i / m_i)(c_i / q_i)$. The coefficient of j is the Peltier coefficient and dividing by T we find that the thermopower is the average of the individual thermopower weighted by the conductivities, $S = (\sigma_1 S_1 + \sigma_2 S_2)/(\sigma_1 + \sigma_2)$.

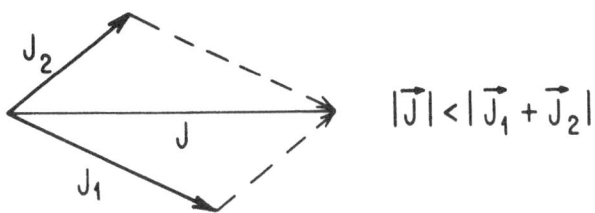

Fig. 7. In the presence of a magnetic field (H) perpendicular to the plane of the page the currents from the two types of particles will be deflected differently as their charges and velocities are inequivalent. The total current is the vector sum and hence always less than the H = 0 current.

The magnetotransport properties are algebraically more difficult for two carriers. In Fig. 7 the configuration of the two currents are shown schematically in the presence of a magnetic field. If either the velocity of the two carriers or their charges are different it will be impossible for a single electric field to cancel out the Lorentz force for both (as happens in the single carrier case), therefore the two currents are no longer parallel to the applied electric field and must be added vectorially. This will always give a total current less than the sum $|J_1| + |J_2|$ and hence the current is diminished in the presence of a magnetic field. This implies a higher resistance or positive magnetoresistance. It is also interesting to note that the single carrier argument against electron-electron scattering breaks down for two carrier systems.

The results of a two carrier Drude model are summarized in Table III.

Evidence for two carrier transport in the quasi-one-dimensional field comes mostly from doping and pressure experiments. In alloys of $TTF_{1-x}TSeF_xTCNQ$ the thermopower shows large changes relating to the reduced conductivity on the doped chains.[16,17] More recent work on doping TMTSF-DMTCNQ has even enabled a qualitative assignment of conductivities and thermopowers to the different carriers on different chains.[18] Hall effect measurements on TTF-TCNQ show a large change under pressure which is related[7] to a change of the relative conductivities on the two chains. Thermopower measurements have confirmed this change[19] (which amounts to a factor of three reduction of S) and together with the earlier Hall measurements indicate that $\sigma TCNQ \sim 5 \ \sigma TTF$ at 300K and atmospheric pressure. This large ratio shows that the TCNQ dominates the transport and hence makes the system look like

Table III. Generalization of the Drude theory for a two carrier system.

σ	$\sigma_1 + \sigma_2$	
S	$\dfrac{\sigma_1 S_1 + \sigma_2 S_2}{\sigma_1 + \sigma_2}$	
κ	$\kappa_1 + \kappa_2$	W. F. law holds
R_H	$\dfrac{R_1 \sigma_1^2 + R_2 \sigma_2^2}{(\sigma_1 + \sigma_2)^2}$	low field
	$\dfrac{1}{n|e|c} \dfrac{\mu_h - \mu_e}{\mu_h + \mu_e}$	high field $n_h = n_e$
$\Delta\rho(H)/\rho_o$	$\dfrac{(\beta_1 - \beta_2)^2\, H^2 \sigma_1 \sigma_2}{(\sigma_1 + \sigma_2)^2}$	low field $\beta_i = e\tau_i/m_i c$
	$\dfrac{\mu_h \mu_e H^2}{c^2}$	high field

a single carrier system. By 8 kbars the ratio is reduced to $\sigma TCNQ - 2\, \sigma TTF$. In contrast to this behavior the single chain conductor Quin $(TCNQ)_2$ has a thermopower reduction of only 30% at 20 kbars.[19]

So far we have only discussed materials which are metallic above $\sim$ 50K. There are materials of great interest in this field which are metallic or semi-metallic down to 1K. The most widely studied is HMTSF-TCNQ. Conductivity,[20,21] Hall effect,[22] magneto-resistance[22] and thermopower[20] all indicate that this material is a semi-metal at low temperatures, certainly under pressure. Again there is agreement in the sign of the carriers between Hall and thermopower measurements where the number of carriers even correspond roughly (from R_H) to the Fermi energy (from S).

For completeness in this section on single particle band transport it is necessary to emphasize that any calculations which hope to fit real data should use the Boltzmann equation results shown in Table IV. The main difference between the latter and

Table IV. Results of the Boltzmann equation treatment of transport.

	σ	S	κ
General	$e^2 \int v_k^2 \tau_k \left(-\frac{\partial f}{\partial \varepsilon}\right) d^3k$	$\frac{e}{T} \dfrac{\int (\varepsilon-\mu) v_k^2 \tau \left(-\frac{\partial f}{\partial \varepsilon}\right) d^3k}{\sigma}$	$\frac{1}{T} \int (\varepsilon-\mu)^2 v_k^2 \tau \left(-\frac{\partial f}{\partial \varepsilon}\right) d^3k$
Metals		$\frac{\pi^2}{3} \frac{k_B^2 T}{e} \left[\frac{\partial \ln \sigma(\varepsilon)}{\partial \varepsilon}\right]_{\varepsilon_F}$	
3-D Free Electron	$ne^2\tau/m$	$\frac{\pi^2}{2} \frac{k}{e} \left(\frac{kT}{\varepsilon_F}\right) + \frac{\pi^2}{3} \frac{k^2 T}{e} \frac{\tau'(\varepsilon_F)}{\tau(\varepsilon_F)}$	σLT (i.e., Wiedemann–Franz law holds if scattering is elastic)
1-D Tight Binding	$\frac{4e^2\tau|t|a \sin(\rho\pi/2)}{\pi\hbar^2}$	$\frac{-\pi^2 k^2 T}{3|e|} \left[\frac{\cos(\pi\rho/2)}{2|t|\sin^2(\pi\rho/2)} + \frac{\tau'(\varepsilon_F)}{\tau(\varepsilon_F)}\right]$	

Drude theory is the use of momentum (k) states with associated
masses, velocities, and scattering times not necessarily equal
for all conducting electrons. One qualitatively new result is
that the thermopower has an additional term related to $\partial\tau(\varepsilon)/\partial\varepsilon$.
This comes from the fact that the current may arise predominantly
from higher or lower energy electrons depending on the energy dependence of τ.

Phonon drag

In discussing our fundamental assumptions in the derivation
of σ we said that the scattering time must relate to processes
which take momentum out of the charge carrier system. Usually
this scattering is from phonons. In Fig. 8 we illustrate this
situation in terms of charged particles (electrons) and uncharged
particles (phonons). The electrons transfer momentum to the
phonons.

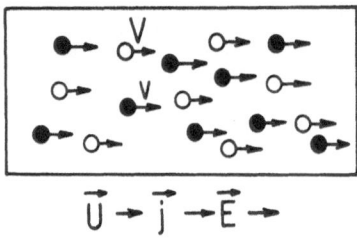

$$\vec{U} \rightarrow \vec{j} \rightarrow \vec{E} \rightarrow$$

Fig. 8. The solid circles represent charged particles (e.g.,
electrons or holes) while the open circles are uncharged particles
(e.g., phonons) with which they collide. The electric field gives
a velocity $\vec{v}$ to the charged particles which drag the uncharged
particles at velocity $\vec{V}$.

The phonons relax back to the electrons with a rate τ_{e-p}^{-1} and relax to zero velocity by scattering off themselves
(in Umklopp processes) and off impurities with a rate τ_{other}^{-1}.
Hence some of the momentum is transferred back to the
electrons, Peierls[23] suggested that the electron-phonon resistivity is reduced by the fraction of momentum which returns to the
electron system. The resistivity becomes:

$$\rho = \rho_o + \rho_{e-p} \left[1 - \frac{\tau_{ep}^{-1}}{\tau_{ep}^{-1} + \tau_{other}^{-1}} \right] \qquad XI$$

where ρ_o and ρ_{e-p} are the resistances due to impurities and phonons respectively.

If the phonons relax to both the electrons and to zero
velocity, they must be "dragged" with a fraction of the electron

velocity given by: $V = v[\tau_{ep}^{-1}/(\tau_{ep}^{-1} + \tau_{other}^{1})]$. This net velocity implies that there will be a heat flow associated with the phonons of the form $U = C_L TV$ where C_L is the total lattice specific heat. The electrical current $j = nev$ can be put into this equation and the Peltier heat and thermopower due to phonon drag calculated as:

$$S_L = \frac{C_L}{ne} \left(\frac{\tau_{ep}^{-1}}{\tau_{ep}^{-1} + \tau_{other}^{-1}} \right) \qquad \text{XII}$$

Traditionally, in normal metals, the phonon drag shows up more strongly in the thermopower than in the resistivity. This results from the large ratio of the lattice specific heat C_L to the electronic specific heat C_e which determines the usual thermopower. At low temperatures S_L increases as $C_L \propto T^3$. At higher temperatures S_L decays as T^{-1}, due to increased phonon-phonon scattering. This produces a peak at $\sim \theta_D/5$.[24] Phonon drag does not appreciably affect the thermal conductivity as there is no net electron velocity in this experiment. The Wiedemann-Franz law does not hold as σ is changed while κ is not. Expected behaviors for phonon drag are shown in Fig. 9.

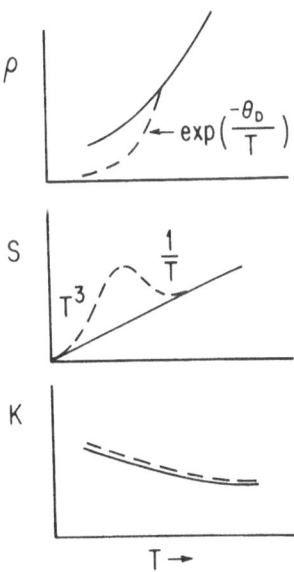

Fig. 9. Typical dependences of the transport coefficients of a metal with strong (dashed line) and weak (solid line) phonon drag effect. The thermopower phonon drag peak occurs in normal metals at $T \sim \theta_D/5$.

Phonon drag has been seen in (SN)$_x$[25] and has been widely discussed by the Jerusalem group[26] in relation to TTF-TCNQ and similar compounds. They have pointed out that in quasi-one-dimensional
materials phonon drag may have different characteristics as one
particular phonon (with wavevector $2k_F$) is responsible for all resistive scattering.

Superconductivity

Aside from transport due to single carriers (as in the Drude
theory), the only other transport mechanism which has been extensively investigated experimentally and theoretically is that found
in superconductors. Rigorously it is incorrect to use the single
particle formalism we have developed to explain any kind of collective transport, we should instead use the Kubo formalism.[27]
However, in order to get some elementary picture we will use the
general concepts from the Drude theory.

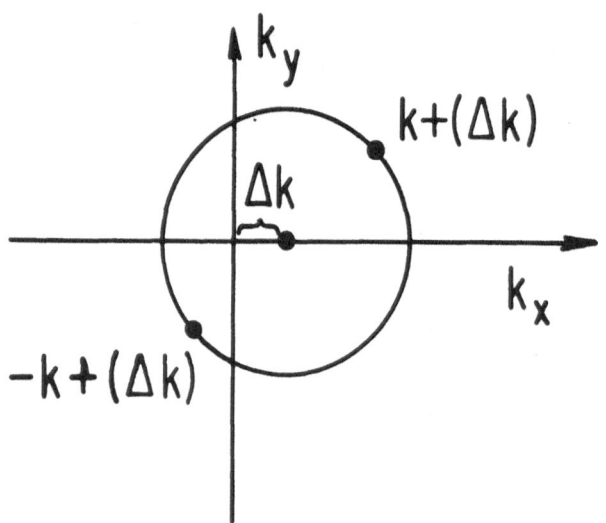

Fig. 10. Superconductivity involves the pairing of electrons
such that all pairs have the same center of mass. In the absence
of a current a pair would be k and -k (with opposite spins). In
this schematic the center of mass has been shifted to indicate
the presence of a current.

The essence of superconductivity is that all electrons gain
energy if they form pairs which have the same center of mass momentum.[28] (The energy gap which is usually found in superconductors is neither necessary nor sufficient to produce superconductivity or zero resistance.) In Fig. 10 the occupation of momentum
states is shown for the situation where a current is flowing. The
usual way that scattering occurs is for a high energy electron on

one side of the Fermi surface to be scattered to a lower energy state on the other side. For a normal electron gas only the kinetic energy is important, and this process gives rise to resistivity. However, for the superconductor which gains potential energy by the particular arrangement of electrons about the center of mass, scattering across the Fermi surface to a lower kinetic energy state increases the potential energy with a net raising of the energy. Thus the only way to reduce the current and lower the energy is for the center of mass to change for all the pairs at the same time. This requires the simultaneous scattering of $\sim 10^{23}$ electrons and gives a scattering rate of $(1/\tau) \sim (1/\tau_e)^{10^{23}}$ where $1/\tau_e$ is the single electron rate. On the time scale of the age of the universe this τ is long, hence we can assume $\tau \sim \infty$ and the conductivity is also infinite.

Only a certain density of electrons are condensed into the superconducting state at temperatures between zero and the superconducting transition (T_c). The conductivity is then the sum of super and normal parts $(\sigma = \sigma_s + \sigma_n)$ which is still infinite. For the superconducting electrons the heat which accompanies a current is zero since the condensed electrons are in their ground state configuration. The thermopower is therefore zero for the super part which dominates the conductivity. The total thermopower is thus zero,[29]

$$S = \frac{\sigma_s S_s + \sigma_n S_n}{\sigma_s + \sigma_n} = \frac{\sigma_n S_n}{\sigma_s} = 0$$

The thermal conductivity is also the sum of normal and super parts. The super part is zero while the normal part is finite $(\kappa = \kappa_n + \kappa_s = \kappa_n)$. The normal component falls, with no discontinuity at T_c, as the density of normal to super electrons decreases as temperature is lowered.[28] The Wiedemann-Franz law should not be applicable as transport is collective not single particle. Indeed the ratio κ/σ is zero and not LT as in Table II.

The magneto-transport is dominated by the Meisner effect which excludes the magnetic field from the bulk of the sample. As a result the Hall coefficient is zero and the resistance remains zero until the critical field which destroys superconductivity returns the resistance to the normal states. These results are summarized in Table V and Fig. 11.

Of the material of interest in the field covered in this conference only two, $(SN)_x$[30] and $NbSe_3$ (under pressure[31] or doped[32]), show superconductivity.

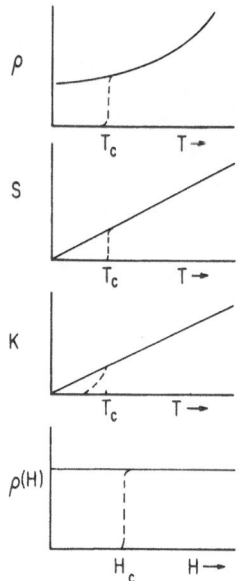

Fig. 11. Transport properties for a normal metal (solid) and for a superconductor (dashed).

Table V.   Transport coefficients for a superconductor.

σ	∞
S	0
κ	Thermal conductivity of normal component ($\kappa/\sigma = 0$, not Wiedemann-Franz)
R_H	0
$\rho(H)$	$\rho_n \theta(H - Hc_2)$

Frohlich Conductivity

One of the most exciting aspects of the quasi-one-dimensional conductors is the possibility of a new form of collective transport unique to systems that undergo charge (or spin) density wave transitions. The Peierls transition[33] which has been extensively discussed throughout this conference produces a charge density wave and an associated periodic distortion of the ionic background as schematized in Fig. 12. If there is no pinning by

impurities or commensurability, or if the CDW is depinned by an
electric field, the CDW can move with the lattice distortion co-
herently moving with it.[34] The translating lattice distortion
at $2k_f$ ('$2k_f$ phonon') looks very similar to the situation which
we encountered before under the title 'Phonon drag.'

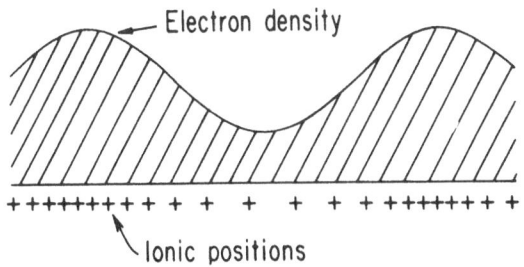

Fig. 12. Charge density wave and periodic ion distortion as result of a Peierls' instability. In the incommensurate case the
CDW has no preferential position and thus is mobile when the distortion moves as well.

      In the case of usual phonon drag, the situation involved
single electrons scattering off thermally excited phonons. In the
case of CDW motion, or Frohlich conductivity, we have a coherent
phonon drag, with a rigid CDW dragging a macroscopically occupied
phonon mode (the lattice distortion). Both incoherent (usual) and
coherent phonon drag result in the reduction of electron-phonon
resistivity (to zero for strong coupling). The difference between
the two is the heat transported by a current and possibly the reduction of scattering off impurities which is predicted for the
Frohlich case.

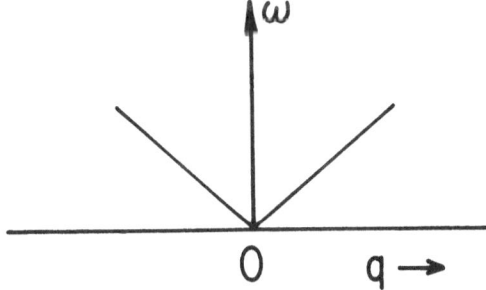

Fig. 13. Dispersion relation for phasons - normal modes of the
CDW. Only the $q = 0$ mode carries current.

      To understand the heat currents for the Frohlich case we must
find the normal modes of the CDW system. The electron density is
given by $\rho_0 \cos[2k_F x - \phi(x)]$. The modes of this system are spatial and time variations of ϕ, or phasons. The phason dispersion
relation is shown in Fig. 13. Whenever we are dealing with a
collective transport phenomenon, it is the $q = 0$ mode which

determines the conductivity. For phasons the $q = 0$ mode has $\omega = 0$ and therefore no energy or heat. The $q \neq 0$ modes are the ones which contribute to the heat content or heat current. This situation is very similar to that which prevails for acoustic phonons. The only phonon which carries momentum is the $q = 0$ mode. All the other modes carry the energy and heat. Under the assumptions that the phasons are eigenmodes of the CDW and are therefore decoupled, there is no heat current to be associated with the electrical current. The Peltier heat and thermopower are zero. Remember that the incoherent phonon drag gives the opposite result of an increased thermopower.

There have been several calculations of the conductivity due to sliding or depinned CDW's, especially in the region above the Peierls' transition temperature.[35] Although the calculation should really be done using the Kubo formalism, it is usually performed phenomenologically using a Drude-like expression with $\sigma_F = N_F e^2 \tau / M_F$ where N_F and M_F are the effective density and mass of the CDW and τ is the lifetime of the $q = 0$ mode due to the creation of electron hole pairs. In the fluctuating region the result is $\sigma_F = (ne^2/m)(4k_B T/\pi\Delta)[\xi(T)/v_F]$.[35] An additional property of CDW conductivity is that it should show nonlinearities (change with applied electric field) since the CDW can be depinned from weak pinning sites.[36]

The thermal conductivity is calculated from all modes except $q = 0$ by a Drude-like expression $\kappa_F = \Sigma_q C_q V_q^2 \tau$. Since this expression has virtually nothing to do with the $q = 0$ mode and the conductivity, the Wiedemann-Franz expression should not hold. Calculations[10] show that the ratio $\kappa_F/\sigma_F \simeq LT_C(k_B T_c/E_F)^2$. This ratio is about 4 orders of magnitude smaller than the value found for single particle transport and again shows that the Wiedemann-Franz ratio is a crucial way of establishing whether the transport is collective or not.

The Hall effect should be substantially the same for Frohlich conductivity as that given in the Drude expression. Unlike the case for superconductors there is no Meissner effect to exclude the magnetic field and the Lorentz force will deflect the CDW until the Hall field builds up to cancel it. The magnetoresistance will, however, have a new term. The spin susceptibility of the Peierls state is lower than that of the metallic state. Since the energy in a magnetic field is lower with a larger susceptibility, the magnetic field tends to weaken the Peierls state.[9] If the CDW fluctuations are conductive this will produce a positive magnetoresistance. If they are resistive, the magnetoresistance is negative. This spin effect is isotropic in applied field direction.

In Table VI the tentative transport coefficients for Frohlich conductors are given.

Table VI. Compilation of tentative properties associated with Frohlich conductivity in the temperature range just above the Peierls' transition temperature.

σ_F	$\dfrac{ne^2}{m}\left(\dfrac{4k_B T}{\pi\Delta}\right)\dfrac{\xi(T)}{V_F}$	(nonlinear)
S	$\dfrac{\sigma_n S_n}{\sigma_F + \sigma_N}$	
κ_F	$\sigma_F L T_c \times \left(\dfrac{k_B T_c}{E_F}\right)^2$	(Wiedemann-Franz law does not hold)
R_H	$1/nec$	
$\Delta\rho(H)/\rho$	$\sim (\mu_B H/\Delta)^2$	

Of the transport experiments performed, those which show characteristics most similar to what is expected for Frohlich conductivity are the large and striking nonlinearities found in NbSe$_3$[37] and the reduction in conductivity of TTF-TCNQ when it is made commensurate by application of pressure.[38] The conductivity of TTF-TCNQ has also been shown to fit the temperature dependence given in Table VI to the same degree with which it can be fit by single particle theories.[39]

This tutorial has presented some simple explanations of the DC transport coefficients for single particle and two forms of collective transport. Care must be taken in applying these simplified models to real systems. Although a great deal can be learned about the transport mechanisms and properties of quasi-one-dimensional conductors from the experiments described here, it is clear that any final analysis must also include the results of a great many other experiments in order to produce a coherent picture of the physics taking place in these unusual compounds.

Acknowledgments for helpful and stimulating discussions are due: A. Alexander, E. Conwell, T. Holstein, D. Jérome, P. Pincus, and M. Weger. Research supported by NSF Grant DMR-79-08560, ONR N00014-76-C-1078, and an A. P. Sloan Foundation Fellowship.

REFERENCES

1. For simple derivations see: N. Ashcroft and N. Mermin, Solid
    State Physics (Holt, Rinehart, & Winston, New York, 1976).
    For more rigorous treatments see: J. M. Ziman, Electrons and
    Phonons (Oxford University Press, London, 1972).
2. P. Drude, Annalen der Physik 1, 566, and 3, 369 (1900).
3. W. Thomson (Lord Kelvin), Proc. Roy. Soc. Edin. 3, 255 (1854);
    L. Onsager, Phys. Rev. 3F, 2265 (1931).
4. J. P. Ferraris, O. O. Cowan, V. V. Walatka, and J. H. Perl-
    stein, J. Amer. Chem. Soc. 95, 948 (1973); L. B. Coleman, M.
    J. Cohen, D. J. Sandman, F. G. Yamagishi, A. F. Garito, and
    A. J. Heeger, Solid State Commun. 12, 1125 (1973).
5. G. A. Thomas et al., Phys. Rev. B 13, 5111 (1976).
6. P. M. Chaikin, J. F. Kwak, T. E. Jones, A. F. Garito, and
    A. J. Heeger, Phys. Rev. Lett. 31, 601 (1973).
7. J. R. Cooper, M. Miljak, G. Delplangue, D. Jérome, M. Weger,
    J. M. Fabre, and L. Giral, J. de Physique 38, 1097 (1977).
8. J. P. Pouget, S. K. Khanna, F. Denoyer, R. Comès, A. F.
    Garito, and A. J. Heeger, Phys. Rev. Lett. 37, 437 (1976).
9. T. Tiedje, J. F. Carolan, A. J. Berlinsky, and L. Weiler,
    Can. J. Phys. 53, 17 (1975).
10. M. B. Salamon, J. W. Bray, G. DePasquali, R. A. Craven, G.
    Stucky, and A. Schultz, Phys. Rev. B 11, 619 (1975).
11. R. Comès, S. M. Shapiro, G. Shirane, A. F. Garito, and A. J.
    Heeger, Phys. Rev. Lett. 32, 1518 (1975).
12. In principle it should be possible to separate the electron
    and lattice contributions to κ by measuring the transverse
    thermal conductivity in a magnetic field, the Righi-Leduc
    effect. See reference 1.
13. J. R. Cooper, Phys. Rev. B 19, 2404 (1979); and D. Jérome,
    J. Phys. Lett. 38, L-489 (1977).
14. M. Weger and J. Friedel, J. de Physique 38, 241 (1977); H.
    Gutfreund and M. Weger, Phys. Rev. B 16, 1753 (1977); H.
    Gutfreund, M. Weger, and M. Kaveh, Solid State Commun. 27,
    53 (1978); also this conference.
15. E. M. Conwell, in: International Conference on Quasi-One-
    Dimensional Conductors, Dubrovnik, Lecture notes in Physics
    96 (Springer-Verlag, Berlin, 1978); and Phys. Rev. Lett. 39,
    777 (1977); also this conference, pp. 213.
16. T. D. Schultz and S. Etemad, Phys. Rev. B 13, 4928 (1976);
    and S. Etemad, Phys. Rev. B 13, 2254 (1976).
17. P. M. Chaikin, J. F. Kwak, R. L. Greene, S. Etemad, and E.
    Engler, Solid State Commun. 19, 1201 (1976).
18. K. Mortensen, C. S. Jacobsen, J. R. Andersen, and K. Bech-
    gaard, in International Conference on Quasi-One-Dimensional
    Conductors, Dubrovnik Lecture Notes in Physics 96 (Springer-
    Verlag, Berlin, 1978).
19. C. Weyl, P. M. Chaikin, and D. Jérome, to be published.

20. A. N. Bloch, D. O. Cowan, K. Bechgaard, R. E. Pyle, R. H. Banks, and T. O. Poehler, Phys. Rev. Lett. 34, 1561 (1975).

21. J. R. Cooper, M. Weger, D. Jérome, D. Lefur, K. Bechgaard, A. N. Bloch, and D. O. Cowan, Sol. State Commun. 19, 749 (1976).

22. J. R. Cooper, M. Weger, C. Delplanque, D. Jérome, K. Bechgaard, J. de Physique Lett. 37, 349 (1976).

23. R. E. Peierls, Ann. Phys. (5) 12, 154 (1932).

24. See for example: Thermoelectric Power of Metals by F. T. Blatt, P. A. Schroeder, C. L. Foiles, and D. Greis (Plenum Press, New York, 1976).

25. L. J. Azevedo, P. M. Chaikin, W. G. Clark, W. W. Fuller, and J. Hammann, J. de Phys., Colloque C6, 39, 446 (1978), LT XV.

26. H. Gutfreund, M. Kaveh, and M. Weger, International Conference on Quasi-One-Dimensional Conductors, Dubrovnik, Lecture Notes in Physics 96 (Springer-Verlag, Berlin, 1978); also see ref. 14 and this conference.

27. R. Kubo, J. Phys. Soc. Japan 12, 1203 (1957).

28. M. Tinkham, Introduction to Superconductivity (McGraw-Hill, New York, 1975).

29. W. F. Viner, in Superconductivity, ed. by R. Parks (Marcel Dekker, New York, 1969).

30. R. L. Greene, G. B. Street, and L. J. Sutter, Phys. Rev. Lett. 34, 577 (1975).

31. P. Monceau, J. Peyrard, J. Richard, and P. Molinie, Phys. Rev. Lett. 39, 161 (1977).

32. W. W. Fuller, P. M. Chaikin, and N. P. Ong, Sol. State Commun. 30, 689 (1979).

33. R. E. Peierls, Quantum Theory of Solids (Clarendon Press, Oxford, 1955).

34. H. Fröhlich, Proc. Roy. Soc. A223, 296 (1954).

35. D. Allender, J. W. Bray, and J. Burdeen, Phys. Rev. B 9, 119 (1974); S. Strässler and G. A. Toombs, Phys. Lett. 46A; M. J. Rice, Sol. State Commun. 16, 1285 (1975).

36. P. A. Lee, T. M. Rice, and P. W. Anderson, Sol. State Commun. 14, 703 (1974).

37. P. Monceau, N. P. Ong, A. M. Portis, A. Meerschaut, and J. Rouxel, Phys. Rev. Lett. 37, 602 (1976); N. P. Ong and P. Monceau, Phys. Rev. B 16, 3443 (1977).

38. A. Andrieux, H. J. Schulz, D. Jérome, and K. Bechgaard, Phys. Rev. Lett. 43, 227 (1979), and J. de Physique Lett. 40, 385 (1979).

39. A. J. Heeger, in Highly Conducting One-Dimensional Solids, ed. by J. Devreese, R. Evrard, and V. van Doren (Plenum Press, New York, 1979).

THEORY OF THE ELECTRICAL CONDUCTIVITY OF ORGANIC METALS

Meir Weger

The Hebrew University, Jerusalem, Israel

A theory for the electrical resistivity of the high-conductivity organic metals, such as TTF-TCNQ, HMTSF-TCNQ, TMTSF-DMTCNQ, TMTSF$_2$PF$_6$, is presented. Some particular experimental features are a T^2 temperature dependence, an extremely strong pressure (i.e. volume) dependence, and a singular frequency dependence. These (and other) features are treated by a phenomenological theory, making use of the *molecular* structure (accounting for example for the very specific lattice anharmonicities) and the *one dimensional* nature of the metalic state, giving rise to phonon-drag at unusually high temperatures, and a quadratic electron phonon coupling involving mainly molecular librations. The phenomenological theory is substantiated by more basic concepts, such as *ab-initio* determinations of the coupling constants, the deloclization in one-dimensional systems, and the role of the on-site Coulomb coupling.

1. INTRODUCTION

The theory of the electrical resistance in metals, due to electron-phonon scattering, essentially follows the classical work of F. Bloch (1). This theory accounts for a linear law $\rho \propto T$ for temperatures $T \gtrsim \Theta_D$. Already the earliest experiments on TTF-TCNQ (2,3) indicated discrepancies from this behavior, and it was soon afterwards shown (4) that $\rho = A + B \times T^n$."A"is a residual resistance, related to defects, which can be made to vanish in extremely good crystals (5). "B" is an inherent quantity, not strongly dependent on defect concentration in good samples. However, B is very strongly pressure dependent. $n \simeq 2.3$ for $60 \text{ K} < T < 300 \text{ K}$, and decreases to $n \simeq 2$ at moderate pressures. Thus, it seems natural (6) to plot ρ *vs*. T^2 (fig. 1).

L. Alcácer (ed.), The Physics and Chemistry of Low Dimensional Solids, 77–100.

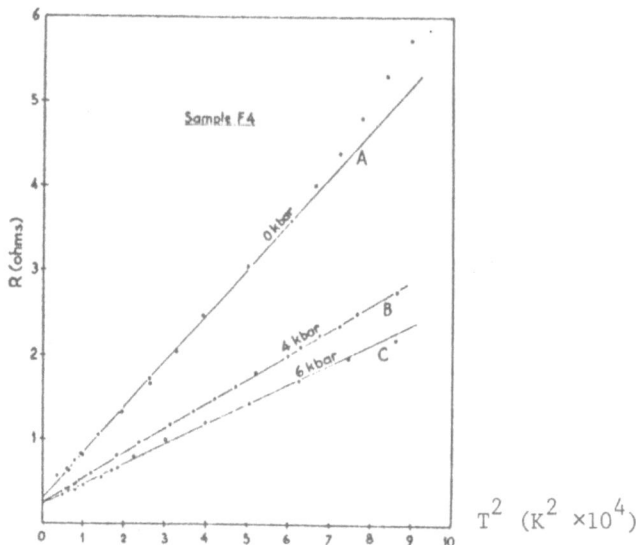

Figure 1. Resistance plotted $vs.\,T^2$ at various pressures
        (ref. 6).

About a dozen theories have been proposed to account for ρ ;
some invoke electron-phonon coupling with various types of phonons
(translations, librations, and intra-molecular modes), and others
invoke electron-electron collisions. Some invoke single-particle
correlations, and others invoke collective effects. Some consider
traveling waves as the basic entity, and others consider localised
standing waves (7,8). All claim to account for the n=2.3 law. Thus,
additional experimental variables are needed to discriminate among
the various theories. Hydrostatic pressure serves as an ideal tool
for this purpose because of its very large effect on the resistivity.
We find that $\partial \ell n \rho / \partial \ell n V \approx 20\%-28\%$ (9), which is about an order of
magnitude larger than for "ordinary" metals. Such a large
quantitative difference implies a *qualitative* difference from
ordinary metals. In this respect, pressure is analogous to the
magnetic field which was an essential variable for proving the
existence of a Fermi-surface in ordinary metals, because of its
huge effect on the resistivity at low temperatures.

When we consider the theory of the transport properties of
TTF-TCNQ, we must take into account the molecular nature of these
materials, as well as the one-dimensional nature of the electronic
states, the (approximate) quantum numbers being (x,k_y,z) [rather
than (k_x,k_y,k_z). y is the chain direction (9)]. Figuratively
speaking, we must always have one foot in the "molecular crystal"
boat, and one foot in the "one-dimensional conductor" boat.

2. THE GRÜNEISEN CONSTANT

Because of the dominant importance of pressure, we must define a variable that characterises its effect. Such a variable is the long-known Grüneisen constant γ. For phonons, γ is defined as $\gamma_j = -\partial \ln \omega_j / \partial \ln V$, where j characterizes the mode. Similar definitions are made for other properties (electronic density of states, etc.). γ_e is determined from the pressure-dependence of the phonon frequencies and the compressibility. For naphtalene and anthracene, which are molecules somewhat similar to TTF and TCNQ in size and shape, $\gamma \approx 3.5$-6 for the external, i.e. the translational and rotational modes (10), and $\gamma_j \ll 1$ for internal modes (bond stretching and bending)(11). For molecular crystals, we must distinguish between the Grüneisen *constants* γ_j and the Grüneisen *function* $\gamma(T) = \Sigma c_j(T) \gamma_j / \Sigma c_j(T)$, where $c_j(T)$ is the specific heat of mode j at temperature T (12). At low temperatures, the internal modes are not excited and $\gamma(T)$ is characteristic of the external modes and thus close to 3-5. At higher temperatures (ambient, say) the very large number of internal modes may make them dominate and $\gamma(T)$ falls, even below 1. The thermal expansion (13), non-linear compressibility (14) etc. are related to $\gamma(T)$, and thus are strongly temperature dependent, much more than in "ordinary" metals. We feel that the Grüneisen constant requires more attention than it has got in the past, in this field.

3. LIBRON THEORY

Two libron modes (per molecule) have the property, that their linear coupling with the electrons vanishes by symmetry. This is because the electron density ψ^2 always transforms into itself under reflection in the bc* plane (i.e. under $x \to -x$), while the libron modes reverse sign under this reflection. (The transverse translon in the x direction, also has this property)(15). Thus, these modes are transverse to all electron states and $(\vec{e} \cdot \vec{k}) = 0$ identically. (Strictly speaking, e and k are orthogonal irreducible representations of the chain symmetry group.) This is because k is always along the chain, for all electrons. (In 3-D, for any phonon, there will always be some electrons for which $(\vec{k} \cdot \vec{e}) \neq 0$). Therefore, in 1-D, these modes couple with the electrons only in second order. If W is the one-electron potential,

$$W = W_o + \tfrac{1}{2} \sum_j (\partial^2 W / \partial \theta_j^2) \theta_j^2 + \ldots \qquad\qquad 1$$

where θ_j is the coordinate characterizing the given libration (or translation), since $\partial W / \partial \theta_j = 0$. The transfer integral along the chain $t_\parallel$ possesses the same property, and we use W to denote it, as well as the one-electron (Madelung) potential. Thus, by the Golden Rule, the scattering rate of an electron by librons is given by:

$$\frac{1}{\tau_{\parallel}} = \frac{2\pi}{\hbar} \frac{1}{2} \sum_j \left[\frac{\partial^2 W}{\partial \theta_j^2} <\theta_j^2> \right]^2 n(\varepsilon_F)$$
                                                                                                    2

as long as the Golden Rule (i.e. the first Born Approximation) applies. (Since $t_{\parallel}$ is affected by $\theta_j - \theta_{j-1}$, and on the average $<(\theta_j - \theta_{j-1})^2> = 2<\theta_j^2>$, a factor of 2 enters when we use $t_{\parallel}$). The resistivity is given by $\rho = (m*/ne^2) 1/\tau_{\parallel}$. As long as $\hbar\omega_j \lesssim k_B T$, i.e. $T \gtrsim \Theta_D$, we can use Boltzmann statistics and $<\theta_j^2> = k_B T/I\omega_j^2$ where I is the appropriate moment of inertia (or mass, for translational modes). Thus, we get: $\rho \propto T^2/\omega^4$.

A particularly striking demonstration of the T^2 law is the recent result of Bechgaard and Jacobsen (17) on TMTSF$_2$PF$_6$ (and related compounds), where the T^2 law holds extremely well over two decades in resistivity (fig. 2). Since TMTSF is heavy (M.W.=448), the Debye temperature should be low (30-40 K), and the use of Boltzmann statistics at these temperatures is reasonable. The extension to Bose-Einstein statistics for $T<\Theta_D$ is trivial.

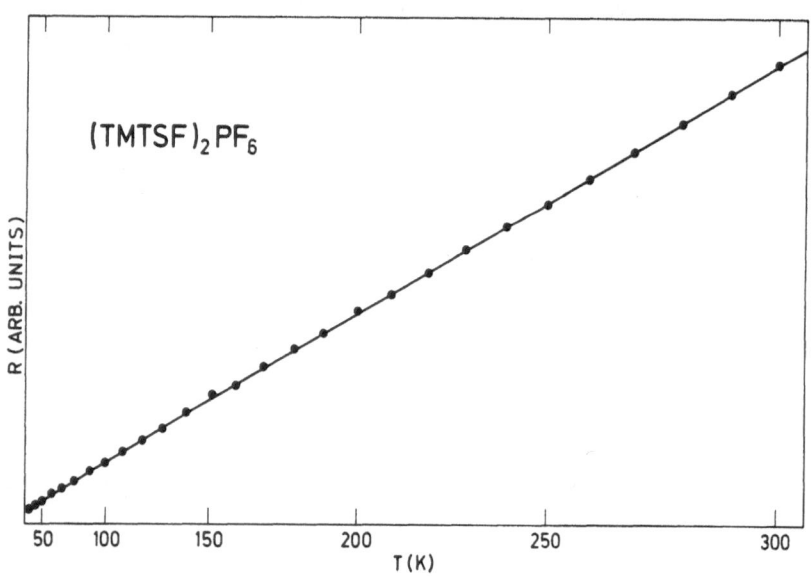

Figure 2. The resistance of TMTSF$_2$PF$_6$ plotted *vs.* T^2.
From Bechgaard and Jacobsen (17).

4. THE PRESSURE DEPENDENCE

The pressure dependence of the electrical conductivity follows from that of the phonon frequencies. Since $\rho \propto \omega^{-4}$, we have:

$$\frac{\partial \ln \rho}{\partial \ln V} = -4 \frac{\partial \ln \omega}{\partial \ln V} = 4 \gamma_e \simeq 16 - 24 \qquad\qquad 3$$

in excellent agreement with experiment. We can easily see how this value of γ_e follows from a Lennard-Jones potential (18) $v = a/r^{12} - b/r^6$. At equilibrium, $\partial v/\partial r = 0$, from which $2a/r_0^{12} = b/r_0^6$. The force constant is given by: $K = \partial^2 v/\partial r^2 = 72a/r_0^{14}$, and its derivative, $-\partial K/\partial r = 21K/r_0$, thus $-\partial \ln K/\partial \ln r = 21$ and $-\partial \ln \omega/\partial \ln r = 10.5$, since $\omega \propto K^{1/2}$. Thus, for phonon modes polarized in the chain direction, $\partial \ln \rho/\partial \ln b = 42$, and since $\delta V/V \simeq 2\delta b/b$ (14), we get $\partial \ln \rho/\partial \ln V \simeq 21$ for such modes. In principle, this result should not be applicable to modes where the atomic motion is not in the chain direction. In the a and c^* directions, the large molecular size modifies the above estimate, since $\delta r/r \simeq \delta a/(a-d_a)$ or $\delta c/(c-d_c)$, where d_a, d_c is the molecular size in the a and c^* directions respectively. This enhancement factor $a/(a-d_a) \simeq c/(c-d_c) \simeq 2$, increases both γ_e and the inverse compressibility in these directions by roughly the same amount, so the effect of a given *stress* in those directions should be comparable to that of a given stress in the b-direction. Therefore, the above reasoning does not depend critically upon the polarization of the phonon modes, and from Zuppiroli's anisotropic stress measurements (19), the modes polarized in the chain direction contribute the most to the resistivity; (probably the TCNQ libration around the long molecular axis). In fig. 3 we plot the experimental

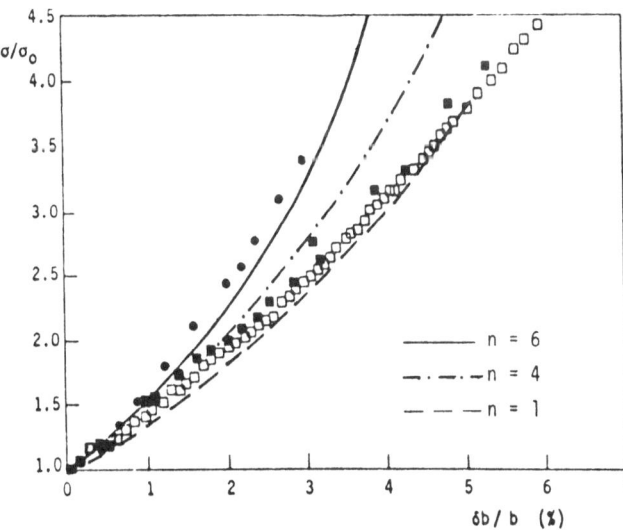

Figure 3. The increase in conductivity plotted *vs.* the decrease in lattice constant (under pressure) for TTF-TCNQ (circles) and TSF-TCNQ (squares). The lines are calculated from libron theory, for a lattice potential given by $v = a/r^{12} - b/r^n$ (ref. 20).

data for TTF-TCNQ and TSF-TCNQ, as well as theoretical curves for
a Lennard-Jones potential $a/r^{12} -b/r^6$, as well as potentials
$a/r^{12} -b/r^n$ for n=4,1 (20). Since the molecules are charged, one
can argue that the Madelung energy contributes a b/r attractive
term, while the bonds possessing a dipolar moment contribute fixed
dipole - induced dipole forces, going like b/r^4. It is seen that
excellent agreement with the TSF-TCNQ data can be obtained. The
effect of the power of the repulsive term is even more drastic, and
Metzger (21) suggested the possibility of a repulsive term with a
power between 9 and 12, also providing excellent agreement with
experiment. However, we wish to point out, that already the "raw"
Lennard-Jones potential provides a rather good agreement with
experiment, *without a single adjustable parameter,* and this good
agreement is even more striking than the perfect fit that can be
obtained with one adjustable parameter. The difference between
TTF-TCNQ and TSF-TCNQ may be due to a soft mode on the TTF chain,
as suggested from the effect of pressure on the Hall constant (22),
on the thermal power (23), and the independent data from the Debye-
Waller factor (24) and Raman scattering (25) which suggest that the
TTF libration in its molecular plane is exceptionally soft. This
interpretation is consistent with that of Conwell (26).

5. THE CONSTANT VOLUME DATA

It was pointed out by Cooper (27) and by Jerome (28) that the
huge effect of pressure on the resistivity gives rise to a qualitatively different behavior of the resistivity at constant volume,
$\rho_V(T)$, from that at constant pressure, $\rho_p(T)$ (fig. 1); they illustrated this by the curve of the resistivity at the 60 K volume as
function of temperature. Just like a curve of $\rho_p(T)$ at a *single*
pressure fails to provide an adequate picture, so does a $\rho_V(T)$
curve at a single volume. Therefore, we show in fig. 4 a set of
curves of $\rho_V(T)$ for several volumes (29). In principle, we can
derive curves at constant volume, or at a constant value of the
lattice parameter b (say), or at a constant value of all three
lattice parameters a, b, and c, (making use of Zuppiroli's uniaxial
stress data (19),) the last requiring a highly anisotropic stress.
The differences between these possibilities are not qualitative,
therefore we follow previous practice and plot data at constant b.
Whether we prefer the constant pressure curves (fig. 1) or the
constant volume ones (fig. 4) is entirely a matter of taste, since
they are *entirely equivalent*; however, we should look at a whole
family of curves, and not at a single one.

The $\rho_V(T)$ curves at the 60 K to 200 K volumes pose no problems,
and have been accounted for quantitatively (20,30) with *only* the
$\rho_{b(60K)}(300K)$ value as adjustable parameter. For the $\rho_p(T)$ curves,
at ambient pressure, the same analysis yields $\rho \propto T^2/(1-0.08 \times T/300)^4$
which fits the n=2.3 law very well (20). Moreover, the exponent n

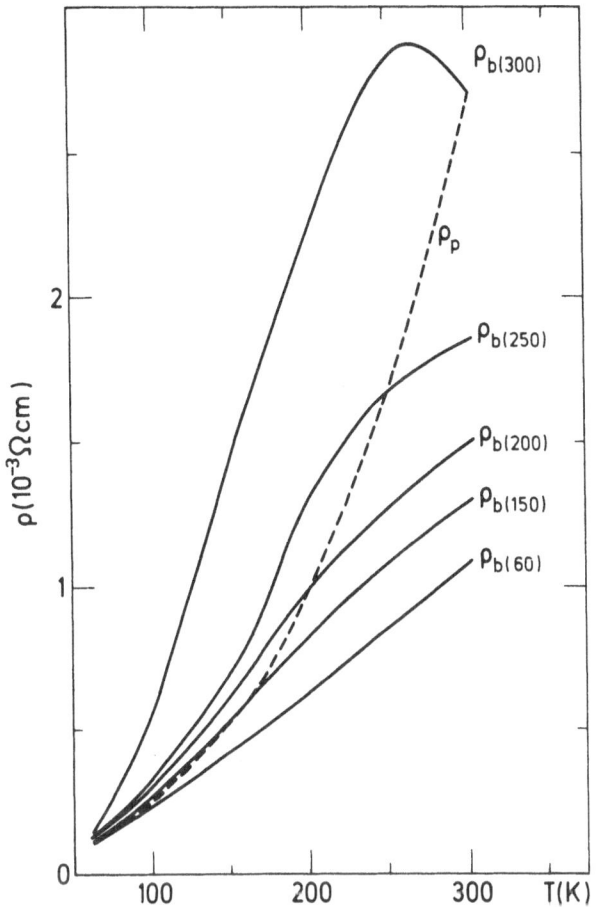

Figure 4. The resistivity of TTF-TCNQ at a constant lattice parameter $b(T_0)$ as function of temperature. T_0 is the temperature at which b attains this value at ambient pressure. The dotted line is the resistivity at ambient pressure. (ref. 29).

that fits this relationship increases somewhat with temperature, in accord with the analysis of Heeger (31). Also, the change of n from 2.3 to 2 under pressure, has been accounted for (29).

Since data on the lattice anharmonicity of TTF-TCNQ is not available, available data on naphtalene and anthracene (32) was used. When we compare TTF-TCNQ at a pressure P to N&A at a pressure P' at which the compressibilities are the same (following the philosophy of Bridgman (33)), we get excellent agreement. When TTF-TCNQ at a pressure P ($\sim$5 kbar) is compared with N&A at the same pressure, agreement is reasonable, while if it is compared with N&A at P=0, agreement is poor, as argued by Zallen and Conwell (34).

This is because the temperature dependence of the phonon frequencies
varies very strongly with pressure,(30) and this "hair splitting"
detail should be considered by theories at the level of precision
that has been attained by now. The argumentation about such details
is an indication of the rapid progress and maturity that our
understanding of the conductivity has reached.

The negative slope of the $\rho_{b(300)}(T)$ curve for $T > 250$ K is
unexpected. The Golden Rule (First Born Approximation) should not
be valid here, since $\hbar/\epsilon_F \tau \simeq 1$ (16); however, this should affect
$\rho_{p=0}(T)$ as well. The same applies to extra relaxation by intramolecular phonons, as suggested by Conwell (35). Phonon anharmonicities can in principle account for such an effect, however, the
degree of anharmonicity required is very large, and requires an
anomalously soft mode. Independent data on the various modes, in
particular the TTF librations (since the translational modes do not
seem to be anomalous, from the neutron diffraction work (36)),
should be available before we can consider this effect to be properly accounted for.

6. THE BLOCH MECHANISM

Classical Bloch theory (1) has been extremely successful in
accounting for the resistivity of metals since 1928. Therefore, it
was natural to try to apply it to organic metals as well, and such
attempts were made since their discovery (13). Some of the
difficulties involved, are:

6.1. Absolute Value.

In order to account for the ambient resistivity of 2 mΩ-cm, and
for $d\rho/dT$, a very large value of the electron-phonon coupling
constant is necessary; $\lambda \simeq 1.3$ was estimated by Heeger (37). This
large value of λ gives rise to a very high Peierls transition
temperature T_p , namely $T_p \simeq 500$ K. If this were indeed the case,
Bloch theory would not be valid in the first place, since $T \ll T_p$.
This argument can be reversed. We can estimate λ from T_p [which
is close to the Mean Field value,(38,9,39)].We get $\lambda = 0.15$-0.2
for T_p=16K-53K. By the Hopfield relation (40) :

$$d\rho/dT = \lambda \ (2\pi k_B/\hbar) \ (4\pi/\epsilon_\infty \omega_p^2) \qquad\qquad 4a$$

where ϵ_∞ and ω_p are derived from the complex dielectric constant

$$\epsilon(\omega)/\epsilon_\infty = 1 - \omega_p^2/(\omega^2 - i\omega/\tau) \qquad\qquad 4b$$

and we get $\rho_{Bloch} \simeq 0.1$-0.2 mΩ-cm at 300 K, and at all temperatures
and for all materials, ρ_{Bloch} is considerably smaller
than the measured ρ .

6.2. Pressure Dependence

The observed variation with pressure, $\partial \ell n\rho / \partial \ell nV \simeq 20$ (formula 3) is too large. For Bloch theory, $\rho \propto T/\omega^2$, and $-\partial \ell n\omega^2 / \partial \ell nV \simeq 10$. The more detailed expression for ρ is: $\rho = (2\pi k_B T/\hbar)(m^*/ne^2)(I^2 n(\varepsilon_F)/2M\omega^2)$. $m^*n(\varepsilon_F)$ decreases somewhat with pressure (27), and Conwell (26) suggested a contribution of 3 from this to $\partial \ell n\rho / \partial \ell nV$. However, the matrix element $I = \langle |\nabla v| \rangle$ increases with pressure, and for an LCAO approximation, $\partial \ell nI / \partial \ell nV = \partial \ell n\varepsilon_F / \partial \ell nV$, since both terms are proportional to the overlap, in the extended Huckel approximation. Thus, the increase of I^2 with pressure cancels the decrease of $m^*n(\varepsilon_F)$. Thus, the contribution of the "electronic" term to the pressure dependence is probably smaller.

6.3. Temperature Dependence

The temperature dependence of the resistivity does not fit. At constant pressure, $\rho_p \propto T^2$ (roughly). At constant volume, $\rho_v \propto T-70$ for T>150 K (fig. 4). Bloch theory predicts $\rho \propto T$ for $T > \Theta_D$ ($\Theta_D \simeq 60K$ here). Also, for T<150 K, ρ_V curves *up*, while for Bloch resistivity it should curve *down* for $T \lesssim \Theta_D$. Conwell (35) suggested a contribution from intra-molecular modes to obtain a temperature dependence which is stronger than in Bloch theory; however, such a contribution should be seen clearly at high pressures ($P \simeq 30$ kbar) where the resistivity due to the external modes is suppressed by an order of magnitude, and experiments (41) at these pressures do not seem to confirm the presence of this mechanism.

7. PHONON DRAG

One of the main "mysteries" of TTF-TCNQ is the strong frequency dependence of the conductivity, which falls off considerably at frequencies of order 10 cm^{-1}. The early Penn data (37) were later confirmed by the excellent measurements of Jacobsen (42). Since this frequency is smaller than the phonon frequencies, and than $k_B T$, it is hard to account for this fall "classically". It was suggested that this frequency dependence is due to phonon drag, $2k_F$ phonons being in equilibrium with the *electrons* (in a stationary, current carrying state) rather than with the other phonons (i.e. the lattice)(16,38,20,43). This is because $2k_F$ phonons are rapidly reabsorbed by the electrons, since *all* electrons at the Fermi surface can participate in the absorption process, in a 1-D system (in contrast with a 3-D one). The possibility of phonon drag was suggested by Peierls in the 30's (44) and it is observed quite frequently in pure metals at helium temperatures (45); but the existence of phonon drag at temperatures higher than Θ_D is somewhat unusual. Detailed estimates of the various relaxation times (20) lead support to this mechanism; moreover, in NbSe$_3$, the continuity of the high-electric-field conductivity at the phase transitions ($\sim$150 K,$\sim$ 60 K) (46) seems to prove the existence of complete phonon drag at those temperatures (47). Additional evidence for phonon drag in TTF-TCNQ is the absence of the Bloch (linear in T) term in the resistivity, its reappearance after irradiation (43); the absence of irradiation effects on the optical conductivity (48). Anomalies in the thermoelectric power (49) seem to be complicated and cannot be accounted-for by phonon drag in a simple way.

A new, important element in the phonon drag picture is the recent observation by Jerome et al (50) of an anomaly in the longitudinal conductivity at 19 kbar, where the CDW becomes commensurate with the lattice. This anomaly is *not* accompanied by a similar anomaly of the transverse conductivity $\sigma_\perp$. Since $\sigma_\perp \propto \tau_\perp^{-1}$, where $\tau_\perp$ is the hopping time between chains, and

$$\tau_\perp^{-1} = \frac{2\pi}{\hbar} |t_\perp|^2 \quad \tau_\parallel/h \qquad\qquad 5$$

where $\tau_\parallel$ is the on-chain scattering time (51), clearly an anomaly in $\tau_\parallel$ should reflect itself in an anomaly in $\sigma_\perp$. However, $\tau_\parallel^{-1} = \tau_{dc}^{-1} + \tau_{2k_F}^{-1}$, where τ_{dc}^{-1} is the scattering rate due to the quadratic-coupling mechanism (section 3), and $\tau_{2k_F}^{-1}$ is the scattering rate by the $2k_F$ phonons. $\sigma_\parallel = ne^2\tau_{dc}/m*$ when complete phonon drag is present, while $\sigma_\parallel = ne^2\tau_\parallel/m*$ when it is completely suppressed. Thus, suppression of phonon drag in the commensurate state, gives rise to an anomaly in $\sigma_\parallel$ without giving rise to one in $\tau_\perp$(and hence, in $\sigma_\perp$). Specifically, suppression of phonon drag here may involve scattering of the electrons (or $2k_F$ phonons) off the $4k_F$ distortion, since $6k_F$ equals a reciprocal lattice vector.†

In addition to the suppression of phonon drag in the commensurate state, the electronic wavefunction will also change (52). Due to the commensurability, the states $\cos(k_F y+\phi)$, with $\phi=0$ (fig. 5, left) and with $\phi = \pi/2$, or $\pi/6$ (fig. 5, right) are not equivalent; in the first, electron-phonon coupling gives rise to triplets of molecules, and in the second - to dublets and singlets. Still, for a harmonic lattice, the states are degenerate. But, when the lattice anharmonicity is taken into account, the states become non-degenerate, the energy difference being

$$\Delta E = \frac{\partial K}{\partial r} <(\delta R_{j+1} - \delta R_j)^3> = 2 \cdot 2\gamma_e \cdot Kb^2 \cdot (\delta R/b)^3 \qquad\qquad 6$$

where δR is the amplitude of the $2k_F$ (or $4k_F$) distortion. In Table 1 we list the values of δR_j, and the harmonic and anharmonic contributions to the lattice energy. For TTF-TCNQ, $2\gamma_e \simeq 20$, $Kb^2 \simeq 10 eV$ $\delta R/b \simeq 0.03$, leading to $\Delta E \simeq 100$ K. This should give rise to an additional increase in the resistivity around and below 100 K. These changes should reflect themselves in small changes in $\sigma_\perp$. The experimental data are not yet sufficiently complete to verify, or refute, the presence of a small anomaly in $\sigma_\perp$ at 19 kbar.

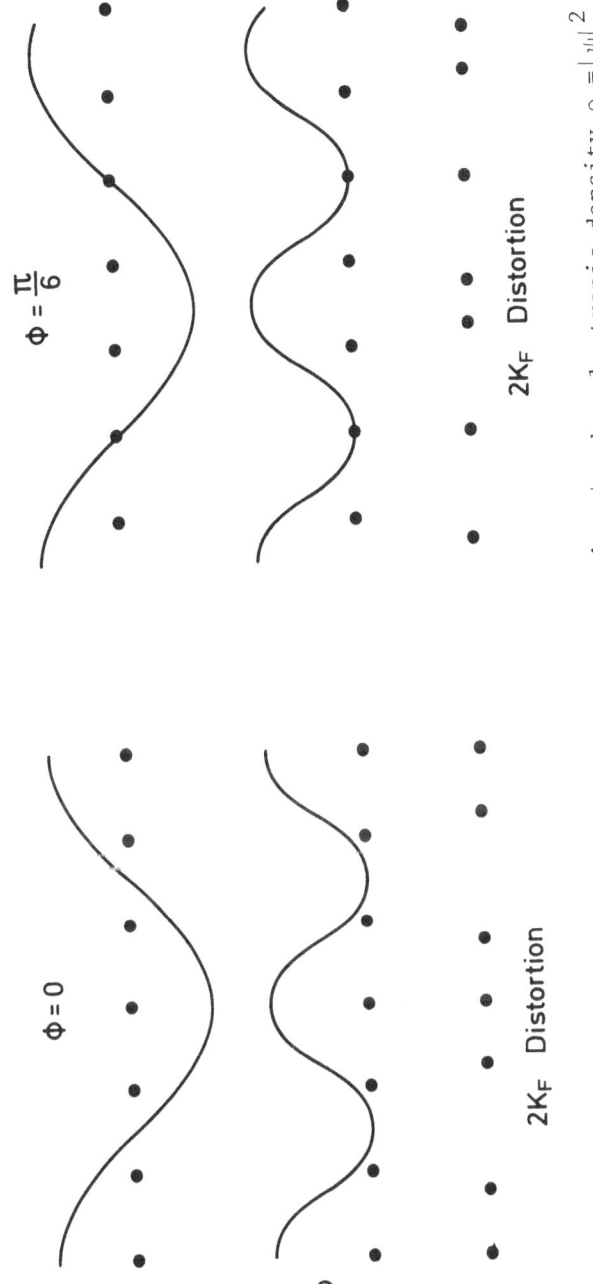

Figure 5. Illustration of the electronic wavefunction ψ, the electronic density $\rho = |\psi|^2$, and the lattice distortion, for the commensurate case of a 1/3 filled band, observed by Jerome in TTF-TCNQ at 19 kbar (50). The standing waves $\cos k_{FY}$ and $\sin k_{FY}$ are no longer degenerate when the lattice anharmonicity is taken into account, and therefore the traveling waves $\exp(+ik_{FY})$ and $\exp(-ik_{FY})$ are no longer eigenstates.

Table 1. The molecular displacements, the changes in molecular separation, and the contributions to the harmonic and anharmonic lattice energy, for the commensurate state (1/3 filled band), for cosine and sine wavefunctions (fig. 5).

	$\phi = 0$				$\phi = \pi/6$			
j	0	1	2	3	0	1	2	3
δR_j	0	1	-1	0	1	0	-1	1
$\delta R_{j+1} - \delta R_j$	1	-2	1	1	-1	-1	2	-1
$(\delta R_{j+1} - \delta R_j)^2$	1	4	1		1	1	4	
$(\delta R_{j+1} - \delta R_j)^3$	1	-8	1		-1	-1	8	
$<(\delta R_{j+1} - \delta R_j)^2>$	2				2			
$<(\delta R_{j+1} - \delta R_j)^3>$	-2				+2			

Note that the phonon drag mechanism can be described by the Boltzmann transport equation, in contrast to Frohlich sliding charge density waves. The validity of the Wiedemann-Franz law was a strong factor leading to the favoring of the former over the later (53). However, as far as the electrical conductivity is concerned, the difference between phonon drag and sliding CDW's is probably not very large (47).

8. ONE DIMENSIONAL LOCALIZATION

In a 1-D system, an infinitesimal potential creates a localized state (54). The present theory assumes from the start travelling wave states. This apparent contradiction poses a dilemma. Friedel pointed out (55), that the resistivity in the 1-D state of HMTSF-TCNQ at 80 K is nearly the same as in the 3-D state at 30 K; thus, it is hard to believe that the mechanisms of resistivity are essentially different in these two states. Still, we must understand *why* this is so. Gogolin et al (56) showed that at finite temperatures phonons may cause delocalization of the electron states, which are localized at 0 K, and give rise to a Drude behavior (characteristic of travelling wave states). However, they require the scattering rate by the phonons to exceed that of the impurities, and under these conditions, the phonons give rise to an appreciable resistance as well, which does not seem to be the case here in the low temperature region (60 K - 100 K). Kaveh et al (57) showed, that for *molecular* crystals, the forward scattering rate by phonons exceeds the backward scattering rate by more than an order of magnitude, because the range of the potential W is large (the molecular size), thus $W(2k_F) \ll W(q \approx 0)$. Thus, the forward-scattering phonons give rise to delocalization, without giving rise to resistivity in the delocalized state. In this picture, there should still be localization at liquid helium temperatures, where even the small-q (forward scattering) phonons are frozen out. There, the 3-D coupling is necessary to ensure delocalization; this 3-D coupling causes a large magnetoresistance (at least in 2-chain-family systems) and thus it is easy to determine in an independent way whether this is the cause for delocalization. In HMTSF-TCNQ (9) and TMTSF-DMTCNQ (58) the states are indeed 3-D below 20-30 K. In TTF-TCNQ above T_p they are definitely not, as determined in an unambiguous way by the measurement of the tunneling rate $\tau_\perp$ by NMR (51). $\tau_\perp^{-1} = 3 \cdot 10^{11} sec^{-1}$ corresponds to a temperature of $1°K$, which is far too small to have an effect at 60 K.

While the approach of Kaveh et al is conservative, and consistent with the Russian work, more radical approaches exist, which claim delocalization at much lower temperatures (59). More experimental work, following the lines of the work by Zuppiroli et al (60) on radiation induced defects, seems to be desired.

The condition $W(2k_F) \ll W(q \approx 0)$ may apply to the potential due to impurities and structural defects as well. Such defects should be extremely inefficient in scattering the electrons in 1-D, since multiple small-q scatterings, which are effective in 3-D, are forbiden in 1-D (fig. 6). This may account for the low

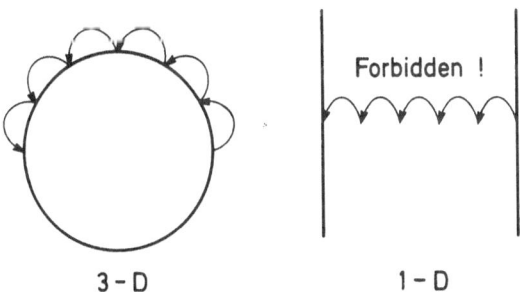

Figure 6. Illustration of the suppression of the residual resistance due to a moderate-range potential, in one dimension.

residual resistance of TMTSF-DMTCNQ, $TMTSF_2PF_6$ (and related compounds), and possibly also $Hg_{3-\delta}AsF_6$ (61), where the residual resistance is less than $1\mu\Omega$-cm in spite of several percent of vacancies in the AsF_6 lattice.

9. THE TRANSVERSE CONDUCTIVITY

A theory of the electrical conductivity of organic metals, must account for the transverse conductivity as well as for the longitudinal one. The Penn group pointed out already in 1974 (62) that $\sigma_{\perp}$ is diffusive; if the hopping time between chains is $\tau_{\perp}$, and the distance is L (L=a/2 and c/2 in the a and c directions), then the diffusion constant is given by $D_{\perp} = L^2/\tau_{\perp}$, and the mobility by $\mu_{\perp} = eD_{\perp}/k_B T$, and the conductivity by:

$$\sigma_{\perp} = [n(\varepsilon_F) k_B T] \ e \ \mu_{\perp} \ = n(\varepsilon_F) \ e^2 \ L^2 \ /\tau_{\perp} \qquad\qquad 7$$

The hopping time was measured directly by NMR (from the cutoff in frequency dependence of T_1)(51) and formula 7 applies very well (fig. 7). The EPR linewidth is also related to $\tau_{\perp}$, since:

$$\Delta H \propto \ (\Delta g^2/\tau_{\|}) \ (\tau_{\|}/\tau_{\perp}) = \Delta g^2/\tau_{\perp} \qquad\qquad 8$$

from the Elliott relation, and from the suppression of the Elliott mechanism by the one-dimensionality (63). This relationship also applies very well (fig. 7).

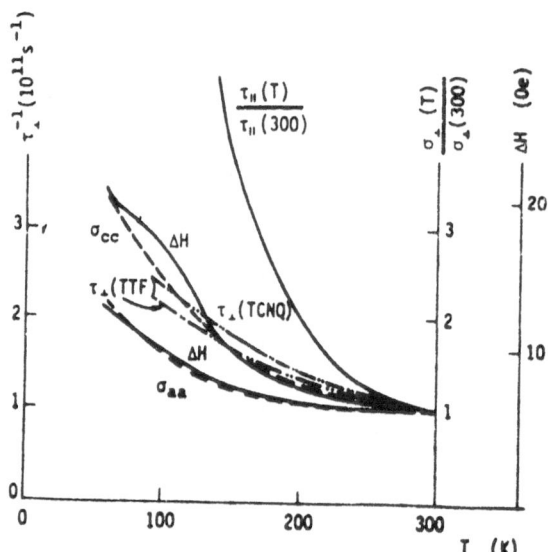

Figure 7. The temperature dependence of the inter-chain hopping time $\tau_{\perp}$ (as measured by NMR), the transverse conductivity, and the EPR linewidth ΔH. The temperature dependence of the longitudinal conductivity (denoted by $\tau_{\|}$ in this figure) is also shown for comparison (ref. 63).

The relationship between $\tau_\perp$ and the on-chain scattering time is given by formula 5 (section 7). However, this is only a rough approximation (see fig. 7). Causes for deviations from this formula are:

9.1. Non Degeneracy.
Non-degeneracy of equal-k donor and acceptor states, as discussed in ref. 63. This is probably the cause for the different temperature dependence of σ_{cc} (donor-donor and acceptor-acceptor hopping) and σ_{aa} (donor-acceptor hopping).

9.2. Phonon Drag.
Emission and absorption of $2k_F$ phonons destroys the phase of the electronic wavefunction, thus contributing to reduce $\tau_\perp^{-1}$, without contributing to $\rho_\parallel$. Therefore we must use for $\tau_\parallel^{-1}$ the value $\tau_{dc}^{-1} + \tau_{2k_F}^{-1}$, rather than τ_{dc}^{-1}, as discussed in section 7.

9.3. Forward Scattering.
Forward scattering (57) also destroys the phase of the electronic wavefunction, thus contributing to reduce $\tau_\perp^{-1}$ without contributing to ρ. The temperature dependence of the forward scattering rate may be different from that of the backward scattering rate, since it attains the value $\epsilon_F/\hbar$ at a lower temperature, and should not increase so fast at higher temperatures (57).

All these 3 factors make $\sigma_\perp$ increase at low temperatures more slowly than $\sigma_\parallel$. Therefore, it is difficult to separate these 3 effects and determine precisely the contribution of each.

Note that formulas 5, 7 yield for the anisotropy:

$$\sigma_\parallel / \sigma_\perp \approx (t_\parallel/t_\perp)^2 (b/L)^2 \qquad\qquad 9$$

where $t_\parallel$, $t_\perp$ are the respective transfer integrals. Thus, the anisotropy is essentially the same as when the conduction process is coherent in the transverse direction as well (3-D state), and essentially the same as when the conduction process is by hopping in the chain direction as well. Therefore, a transition from a 1-D state to a 3-D state does not involve a radical change in the anisotropy; and more involved properties (magnetoresistance, Hall effect, diamagnetic susceptibility) should be investigated (9).

Thus, we may summarize, that an *approximate* theory of the transverse conductivity is almost trivial, when the longitudinal conductivity is known, but a *precise* theory is complicated and involves many factors; the precision with which $\sigma_\parallel$ is accounted for theoretically, is probably better now than that of $\sigma_\perp$.

10. THE ROLE OF U

Up to now, we did not mention the role of the on-site Coulomb repulsion U. This repulsion has been invoked to account for the enhanced susceptibility (64,65) and the $4k_F$ reflections (66,67). There is no question that such a repulsion must exist; the question is just what is its magnitude. The "Big U" school argues that $U \simeq 4t_{||}$ (i.e. the bandwidth), while the "small u" school argues that $U \ll 4t_{||}$. The transport properties support the small-u school; up to now, no suggestion (even on a level of a "hand waving argument") was made on how to account for the semi-metalic state of HMTSF-TCNQ, with an effective ε_F of a few meV (9,68), by "Big U". The argument was made (66,67) that the $4k_F$ reflections are an incipient Wigner condensation, which occurs when $U \simeq 4t_{||}$. The problem is, that a Wigner condensation is a 3-D phenomenon. The Coulomb repulsions between neighboring molecules, V_1, are *not* anisotropic; the distance between atoms on neighboring molecules is about 3 Å *both* along the chains, and between chains. On the other hand, the $4k_F$ reflections are very strongly 1-D over a very wide temperature range (300K → 50K). This property requires a re-evaluation of the theory for these reflections.

A "small u" theory for 1-D systems was proposed by Overhauser in 1960 (69). There it was shown that the SDW has a lower energy than the metalic state at OK for U as small as we please; however, the gap of the SDW state is essentially given by a BCS like relationship $\Delta_{SDW} \simeq \varepsilon_F \exp(-1/\mu)$, where $\mu = U/4t_{||}$. Moreover, the charge of the Overhauser state is perfectly uniform, thus there is no electrical coupling between chains (and magnetic coupling is too weak to be significant). Based on this work, it was suggested just before this conference (52) that the 1-D nature of the Overhauser state suppresses the real transition temperature well below the mean-field temperature, $k_B T_{SDW}^{MF} \simeq \Delta_{SDW}/1.75$ towards OK, and thus we have a fluctuating state over a very wide temperature range, and the $4k_F$ reflections are a manifestation of this fluctuating state. Thus, we may have $T_{SDW}^{MF} \simeq 160$ K, $T_{CDW}^{MF} \simeq 80$ K, $T_{CDW} = T_P = 53$ K, $T_{SDW} \simeq 53$ K (fig. 8). This accounts for the observation of the $4k_F$ reflections at high temperatures, where the $2k_F$ reflections (which are a manifestation of the CDW state) are not observed. To obtain the $4k_F$ reflections, an electron-lattice coupling term was added to Overhauser's Hamiltonian.

The difference between "Big U" and "small u" theories is, that the first is a strong-coupling theory, while the second is a weak-coupling one. If we ask:"At what value of U is the magnitude of the modulation of the electronic density, with wavevector $4k_F$, comparable to the electronic density itself", the answer is $U \simeq 4t_{||}$, i.e. "Big U". However, if we ask:"at what value of U, is the magnitude of the modulation of order $\delta R_{4k_F}/b$", where δR_{4k_F} is the amplitude of the molecular displacement with wavevector $4k_F$, the answer is $U \ll 4t_{||}$, i.e. "small u".

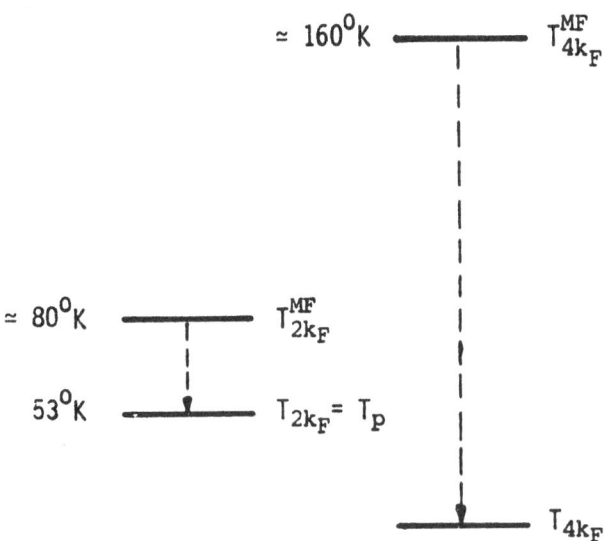

Figure 8. Illustration of the large suppression
of the SDW transition, which gives the $4k_F$ reflections, compared with the CDW one (giving the
$2k_F$ reflections), due to the extreme one-dimensional nature of the former (ref. 52).

Moreover, we should note that the U we talk about, is a
"dressed", i.e. screened U, and moreover only the short-range
component of it, i.e. $U = U_0 - 2V_1$, where U_0 is the on-site
interaction and V_1 the nearest-neighbor (same chain) interaction.
This is *not* the quantity that enters the cohesive (i.e. Madelung)
energy, for example (21,70). $T_{SDW}^{MF} \simeq 160$ K implies $U \simeq 0.1$ eV,
while the "bare" U can be much, much bigger.

Another feature of Overhauser's theory is that the spin
susceptibility $\chi_s(q)$ is affected *only* near $q = 2k_F$, since the
coupling is weak; so we should not expect big anomalies in the
measured susceptibility $\chi_s(0)$ near $T_{4k_F}^{MF}$. The enhancement of
the susceptibility observed in TTF-TCNQ, and some other materials
(71) at high temperatures, may be accounted for by polaron theory
(20); any localization giving rise eventually to a Curie law,
whether the localization is due to U or to a drastic reduction of
the electronic mean-free-path by phonon scattering.

It may also be instructive to point out the relation of this
theory to Kommandeur's work (72). What he calls a "one electron
bond", we call an Overhauser SDW state, since the maxima in the
electron density are due to single electrons (spin-up or spin-down).
For a quarter-filled band, this indeed gives rise to dimerization(52).

What he calls a "two electron bond", we call a Peierls-Frohlich
CDW state, since the maxima in the electronic density are due to
simultaneous occupation by spin-up and spin-down electrons; for the
quarter filled band, this indeed gives rise to tetramerization (52).

11. THE COUPLING CONSTANT

As noted in section 5, the "libron" theory of the resistivity
contains one adjustable parameter - the value of the resistivity at
ambient temperature and pressure. Clearly, this quantity must be
calculated for the theory to be complete. The quantity involved here
is $\partial^2 W/\partial\theta^2$ (section 3). When we take for W the transfer integral,
$t_{||}$, the coupling constant is rather weak; $\partial^2 t_{||}/\partial\theta^2 \simeq t_{||}$ (since for
a p_z state, $\psi \propto \cos\theta$), and $\partial^2 t/\partial\theta^2 <\theta^2> \simeq t\times3\cdot10^{-3} \simeq 3$ meV.
However, the 1-electron potential gives rise to a much larger
contribution, namely

$$\partial^2 W/\partial\theta^2 \ <\theta^2> \ \simeq \ W \ (R_1/\Delta R)^2 \ <\theta^2> \ \simeq \ 30 \ \text{meV} \qquad\qquad 10$$

where $W \simeq Ze^2/R_1 \simeq 3$ eV is the potential, ΔR is the distance
between neighbouring atoms (belonging to different molecules),i.e.
$\Delta R \simeq 3$ Å, and R_1 is half the length of the molecule, $R_1 \simeq 5$ Å. Thus
we get a coupling constant of about 30 meV for one mode. Since we
have 6 modes per molecule (3 translons and 3 librons), and, moreover,
the movements of the neighboring molecules (there are 12 nearest
neighbors) also contribute to the coupling, we obtain a total
coupling constant of about 100 meV, which accounts for the observed
resistivity.(Probably this value is trustworthy to within a factor
of 2).

The key question here is, why is this quadratic coupling
constant so big, namely big enough to overwhelm the linear coupling
(section 6) at ambient conditions, while in "ordinary" metals this
is not the case. When we make a rough comparison between the
quadratic and linear coupling constants, we get a ratio:

$$\frac{\text{Quadratic}}{\text{Linear}} \simeq p \left(\frac{\omega_{LA}}{\omega_{Lib}}\right)^2 \left(\frac{\text{size of Molecule}}{\substack{\text{intermolecular.}\\ \text{atomic separation}}}\right)^2 \{Q^2<\delta r^2>\} \qquad 11$$

where p (p=6) is the number of modes per molecule, ω_{LA} is the
frequency of the LA mode, and ω_{Lib} is an average frequency of the
librational (and translational, transverse and longitudinal) phonons
$[(\omega_{LA}/\omega_{Lib})^2 \simeq 2]$; Q is the Slater parameter of the wavefunction
$[\exp(-Qr)]$, and $<\delta r^2>$ is the amplitude of the thermal motion.
The quantity $Q^2<\delta r^2> \simeq 1/6$ is small, and is the basis of Bloch's
statement (1) that the quadratic term is a small correction;
(although in soft organic crystals this term is larger than in
aluminum, say). A factor that has not yet received attention, is

the role of the size of the molecule. In a localized picture, we
can distinguish between the electron-phonon coupling arising from
the motion of the molecule on which the electron sits, and that of
the neighboring molecule. The first gives rise to quadratic coupling
(by symmetry). For atoms, the motion of a neighbor affects the
potential on the *whole* atom, since the size of the electronic
wavefunction is smaller (or at most equal) to the atomic separation.
Therefore, the electron-phonon coupling due to motion of the
neighbor is big. For molecules, the movement of a neighboring
molecule (and there are 12 of them, 8 of which are illustrated in
fig. 9) affects the potential of only a small part of the large
molecule, namely of the order of the interatomic separation. This
is the origin of the third term in formula 11. This term is of

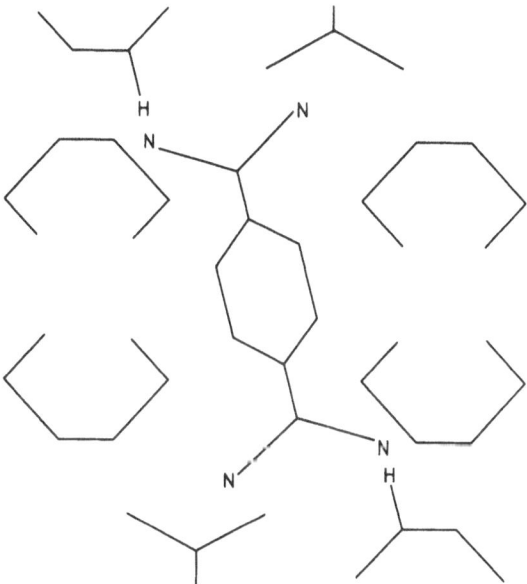

Figure 9. Illustration of the contributions to the change
of the one-electron potential due to a libration of a TCNQ
molecule in TTF-TCNQ. Essentially all 12 neighbors (8 of
which are illustrated) contribute to the change of the
potential on the TCNQ nitrogens (where most of the charge
is). By symmetry, the change is *quadratic* in the libration
angle. The movement of the TCNQ affects, however, the
potential on only a *small part* of each neighboring molecule.
This change is *linear* in the libration angle.

order 10, thus we get a quadratic coupling contributing about 20
times more to the resistivity than the linear coupling.

In the above estimate, we assume that all modes (librations, translations) contribute equally. More detailed estimates show that the librons contribute more. The essential feature here is the importance of the *size* and *shape* of the molecules. TCNQ and TTF molecules are long, narrow, and flat. Therefore, the amplitude of libration around the long molecular axis is very large; and since the π-wavefunctions are essentially planar, the coupling of this libration with the electrons is very strong, and this is the reason why the quadratic coupling dominates. Different molecules, such as TTT or TST (73), are broad, therefore the amplitude of their librations is much smaller, and the quadratic coupling need not dominate the linear coupling to such an extent.

Clearly, this order-of-magnitude estimate must be corroborated by a detailed numerical calculation of the coupling constants.

12. WHY ARE SOME ORGANIC METALS SUCH GOOD CONDUCTORS ?

The survey of the conductivity theory presented here is rather complex; moreover, much important work is not discussed here at all. We find that some organic metals are excellent conductors, attaining conductivities in excess of $10^4 (\Omega\text{-cm})^{-1}$ (TTF-TCNQ, HMTSF-TCNQ, HMTSF-TNAP) in the liquid nitrogen region, and in excess of $10^5 (\Omega\text{-cm})^{-1}$ (TMTSF-DMTCNQ, $TMTSF_2PF_6$) below 20 K. Therefore we may ask whether there is some general underlying feature that makes (some) one dimensional materials such good conductors. A partial answer to this is the *restriction of phase space*. The quadratic coupling of the libron, and absence of linear coupling, arises from a restriction in the direction of the electronic wavevector k to be along the chain (section 3). Phonon drag arises from the restriction of phase space available for phonons interacting linearly with the electrons ($q_\parallel$ is restricted to $2k_F$)(section 7). The low residual resistance is due to inavailability of phase space for multiple scattering (fig. 6). Similarly, the stability of solitons may be traced to a similar cause. On the opposite side, localization by defects and impurities is also strong in 1-D because there are less channels for the electron to leak away. The Peierls transition can also be regarded as such an effect. Therefore, 1-D systems seem to tend to extremes - either be very good conductors, when the purity is reasonable and the linear electron-phonon coupling is not too strong; or be insulators, if this is not the case. 3-D metals seem to be more compromising, because of the larger phase space available for the various processes.

13. CONCLUSION

Various conductivity mechanisms have been proposed for the high-conductivity organic metals. First, the *classical Bloch* mechanism, based on the *annahme* that all phonons are in equilibrium with the *lattice*, and applying to conditions where the quadratic coupling need not be entirely absent, but is considerably weaker than the linear coupling. Second, there is the *Boltzmann Phonon Drag* mechanism, namely a process treated by the Boltzmann transport equation, which assumes no correlation neither among electrons nor among phonons, the $2k_F$ ones in particular. Third, there may be *Frohlich Sliding CDW's*. This mechanism is not as well defined as the preceding two, and the thermal conductivity and TEP expected in this state are not understood yet. Fourth, there is *BCS Superconductivity*, fluctuating or permanent. All three last mechanisms give rise to paraconductivity, therefore when we talk about paraconductivity, we should be specific about which mechanism we imply.

Different materials display different mechanisms, and even the same material may change from one region to another when the temperature, pressure, doping, or other conditions are varied. There is not yet complete concensus as to where TTF-TCNQ at 100 K (say) is; for TMTSF-DMTCNQ at 20 K and 12 kbar, both fluctuating BCS (58) and classical Bloch (74) processes have been suggested. There has been much progress in the last year both experimentally and theoretically, and the field still seems to be very fruitful.

Acknowledgement

This talk is concerned mainly with work done jointly with H. Gutfreund and M. Kaveh in Jerusalem. The author benefitted greatly from a stay as Nordita guest at the II. C. Ørsted Institute in Copenhagen, and collaboration with K. Bechgaard and K. Carneiro.

References

1. F. Bloch, Z. Physik 52(1928)555.
2. J.P. Ferraris, D.O. Cowan, V.V. Walatka, J.E. Perlstein, J. Am Chem. Soc. 95(1973)948.
3. L.B. Coleman, M.J. Cohen, D.J. Sandeman, F.G. Yamagishi, A.F. Garito, A.J. Heeger, Solid State Commun. 12(1973)1125.
4. R.P. Groff, A. Suna, and R.F. Merrifield, Phys. Rev. Lett. 33(1974)418.
5. J.P. Ferraris and T.F. Finnegan, Solid State Commun. 18(1976) 1169.
6. J.R. Cooper, D. Jerome, M. Weger and S. Etemad, J. Physique Lett. 36(1975)L-219.
7. Organic Conductors and Semiconductors, Conference Proceedings Siofok Hungary 1976, Lecture Notes in Physics 65 (Springer Verlag) 1977.

8.   Quasi One Dimensional Conductors, Conference Preceedings,
      Dubrovnik 1978, Lecture Notes in Physics 96, S. Barisic,
      A. Bjelis, J.R. Cooper and B. Leontic Eds.(Springer) 1979.
9.   D. Jerome and M. Weger in Chemical Physics of One Dimensional
      Metals, H.J. Keller Ed. NATO ASI Series B25, Plenum Press N.Y.
      1977. M. Weger, Bull. E.P.S. News 9(1978)7/8 p. 7.
10.   D.A. Daws, L. Hso, S.S. Mitra, O. Brafman, M. Hayek, W.B.
      Daniels Chem. Phys. Lett. 22(1973)595.
11.   R. Zallen and M.L. Slade, Phys. Rev. B18(1979)5775.
12.   T.G. Gibbons, J. Chem. Phys. 60(1974)1094.
13.   D.E. Schafer, G.A. Thomas and F. Wudl, Phys. Rev. B12(1975)
      5532.
14.   D. Debray, R. Millet, D. Jerome, S. Barisic, L. Giral and
      J.M. Fabre, J. Physique Lett. 38(1977)L-227.
15.   M. Weger and J. Friedel, J. Physique 38(1977)241,881.
16.   H. Gutfreund and M. Weger, Phys. Rev. B16(1977)1753. M. Weger
      and H. Gutfreund, Comments on Solid State Phys. 8(1978)135.
17.   K. Bechgaard and C.S. Jacobsen, this Conference, and to be
      published.
18.   M. Kaveh, H. Gutfreund and M. Weger, Solid State Commun.
      27(1978)53.
19.   S. Bouffard, A. Bittar and L. Zuppiroli, p. 103 of ref. 8.
20.   M. Kaveh, H. Gutfreund and M. Weger, p. 105 of ref. 8.
21.   R.M. Metzger, J. Chem Phys. (in press) and this Conference,
      pp. 233.
22.   J.R. Cooper, M. Miljak, G. Delplanque, D. Jerome, M. Weger,
      J.M. Fabre and L. Giral, J. Physique 38(1977)1097.
23.   C. Weyl, P.M. Chaikin, A. Andrieux and D. Jerome (to be
      published).
24.   A.J. Schultz, G.D. Stucky, R.H. Blessing, P. Coppens,
      J. Am Chem. Soc. 98(1976)3194.
25.   H. Kuzmany and H.J. Stoltz, J. Phys. C 10(1977)2241.
26.   E.M. Conwell, this Conference, pp. 213.
27.   J.R. Cooper, Phys. Rev. B19(1979)2404.
28.   R.H. Friend, M. Miljak, D. Jerome, D.L. Decker and D. Debray
      J. Physique Lett 39(1978)L-134.
29.   M. Kaveh and M. Weger, to be published.
30.   M. Weger, M. Kaveh and H. Gutfreund, J. Chem. Phys. (in press).
31.   A.J. Heeger in One Dimensional Highly Conducting Solids,
      Devreese Ed. Plenum Press N.Y. 1979.
32.   M. Nicol, M. Vernon and J.T. Woo, J. Chem. Phys. 63(1975)1992.
33.   P.W. Bridgman, Collected Experimental Papers, Harvard University
      Press 1964, Volume VI, pp. 3620, 3926.
34.   E.M. Conwell and R. Zallen, Solid State Commun. (in press).
35.   E.M. Conwell, Phys. Rev. Lett. 39(1977)777.
36.   S.M. Shapiro, G. Shirane, A.F. Garito and A.J. Heeger,
      Phys. Rev. B15(1977)2413.
37.   A.J. Heeger and A.F. Garito in Low Dimensional Cooperative
      Phenomena, H.J. Keller Ed. NATO ASI Series 9, Plenum Press 1974.
38.   J. Bardeen, in the book of ref. 31.

39. J.B. Nielsen and K. Karneiro, P. 238 of ref. 8, and to be published.
40. J.J. Hopfield, Comments on Solid State Physics 3(1970)48.
41. C.W. Chu, J.M. Harper, T.H. Geballe and R.L. Greene, Phys. Rev. Lett. 31(1973)1431. R.H. Friend, D. Jerome, J.M. Fabre, L. Giral and K. Bechgaard, J. Phys. C 11(1978)263.
42. C.S. Jacobsen, p. 223 in ref. 8.
43. M. Kaveh, H. Gutfreund and M. Weger, Phys. Rev. B18(1978)7171.
44. R.E. Peierls, Quantum Theory of Solids, Clarendon Press Oxford 1955.
45. P.M. Chaikin, this Conference, pp. 53.
46. P. Monceau, N.P. Ong, A.M. Portis, A. Meerschant and J. Rouxel Phys. Rev. Lett. 37(1976)602.
47. A.J. Heeger, M. Weger and M. Kaveh, p. 316 of ref. 8.
48. W.J. Gunning and A.J. Heeger, to be published. M. Kaveh and M. Weger, Phys. Rev. B (sept. 1979).
49. P.M. Chaikin, R.L. Greene and E.M. Engler, Solid State Commun. 25(1978)1009.
50. A. Andrieux, H.J. Schultz, D. Jerome and K. Bechgaard, J. Physique Lett. 40(1979)L-385.
51. G. Soda, D. Jerome, M. Weger, J. Alizon, J. Gallice, H. Robert, J.M. Fabre and L. Giral, J. Physique 38(1977)931.
52. M. Weger and H. Gutfreund, Solid State Commun. (in press).
53. M.B. Salomon, J.W. Bray, G. de Pasquali, R.A. Craven, G. Stucky and A. Schultz, Phys. Rev. B11(1975)619. M. Kaveh, H. Gutfreund and M. Weger, Phys. Rev. B 20(1979)543.
54. N.F. Mott and W.D. Twose, Adv. Phys. 10(1961)107.
55. J. Friedel, Dubrovnik Conf. and private communication.
56. A.A. Gogolin, V.I. Melnikov and E.I. Rashba, JETP 42(1976)168.
57. M. Kaveh, M. Weger and H. Gutfreund, Solid State Commun. 31 (1979)83.
58. A. Andrieux, C. Durore, D. Jerome and K. Bechgaard, Comptes Rendus B (May 1979) and J. Physique (in press).
59. P. Wolfle and W.Gotze , Solid State Commun. 30(1979)
60. L. Zuppiroli, J. Arionceau, M. Weger, K: Bechgaard and C.Weyl J. Physique Lett. 39(1978)L-170.
61. C.K. Chiang, R. Spal, A. Denensten, N.D. Miro and A.G. McDiarmid, Solid State Commun. 22(1977)293.
62. A.J. Heeger in the book of ref. 9.
63. M. Weger, J. Physique 39(1978)C6-1456.
64. J.B. Torrance, in the book of ref. 9.
65. L.G. Caron, M. Miljak and D. Jerome, J. Physique 39(1978) 1355.
66. V.J. Emery, Phys. Rev. Lett. 37(1976)107.
67. J. Kondo and K. Yamaji, J. Phys. Soc. Japan 43(1977)424. J. Hubbard, Phys. Rev. B17(1978)494.
68. M. Weger, Solid State Commun. 19(1976)1149.
69. A.W. Overhauser, Phys. Rev. Lett. 9(1960)462.
70. A.N. Bloch, this Conference.
71. P. Delhaes and S. Flandrois, this Conference.

72. J. Kommandeur, this Conference, pp. 197. S. Huizinga, J. Kom-
     mandeur, G.A. Sawatzky, B.T. Thole, K. Kopinga, W.J.M. de
     Jonge, and J. Roos, Phys. Rev. B19(1979)4723.
73. V.F. Kaminskii, M.L. Khidekel, R.B. Ljubovskii, J.J. Shchegolev,
     R.D. Shibaeva, E.B. Yagabskii, A.V. Zvarkina and G.L. Zuereva,
     Phisica Status Solidi A44(1977)77.
74. U. Hardebusch, W. Gerhardt, J.S. Schilling, K. Bechgaard,
     M. Weger, M. Miljak and J.R. Cooper, Solid State Commun.
     (in press).

NEUTRON SCATTERING STUDY OF TTF-TCNQ UNDER HYDROSTATIC PRESSURE UP TO 5 Kbar

S. Megtert, R. Comès
Laboratoire de Physique des Solides associé au CNRS
Université Paris-Sud, Bâtiment 510
91405   ORSAY   (France)

R. Pynn and C. Vettier
I.L.L. - B.P. 156
38042   GRENOBLE-CEDEX   (France)

A.F. Garito
University of Pennsylvania
PHILADELPHIA, Penn. 19104, (USA)

I.   INTRODUCTION

The pressure dependence of the three phase transitions of TTF-TCNQ has been recently mapped from conductivity measurements (1) as shown in fig 1. This phase diagram reveals two pressure ranges of interest :

a) up to about 15 Kbar where the three phase transitions, observed at ambient pressure, merge into one : in this range, the dominant role seems to be played by the changes induced by pressure on the interchain   coupling ;

b) between 17 and 21 kbar where the single phase transition is of first order while it is of second order on each side, with a sharp minimum in the longitudinal conductivity (2) : in this range, the dominant role seems to be played by changes induced by pressure in the conducting chains themselves.

We shall describe below the present experimental status of TTF-TCNQ as deduced from structural studies under hydrostatic pressure performed by neutron scattering at the Institute Laue-Langevin. The experimental conditions are the same as described in an earlier report (3).

L. Alcácer (ed.), The Physics and Chemistry of Low Dimensional Solids, 101–112.

Figure 1

The phase diagram of TTF-TCNQ
as deduced from high pressure
transport studies (from Friend
et al (1)).

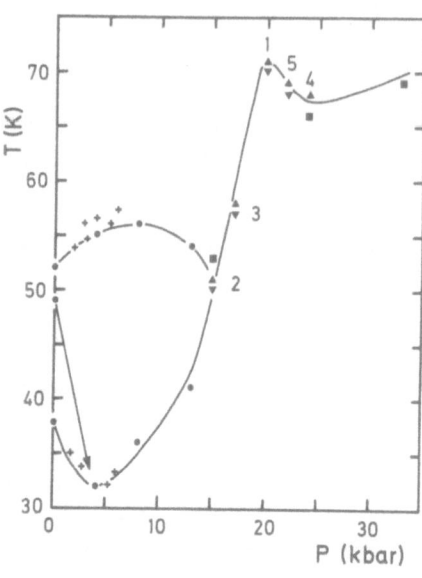

Figure 2

Gradual change with decreasing
temperature of the wave vector
corresponding to $4k_F$ scattering
in TTF-TCNQ. Since $4k_F = 1-0.45b^*$
at room temperature, this
change corresponds to an
increasing charge transfer when
the lattice contracts with
decreasing temperature (from
Khanna et al (4)).

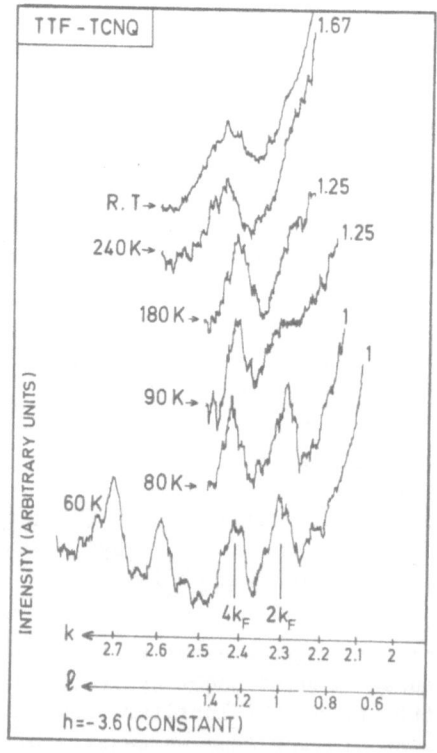

II. PRESSURE INDUCED CHARGE TRANSFER CHANGE

The first order character of the Peierls transition between 17
and 21 Kbar was interpreted as a direct consequence of commensurability between the $2k_F$ charge density wave (CDW) and the
underlying lattice (1, 2). This assumption was partially supported by X-rays measurements on the $4k_F$ precursor scattering
(4, 5) ; as shown in fig 2, this scattering is indeed temperature dependent and its variation corresponds to an increase of
the charge transfer as the lattice contracts. Starting from
$2k_F = 0.295$ b* at ambient pressure and low temperature, a
small increase induced by pressure could lead $2k_F$ to cross
the commensurate value 1/3 b*,locking the CDW to the lattice.
This is expected to occur in the pressure range 17 to 21 Kbar
where the Peierls transition is of first order. A further
increase of the charge transfer at higher pressure (above
21 Kbar) would overcome the pinning energy to the lattice and
restore the incommensurate CDW and therefore the second order
character of the Peierls transition.(1).

Elastic neutron scattering measurements have indeed
revealed the increase of the $2k_F$ wave vector from 0.295 b* at
ambiant pressure to 0.308 b* at 5Kbar (3) (which corresponds to
a charge transfer increase from 0.59 electrons to 0.616 electrons). Measurements at higher pressure have   to date not
been possible, but this increase clearly extrapolates to about
1/3 b* around 17 Kbar. Figure 3 which includes additional data
obtained under the same experimenal conditions as in the
earlier report (3) illustrates this increase of $2k_F$ as a function of pressure. Figure 4 shows some typical scans, through
satellites of the modulated phases and in chain direction,
from which the change of $2k_F$ is inferred.

The structural study, even with its present upper limit of
5Kbar, brings consequently strong support to the explanations
of the anomalous transport properties around 19 Kbar in terms
of commensurability effects.

III. THE PHASE DIAGRAM OF TTF-TCNQ UP TO 5 Kbar

The other point of interest is of course the study of the
effect of pressure on the interchain coupling which is revealed by the pressure dependence of the three phase transitions
observed at ambient pressure ($T_H = 54$ K, $T_M = 48$ K, $T_L = 38$ K).

These transitions are presently understood as arising from the
successive Peierls distortions on the TCNQ chains ($T_H = 54$ K)
and on the TTF chains ($T_M = 48$ K), followed by an
intermediate phase between 48 and 38 K. This intermediate phase
is due to the competition between the coupling of identical

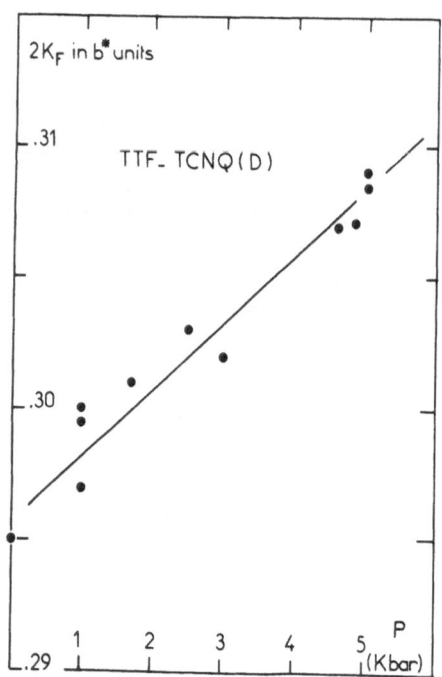

Figure 3

Increase of the $2k_F$ wave vector as a function of applied pressure ; deduced from the longitudinal wave vector component of the $(0.25\ a^{*},\ (1-2k_F)b^{*},\ 0c^{*})$ and $(0.5a^{*},(1-2k_F)b^{*},0c^{*})$ satellite reflexions of the low temperature modulated phases.

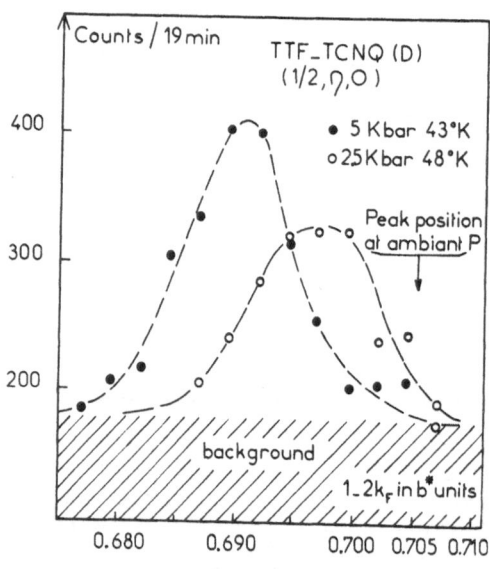

Figure 4

Typical scans showing the shifting in b^{*} direction of the satellite peak position at 2.5 and 5Kbar and in the phase with transverse 2a multiplicity.

chains and unlike chains which drives the transverse periodicity from its 2a value (between 54 and 48 K) towards the 4a
value where it locks-in discontinuously at 38 K (6, 7).

Another aspect which will be more directly relevant to the
present neutron scattering investigation, is that it.was clearly shown that the 54 K transition corresponds to charge density
waves mainly polarised in the transverse c^* direction, in contrast with the 49 K transition which corresponds to a longitudinal distortion in chain direction (4, 8). Below we shall
describe results which have been obtained in the distorted
state of TTF-TCNQ, but exclusively on a satellite with the
scattering vector

$$\vec{Q}_L = (\xi\vec{a}^*, (1 - 2k_F)\vec{b}^*, 0\vec{c}^*)$$

The intensity of such a satellite, arising from a displa-
cive modulation, is proportional to $(\vec{Q}_L.\vec{e})^2$, where $\vec{e}$ is the
polarisation of the lattice distortion. This means that we can
only measure   the pressure dependence of the longitudinally
polarised distortion. In other words, at ambient pressure, our
measurements would only concern the phase below T_M = 48 K,
because such satellites have zero intensity above this
temperature.

Figure 5 shows some typical scans in a^* direction through
the Q_L satellite, at 43 K, and as a function of increasing
pressure. At ambient pressure, this temperature corresponds to
the intermediate phase at X (T)a x $2\pi/2k_F$ b x c. With increasing
pressure the scans of figure 5 show that the transition between
the intermediate phase (X(T, P) multiplicity) and the 2a multiplicity phase takes place between 1.5 and 2.5 Kbar. At this
last pressure, the satellite has reached the transverse zone
boundary at 0.5 a^*. This confirms, regarding the multiplicity
in a^* direction, the decrease with increasing pressure of the
T_M transition as deduced from transport studies (fig 1). Refe-
ring to the details given above about the polarisation of the
distortion it is however a surprise to observe the $\vec{Q}_L$ satellite
at 0.5 a^*(2a multiplicity)in a phase which at ambient pressure
corresponds to a purely transverse distortion which cannot contribute to the Q_L satellite intensity .We shall come back to
this point   below, and first follow the phase diagram
from the point of view of multiplicity along a.

Figure 6 corresponds to the temperature dependence of the
Q_L satellite at 2.5 Kbar, it shows that, at this pressure, the
T_M transition has merged with the lock-in T_L transition. At
2.5 Kbar the intermediate phase with the multiplicity x(T, P.)a
has disappeared, and the transverse multiplicity changes
directly and discontinuously from 4a below 34 K to 2a above 35K.
This is again in fair agreement with the phase diagram of
figure 1.

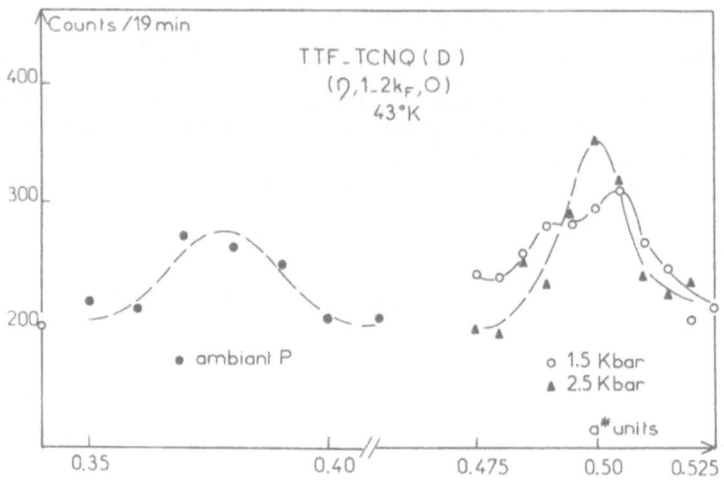

Figure 5

Scans along the a* direction through the low temperature satellites at constant temperature 43°K and increasing pressure. The
phase transition between the x(T, P)a* and the 2a transverse
multiplicity occurs between 1.5 and 2.5 Kbar.

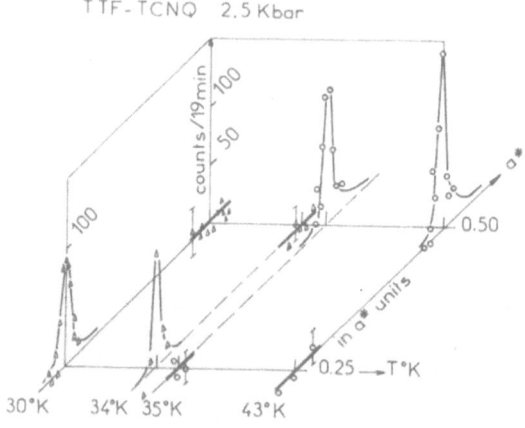

Figure 6

Series of scans in a* direction at 2.5Kbar and with increasing
temperature. The satellite peak position shifts discontinuously
from 0.5 a* below and at 34°K to 0.5 a* at and above 35°K.

If the Q_L satellite intensity is now followed at constant temperature (10°K) as a function of increasing pressure (fig 7) an unexpected feature is observed between 3 and 4.6 Kbar. Below and at 3 Kbar, as observed at ambiant pressure the transverse multiplicity is 4a. At 4.6 Kbar, the intensity at the corresponding position (0.25 $a^{::}$, (1 - $2k_F$)$b^{::}$, $0c^{::}$) has dropped to zero, and the satellite is found at the position (0.5 $a^{::}$, (1 - $2k_F$)$b^{::}$, $0c^{::}$) where no intensity was detected at 3 Kbar. Figure 8 shows typical scans through these satellites at respectively 3 and 4.6 Kbar. This shows that at low temperature and with increasing pressure, a structural modification, not revealed from the transport studies, brings the transverse multiplicity from 4a back to 2a. Subsequent measurements performed at Brookhaven have even revealed that two phase transitions occur in this pressure range with   an intermediate doubly incommensurate phase (9).

Regarding the high pressure 2a multiplicity phase, figure 10 illustrates the temperature dependence of the Q_L satellite intensity in the 4.6 - 5 Kbar pressure range. The satellite intensity drops to zero at the upper phase boundary (fig 1) but no detectable change is observed at the lower phase boundary line inferred from the transport studies. As mentioned several times above, the longitudinally polarised distortion component is the only one to contribute to the Q_L satellite intensity.

IV. DISCUSSION

The direct structural evidence of the pressure dependence of the value of the $2k_F$ wave vector of TTF-TCNQ, deduced from the present neutron scattering studies, gives clearly strong support to the idea that   commensurability effects of the charge density waves are responsible for the particular transport properties observed around 19 Kbar.

The phase diagram as deduced from the type of multiplicity of the low temperature distorted phases and shown in figure 11 reveals however important differences compared to the diagram of figure 1. Besides the additional structural change observed at low temperature in the 3 to 5 Kbar range (hatched area of fig 12), and investigated in more detail in ref 9, only one phase transition is observed at 5 Kbar at about 57 K. Some implications of this earlier results were already discussed by Megtert et al (3) who put forward two possible explanations :

i)   at 5 Kbar, $2k_F$ of TTF-TCNQ is 0.308 $b^{::}$. This is very close to the value of 0.315 $b^{::}$ observed in TSeF-TCNQ at ambiant pressure where only one phase transition is observed (10,11). TTF-TCNQ at 5 Kbar could therefore be similar to TSeF-TCNQ at ambiant pressure. More subtle additional effects as revealed by recent EPR work on TSeF-TCNQ (12) might be difficult

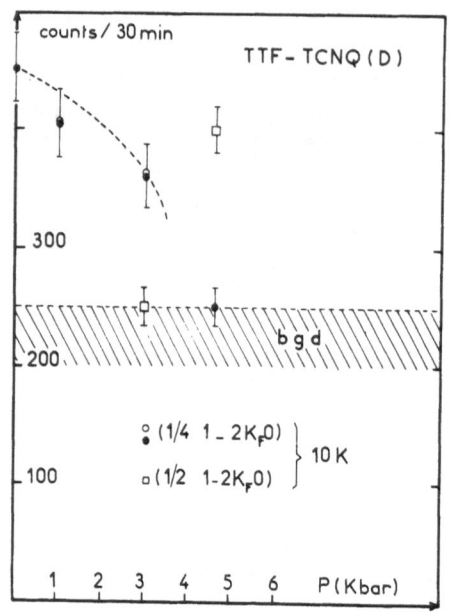

Figure 7

Peak intensity of the $(0.25,(1-2k_F),0)$ and $(0.5,(1-2k_F),0)$ satellites as a function of pressure revealing an additional phase transition at 10 K between the 4a and 2a multiplicity phases. Circles correspond to scans through the 0.25a⸫ (4a) and squares to 0.5a⸫ (2a) positions. (From ref (3)).

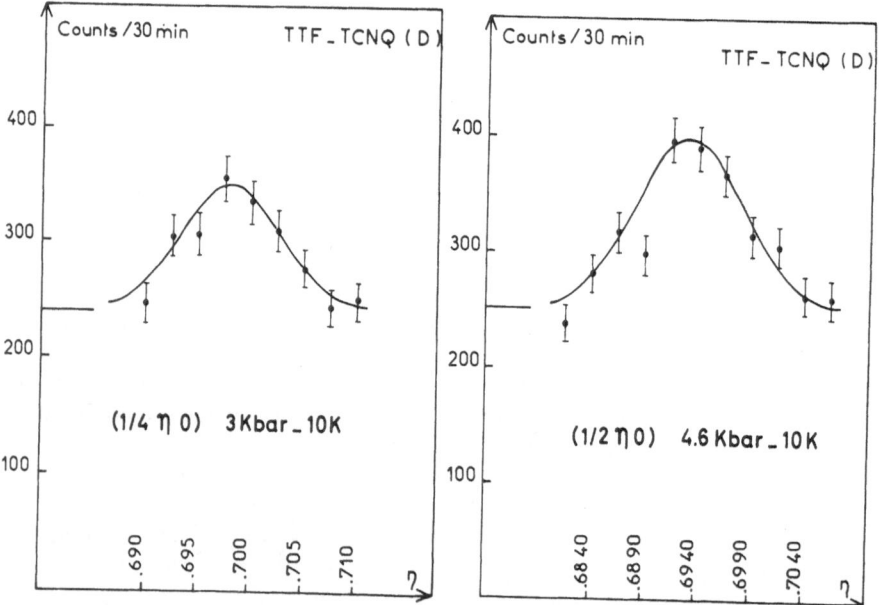

Fig 8 - Typical scans through the $(0.25, (1 - 2k_F), 0)$ satellite at 3kbar and the $(0.5, (1- 2k_F), 0)$ satellite at 4.6Kbar. (From ref (3)).

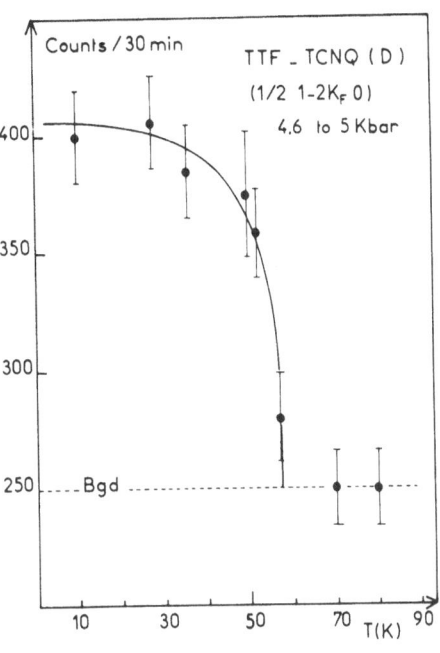

Figure 10

Temperature dependence of the
(0.5, 1-2k$_F$, 0) satellite

observed above 4.6Kbar. Note
that the intensity drops to
zero at the upper transition
of the phase diagram deduced
from transport studies (fig 1).
No significant change is observed at the conjectured low temperature phase separation line
from figure 1. (From ref (3)).

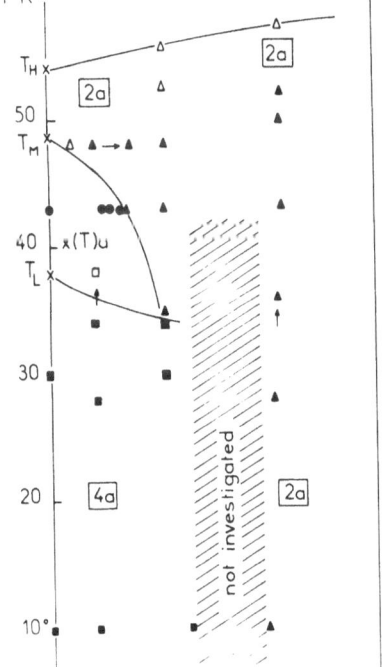

Figure 11

Phase diagram up to 5Kbar as
revealed by the present investigation. Triangles, circles and
squares correspond respectively
to data for the 2a, x(T, P)a and
4a transverse multiplicities. Open
symbols are for zero intensity or
very weak intensity (on the phase
separation lines). As shown by
subsequent measurements by Fincher
et al (9), two phase transitions
occur at low temperature with
increasing pressure in the hatched
area, with again an intermediate
phase corresponding to the incommensurate transverse multiplicity
x(T,P)a.

to detect by structural studies ;

ii) at 5 Kbar, the upper phase transition could involve only the TTF chains and the longitudinally polarised charge density waves, and the lower transition only the TCNQ chains and the transverse polarised distortion not detectable on the Q_L satellite. In other words the opposite situation from that observed at ambiant pressure.

While these two explanations seem mutually exclusive, the extraordinary situation is that they both received partial experimental support but in slightly different pressure ranges.

The newer structural studies seem to rule out the second possibility (ii) : it implies indeed that the 48 K transition (TTF chains and longitudinal CDW) should increase with increasing pressure although we have shown that from the point of view of multiplicity along a, the 48 K transition decreases, which is in agreement with the transport studies. Further, the measurements by Finscher et al (9) at 5 Kbar rule out the formation of a transverse distortion component at the lower phase boundary inferred from the transport measurements. Together with the present results it shows unambiguously that only one clear phase transition takes place at 5 Kbar. Apparently this leaves but the first possibility (i).

Thermopower studies under high pressure performed by Weyl et al (13) show that around 8 Kbar two transitions take place with the donor TTF chains undergoing a distortion at the upper phase boundary, while the acceptor TCNQ stacks only distort at the lower boundary. This study also shows that the pressure range around 5 Kbar discussed above, corresponds precisely to the boundary region between the ambiant type sequence (first TCNQ then TTF at lower T) and a reversed higer pressure type sequence. These results tend to support the hypothesis (ii).

V. CONCLUSION

The problem is now to resolve the dilema concerning the 48 K transition. We have already mentioned in section III the unexpected presence under pressure and in the 2a multiplicity phase, of a sharp satellite corresponding to a longitudinally polarised distortion. Figure 12 gives more data in this respect. Figure 12a shows that the Q_L satellite at 48 K and 0.5 Kbar is broad, but sharpens rapidly with increasing pressure; at 1 Kbar it is already resolution limited. If one plots the peak intensity of this satellite versus pressure, the curve of figure 12b is obtained. Figure 12b seems to reveal another "phase transition" at about 1 Kbar, corresponding to the onset of the longitudinal distortion.

    These results lead to the suggestion that there are two
"phase" separation lines coming out from about 48 K. One
corresponds to the transverse multiplicity change (2a $\not\gtrless$ x(T,P)a)
and decreases in temperature for increasing pressure ; the
other one corresponds to the formation of the longitudinal distortion without change in multiplicity (2a $\not\gtrless$ 2a) and would
increase in temperature with increasing pressure.

    While this suggestion needs more detailed experimental
investigation, if confirmed, it could completely modify the
understanding of the successive phase transitions in TTF-TCNQ,
and eventually shed new light on the contradiction raised
above in the discussion section. More complex phase diagrams
than usually expected have indeed already been conjectured (14).

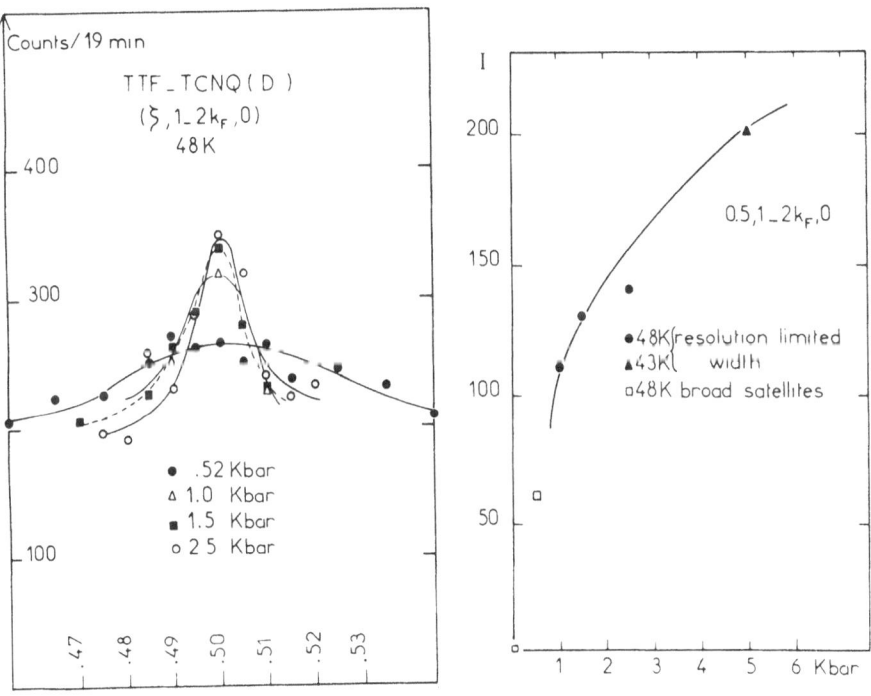

Figure 1 2 : (a) scans in a* direction and at 48K with increasing
pressure, through the (0.5, 1-2k$_F$),0) satellite. Note the increasing sharpness in a phase where the transverse multiplicity
is constant (= 2a) ; (b) peak intensity increase of the same
satellite with increasing pressure.

REFERENCES

(1) Friend, R.H., Miljak, M. and Jerome, D. : 1978, Phys. Rev.
     Lett., 40, p. 1048.
(2) Andrieux, A., Schulz, H.J., Jerome, D. : 1979, Phys. Rev.
     Lett., 43, p. 227 and this conference.
(3) Megtert, S., Comes, R., Vettier, C., Pynn, G., Garito,A.F.:
                1979, Sol. St. Comm.., 31, p. 977.
(4) Khanna, S.K., Pouget, J.P., Comes, R., Garito, A.F.,
     Heeger, A.J. : 1977, Phys. Rev., B 16, p. 1468.
(5) Pouget, J.P., Khanna, S.K., Denoyer, F., Comes, R.,
     Garito, A.F., Heeger, A.J. : 1976, Phys.Rev. Lett., 37,
     p. 437.
     Kagoshima, S., Ishiguro, T., Anzai, H. : 1976, J.Phys.Soc.
     Jap., 41, p. 2061.
(6) Comes, R., Shapiro, S.M., Shirane, G., Garito, A.F.,
     Heeger, A.J. : 1975, Phys.Rev.Lett., 35, p. 1518.
(7) Bak, P., and Emery, V.J. : 1976, Phys.Rev.Lett., 36,p.978.
(8) Pouget, J.P., Shapiro, S.M., Shirane, G., Garito, A.F.,
     Heeger, A.J. : 1979, Phys. Rev., B 19, p. 1792.
(9) Fincher, C. et al : to be published.
(10) Weyl, C., Engler, E.M., Bechgaard, K., Jehanno, G.,
     Etemad, S. : 1976, Sol. St. Comm., 10, p. 925.
(11) Kagoshima, S. : 1979, Sol. St. Comm., 28, p. 485.
(12) Schultz, T.D., Tomkiewicz, Y., : 1979, Bulletin APS March
     Meeting, vol 24, n° 3, p. 231.
(13) Weyl, C., Chaikin, P. : to be published in this Conference.
(14) Hartzstein, C., Zevin, V.,, Weger, M. : to be published.
     A primilary report has been published in Quasi One Dimen-
     sional Conductors, Lecture Notes in Physics, (Springer-
     Verlag), 76, 1979.

PRELIMINARY X-RAY DIFFUSE SCATTERING STUDY OF TMTSF-DMTCNQ

J.P. Pouget[+], R. Comès[+], K. Bechgaard[::]

[+] Laboratoire de Physique des Solides associé au CNRS
Université de Paris-Sud, Bâtiment 510
91405   ORSAY   (France)

[::] H.C. Oersted Institute, Universitetsparken 5,
D.K. 2100   COPENHAGEN   (Denmark)

I. INTRODUCTION

Since the discovery of the genuine $4k_F$ scattering in TTF-TCNQ, a variety of one dimensional conductors have been investigated by X-ray diffuse scattering (TSeF-TCNQ, HMTTF-TCNQ, HMTSeF-TCNQ, NMP-TCNQ, ...) (1). Among them, only HMTTF-TCNQ, in addition to the usual $2k_F$ scattering, has revealed unambiguous $4k_F$ effects both in the form of 1-D precursor scattering and low temperature satellites (1). But at the difference of TTF-TCNQ, the $2k_F$ scattering of HMTTF-TCNQ is dominant at all temperatures leaving, although unlikely, open a possible explanation of the $4k_F$ phenomena in this compound in term of second order effects.

NMP-TCNQ, the properties of which were first explained with large U models (2), was believed to be a potential candidate for such $4k_F$ effects. Removing several assignment ambiguities about the diffuse scattering in pure NMP-TCNQ (3), new results on phenazine doped materials ($NMP_{1-x}Phen_x$-TCNQ) rule out the existence of $4k_F$ scattering in this material (4). Assuming that the only observed scattering around $q_2 \simeq 1/3$ a:: in the alloy arises from the unperturbed TCNQ chains, and that the charge transfer from NMP to TCNQ decreases with the number of electrons available on the donor stacks (1-x per NMP molecules), results summarized in table 1 and figure 1, lead to the conclusion that the two kinds of scattering observed in pure NMP-TCNQ,

113

Table I

NMP_x - $Phen_{1-x}$ TCNQ			
x	0	0.14	0.214
q_2	$1/3\pm0.01$	0.30 ± 0.01	0.28 ± 0.01
q_1	$1/6\pm0.01$	(?)	(?)
C.T. (TCNQ)	2/3e	0.60e	0.56e

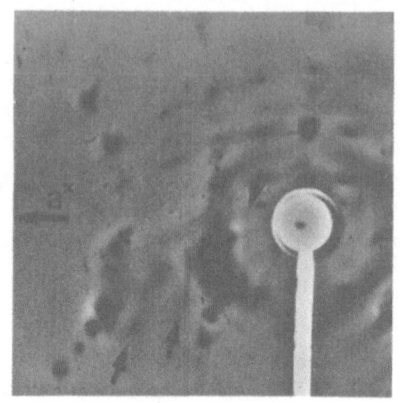

<u>Figure 1</u> - X ray diffuse scattering pattern from $NMP_{0.786}$-$Phen_{0.214}$TCNQ taken at 25K, showing the q_2 = 0.28a* scattering (black arrows). No clear q_1 scattering can be detected. Ascribing this scattering to a $2k_F$ anomaly on the "unperturbed" TCNQ stacks, one gets a 0.56e charge transfer per TCNQ molecule smaller than that 0.66e of pure NMP-TCNQ. Exposure time 116 hours.

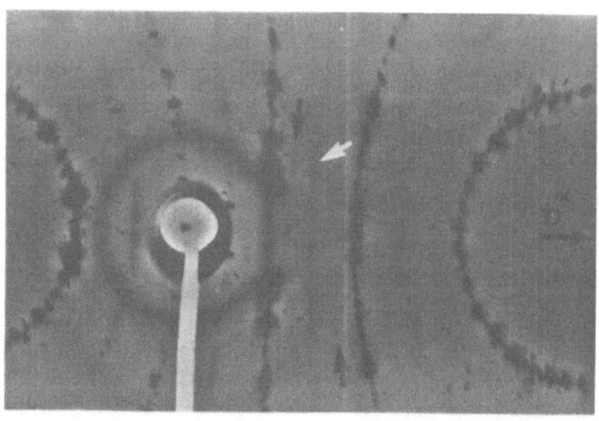

<u>Fig 2</u> - X-ray diffuse scattering pattern from Qn(TCNQ)$_2$ taken at 25K, showing the nearly temperature independent and stronger q_2 = 0.5b* scattering (white arrow) and the weaker q_1 = 0.25b* scattering (black arrow). Exposure time 160 hours.

at $q_1 = 1/6$ a* and $q_2 = 1/3$ a*, must be assigned respectively
to $2k_F^{NMP}$ from the NMP chains and $2k_F^{TCNQ}$ from the TCNQ chains.

Other potential candidates for $4k_F$ anomalies coming from
strong Coulomb correlations, seem to be 1/4 filled band conductors namely 1:2 TCNQ salts such as NEM-(TCNQ)$_2$ (5) or
Qn(TCNQ)$_2$ (6). X-ray diffuse scattering patterns obtained from
various Qn(TCNQ)$_2$ samples show a quasi-one dimensional scattering observable up to room temperature at the wave vector
$q_2 = 0.5$ b* in chain direction, and a weaker scattering only
detectable at low temperature at the wave vector $q_1 = 0.25$ b*
(fig.2) (7). This scattering is interpretable with the assignments $q_2 = 0.5$ a$^* = 4k_F$ and $q_1 = 0.25$ a$^* = 2k_F$; these values
give a 1/2 electron charge transfer to the TCNQ stacks which was
earlier inferred from other experiments (8). While this assignment may be correct, the present observation of transverse and
nearly temperature independent short range correlation in the
q_2 scattering, for the 300 K - 25 K temperature range studied,
cannot rule out an alternative origin of this scattering due to
orientational short range ordering of the quinolinium molecules
(9).

A more recently synthetized one dimensional conductor
TMTSF-DMTCNQ (10), which arises considerable interest because of
its very high conductivity under pressure(11), provides however
the best example of unambiguous $4k_F$ effects since the original
observation on TTF-TCNQ(12). Preliminary results on this
compound will be described below.

II.   EVIDENCE OF $2k_F$ and $4k_F$ SCATTERINGS IN TMTSF-DMTCNQ

The triclinic structure of TMTSF-DMTCNQ (13), schematically
shown in figures 3 to 5, shows first that the 1-D conductor is
three dimensionally ordered regarding the average reference
structure. This avoids the complications of possible orientation-
nal ordering as in NMP-TCNQ or eventually in Qn(TCNQ)$_2$. Microdensitometer reading of an X-ray pattern taken at 100°K reveals
diffuse scattering in reciprocal sheets perpendicular to the
chain axis at $q_2 = 0.5$ a* and $q_1 = 0.25$ a*(fig.6), which will be
assigned respectively to $4k_F$ and $2k_F$ anomalies below. These values correspond surprisingly to a charge transfer of 1/2 electron
per molecule and a 1/4 filled band again, and to commensurate
charge density waves (at ambiant pressure).

III.   THE $2k_F$ SCATTERING and THE 42 K PHASE TRANSITION

Figure 7 shows an X-ray diffuse scattering pattern taken at
25 K, that is to say below the phase transition temperature of
42 K determined from transport (14) and magnetic (15) studies.
Clearly defined satellite reflections with a wave vector compo-

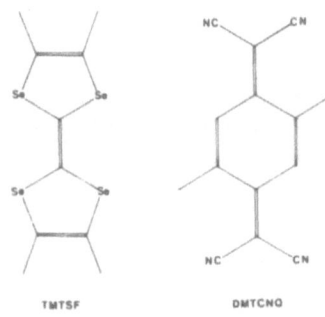

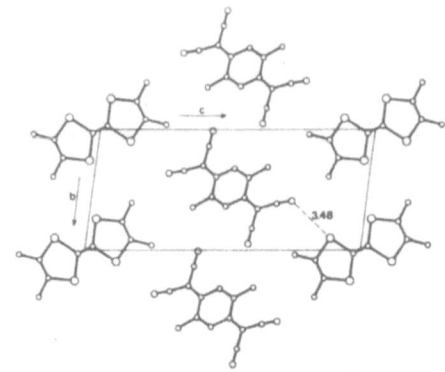

Fig.3 - Donor and acceptor molecules(from ref (13)).

Fig.4 - Crystal packing in TMTSF-DMTCNQ, viewed along a" (from ref (13)).

Fig.5 - Crystal packing in TMTSF-DMTCNQ, viewed along b" (from ref (13)).

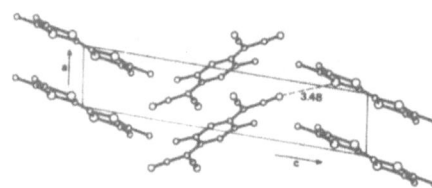

Figure 6

Microdensitometer readings from a pattern of TMTSF-DMTCNQ taken at 100°K. $2k_F$ scattering is observed at the wave vector $0.25a"$ and $4k_F$ scattering at the wave vector $0.50a"$. This implies a charge transfer of 0.5 electrons and a 1/4 filled TCNQ band.

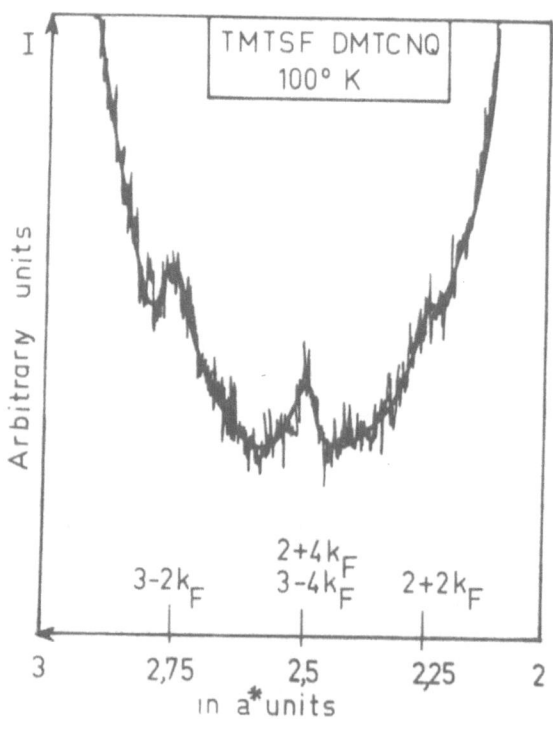

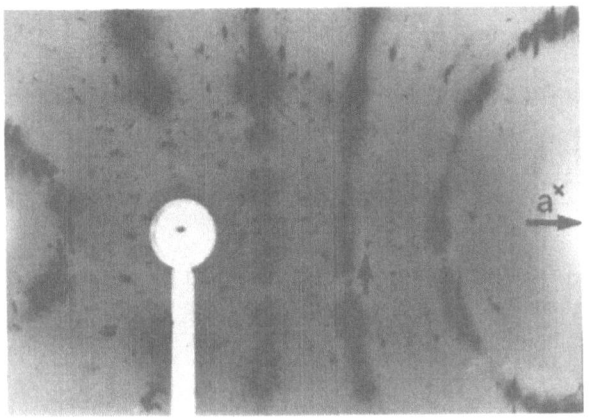

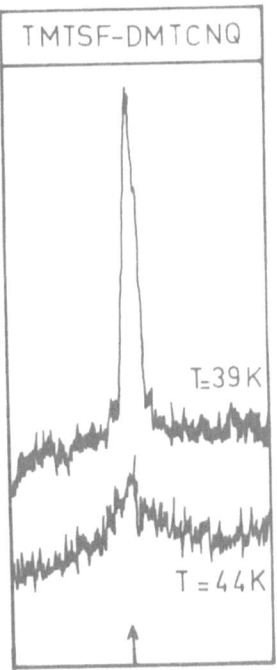

Figure 7 - X-ray diffuse scattering pattern from TMTSF-DMTCNQ taken at 25K. $2k_F$ (arrow) as well as $4k_F$ satellites are observed. Exposure time : 4 hours.

Figure 8 - Microdensitometer readings in the $2k_F$ sheet, and through the satellite shown by the arrow in figure 7, from patterns taken at 39K and 44 K and showing the broadening above the 42K phase transition.

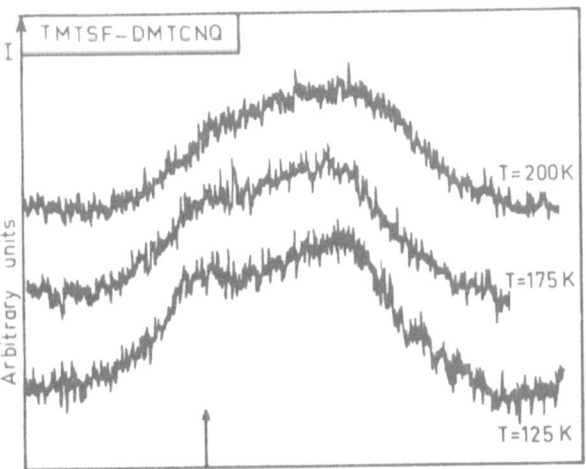

Figure 9 - Microdensitometer readings in the $2k_F$ sheet revealing short range transverse order up to about 175 K. The arrow locates the position of a satellite observed below 42K, and corresponds to the arrow in figure 8.

nent in chain direction of 0.25 a^* ($2k_F$) are observable on
figure 7 (arrow) implying long range 3-D order of the charge
density waves. The transverse wave vector components of these
satellites could not be determined from the present study,
mainly because the crystal quality of the first available samples was poor, as shown by the numerous in layer reflexions and
streaking. Microdensitometer scans through the $2k_F$ satellite,
shown by the arrow on figure 7, are presented figure 8 for
T = 39 K and T = 44 K. They clearly show the broadening resulting from the loss of the long range transverse interchain
order just above the 42 K phase transition (scan at 44 K on
figure 8). Noticeable on the microdensitometer scans shown in
figure 9 is that some transverse short range order between the
chains, remains detectable up to 175 K. It is only around 200 K
that the $2k_F$ scattering has a quasi one dimensional nature,
temperature above which it vanishes rapidly in the background
of usual thermal scattering (it was detected up to 225 K).

IV. $4k_F$ SCATTERING AND A $4k_F$ TRANSITION (?)

In the following we shall refer to observations made in
the areas of patterns which are free from higher order $\lambda/2$ contamination reflexions which can also arise at the 0.5 a^* wave
vector ; that its to say regions between the h = 2 and h = 3
layers of the patterns of figure 10.

Scattering which is well characterized as one dimensional
in nature is already observable at room temperature at the
wave vector 0.5 a^* ($4k_F$) in the pattern (a) from figure 10. It
remains clearly one dimensional down to about 150 K, with a very
slow intensity increase for decreasing temperatures, recalling
the case of the $4k_F$ scattering from TTF-TCNQ ; this increase is
of course enhanced when the correction is made for the decreasing phonon population factor, assuming a dynamical origin of
this scattering. Around 150 K, well defined reflexions appear
on top of a still observable, although weaker 1-D scattering,
in the $4k_F$ reciprocal sheets. The intensity of these reflexions
increases with decreasing temperature and the underlying 1-D
scattering gradually decreases and becomes hardly detectable
below 75 K. This may indicate a 3-D ordering of the $4k_F$ charge
density waves well above 42 K.

V.   CONCLUSION

In the long search for another compound besides TTF-TCNQ
revealing unambiguously genuine $4k_F$ scattering, the preliminary
results described above show that TMTSF-DMTCNQ is the adequate
material. It is a well ordered crystal, with high temperature
one dimensional $4k_F$ scattering, and a $2k_F$ scattering developing
at lower temperatures. The structural study confirms the 42 K

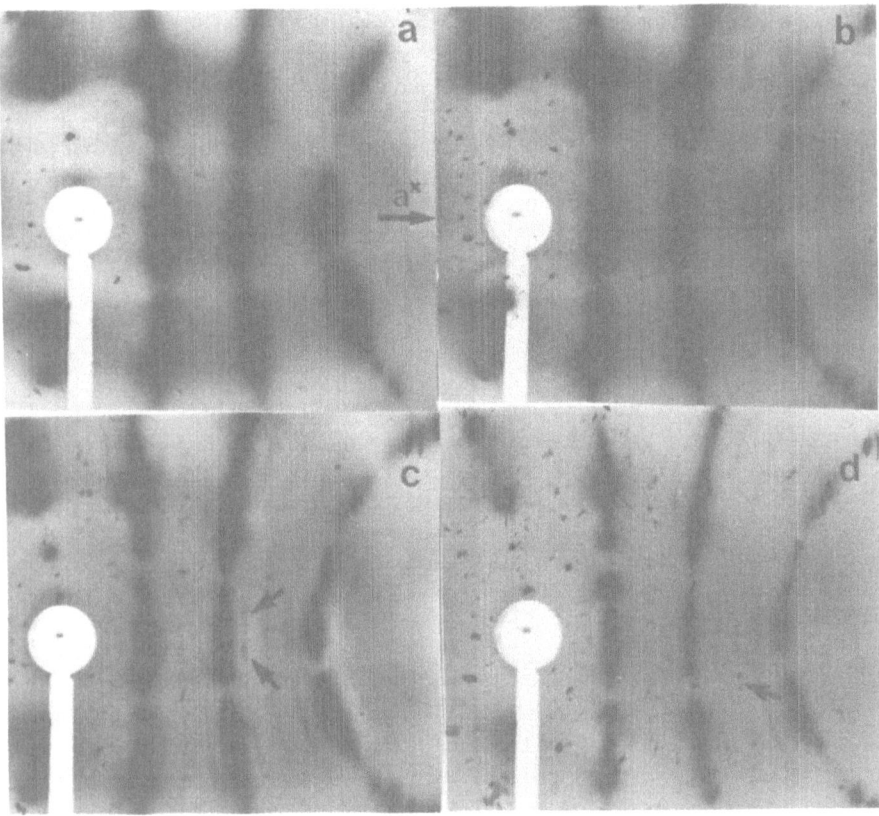

Figure 10 - X-ray diffuse scattering patterns from TMTSF-
DMTCNQ in the conducting phase (T > 42K) showing $2k_F$ and $4k_F$
scatterings at 295K(a), 175K (b), 100K (c), 44K (d). Note that
1D $4k_F$ scattering is visible up to room temperature and that
$4k_F$ satellites (arrow in pattern (d)) are visible well above
the 42K phase transition (between the h = 2 and h = 3 layers,
high order $\lambda/2$ contamination seems unlikely since not observed
at 175K (b) and 295K (a)). Note also that the $2k_F$ scattering
has emerged from the thermal diffuse scattering intensity at
175K (b), and shows marked interchain correlations far from the
42K phase transition (arrows in pattern (c)). These patterns
show clearly that the variation in the reciprocal space of the
intensity of the 1D scattering within each $2k_F$ and $4k_F$ sheets
follows relatively well that of the nearest main Bragg reflexions.
This indicates, as earlier noticed in other TCNQ salts (1), that
the main origin of the $2k_F$ and $4k_F$ scatterings is due to translations of rigid molecular units. Exposure time : (a) 1 1/2hour,
(b) 2 hours, (c) 3 hours, (d) 2 hours.

phase transition as related to the ordering of the $2k_F$ charge density waves. Another temperature range of interest seems to be around 150 K where a change of slope in the temperature dependence of the thermoelectric power was observed earlier (14). It is around this temperature that short range transverse order builds up significantly for the $2k_F$ modulation waves ; it is also around this temperature and in any case well above the 42 K $(2k_F)$ phase transition, that a $4k_F$ transition may take place. This possible $4k_F$ transition in TMTSF-DMTCNQ should however be considered with great caution, higher order $\lambda/2$ contamination reflexions, which could be enhanced by the decreasing Debye Waller factor at low temperatures, cannot be completely ruled out, and such reflexions could also be located in the ·0.5 a∗ layer.

We are very grateful to our colleagues A.J. EPSTEIN for the phenazine doped NMP crystals and to K. HOLCZER and H. STRZELECKA for the Quinolidium (TCNQ)$_2$ samples from which the patterns of figures 1 and 2 were taken, and to P.M. CHAIKIN and D. JEROME for numerous discussions.

REFERENCES

(1) See for example Pouget, J.P., Megtert, S. and Comès, R. : 1979, One dimensional conductors "Lecture Notes in Physics" (Springer Verlag) 95, p. 14.
(2) Epstein, A.J., Etemad, S., Garito, A.F. and Heeger, A.J. : 1972, Phys. Rev.. B5, p. 952.
(3) Pouget, J.P., Megtert, S., Comès, R. and Epstein, A.J. : Phys. Rev. B (in press).
(4) Pouget, J.P., Comès, R. and Epstein, A.J. : to be published.
(5) Sawatzky, G.A., Huizinga, S. and Kommandeur, J. : 1979, in ref. (1), 96, p. 34.
(6) Chaikin, P.M., Kwak, J.F. and Epstein,A.J, : 1979, Phys. Rev. Lett., 42, p. 1178.
(7) Pouget, J.P., Comès, R., Holczer, K. and Strzelecka, H. : to be published.
(8) Clark, W.G., Hamman, J., Sanny, J. and Tippie, L.C. : 1979, in ref. (1), 96, p. 255.
Devreux, F. : 1979, thesis, Grenoble (France), unpublished.
(9) In this picture the position and width in b∗ of the q_2 scattering give a local order (correlation length $\xi_{//} \sim 50A$) of quinolinium molecules in which the b parameter is doubled.
(10) Andersen, J.R., Jacobsen, C.S., Rindorf, G., Soling, M. and Bechgaard, K. : 1975, J. Chem. Soc. Chem. Comm., p. 883.

(11) Andrieux, A., Duroure, C., Jerome, D. and Bechgaard, K. :
      1979, Journ. de Phys. Lettre 40, p. L381.
(12) Pouget, J.P., Khanna, S.K., Denoyer, F., Comès, R.,
      Garito, A.F. and Heeger, A.J. : 1976, Phys. Rev. Lett 37,
      p. 437.
(13) Andersen, J.R., Bechgaard, K., Jacobsen, C.S.,
      Rindorf, G., Soling, H. and Thorup, N. : 1978, Acta
      Cryst. B34, p. 1901.
(14) Jacobsen, C.S., Mortensen, K., Andersen, J.R. and
      Bechgaard, K. : 1978, Phys. Rev. B18, p. 905.
(15) Tomkiewicz, Y., Andersen, J.R. and Taranko, A.R. : 1978,
      Phys. Rev. B17, p. 1579.

FLUCTUATING COLLECTIVE CONDUCTIVITY AND SINGLE-PARTICLE CONDUCTI-
VITY IN 1-D ORGANIC CONDUCTORS

D. Jérome

Laboratoire de Physique des Solides, Université Paris-
Sud, 91405 Orsay (France)

This article is devoted to a presentation of recent results in
the field of Quasi-One-Dimensional conductors, obtained with the
help of high pressure studies.

Section I presents a reinvestigation of the TTF-TCNQ phase diagram based on data of conductivity and thermoelectric power.
Section II discusses the peaking of the phase transition temperature in a narrow pressure domain around 19 kbar which suggests the
existence of commensurability occuring between the wavelength of
the charge density wave (CDW) and the underlying lattice. Section
III relates the loss of longitudinal conductivity observed in the
metallic regime, within the pressure domain 17-21 kbar. In Section
IV, the lack of anomaly of transverse conductivity in the commensurability domain (17-21 kbar) is interpreted as establishing a
fluctuating collective origin for most of the TTF-TCNQ metallic
conduction at low temperature. In Section V we have undertaken a
decomposition of the metallic conductivity of TTF-TCNQ into two
components, namely a fluctuating collective contribution, $\sigma^F_{//}$, and
a single-particle contribution, $\sigma^{sp}_{//}$; the latter component characterized by its large pressure coefficient. A comparison between
$\sigma^{sp}_{//}$ and other Fermi level properties is also included in Section V.
Therefore, we propose a possible interpretation of the single-particle electron scattering time in terms of first order electron-
phonon scatterings, enhanced by the electron-electron interactions.

I. TTF-TCNQ PHASE DIAGRAM

The succession of phase transitions occurring at low temperature
in TTF-TCNQ is a feature unique in the case of two-chain 1-D con-

L. Alcácer (ed.), The Physics and Chemistry of Low Dimensional Solids, 123–142.
Copyright © 1980 by D. Reidel Publishing Company.

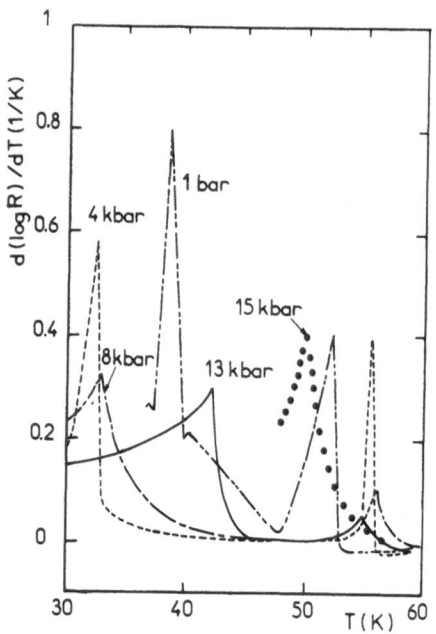

Fig. 1 : A plot of dlogR/dT
versus T, at various
pressures. At 15 kbar
a single peaking is
observed.

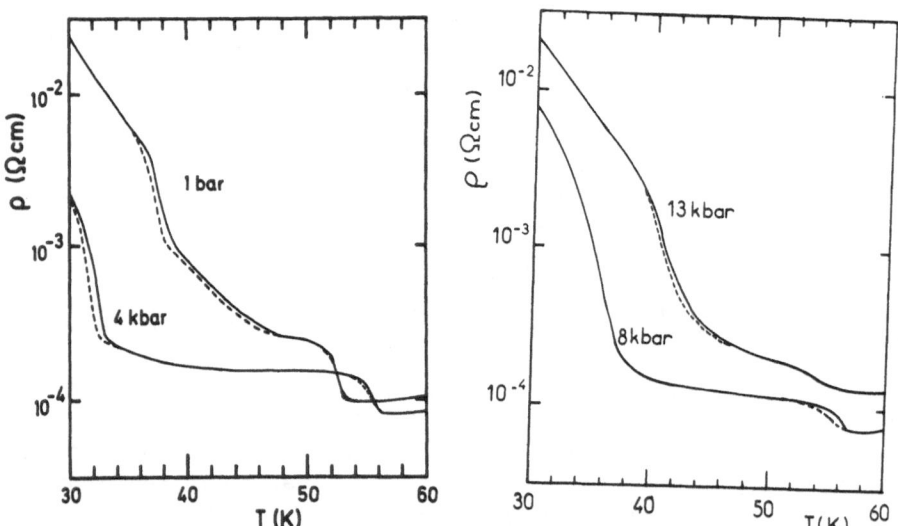

Fig. 2 : Behaviour of the resistivity between T_H and T_L.

ductors, when periodic lattice distortions are driven by the Peierls mechanism (1).

The onset of 3-D ordering is detected directly either by X-ray diffuse scattering (2) or by neutron scattering experiments (3). Moreover, significant anomalies in transport and thermodynamic properties accompany the phase transitions. As far as resistivity is concerned, a maximum of $\partial \log \rho / \partial T$ at a temperature T_c is usually considered as the sign of a phase transition (4). Whenever the resistivity of the distorted state is written

$$\rho(T) = \rho_{min} \exp \frac{\Delta(T)}{kT} \tag{1}$$

a peaking of $\partial \log \rho / \partial T$ in (1) is related to a very large value of $d\Delta/dT$ at $T = T_c$.
The behaviour of $\partial \log \rho / \partial T$ in TTF-TCNQ at different pressures is displayed on figure 1.
For pressures lower than $\sim$ 15 kbar the peaking of $\partial \log \rho / \partial T$ occurs at two distinct temperatures : T_H (high) (5) and T_L (low) (6). It is clear however that the strongest peaking is always related to T_L and above 15 kbar the two transitions merge into a single one. Since both chains, TTF and TCNQ, are likely to contribute to the conduction, the change of conductivity at T_H or T_L cannot point to the nature of the chain on which a distortion sets in. TTF-TCNQ remains a fairly good conductor between T_H and T_L and only half of the charge carriers are lost at T_H, figure 2.
Fortunately, measurements of the thermoelectric power coefficient under pressure yield additional informations (7). The absolute TEP coefficient of a two-chain conductor is expressed in terms of the individual TEP coefficients of each chain S_Q and S_F by (8)

$$S = \frac{\sigma_Q S_Q + \sigma_F S_F}{\sigma_Q + \sigma_F} \tag{2}$$

For a single one-dimensional band the absolute TEP coefficient is given by

$$S = - \frac{\pi^2 k_B^2 T}{3/e/} \left(\frac{\cos \pi \rho/2}{2|t_{//}| \sin^2 \pi \rho/2} + \frac{\tau'(\varepsilon)}{\tau(\varepsilon)} \right)_{E_F} \tag{3}$$

where $\tau(\varepsilon)$ is the energy dependent electron scattering time, ρ the amount of charge transfer and $t_{//}$ the longitudinal transfer integral. When the band structure contribution to the TEP dominates in equ(3), the sign of the TEP is negative for $\rho < 1$ and positive for $\rho > 1$. Since the X-ray results (9) indicate a charge transfer of 0.55 electron under ambient conditions, we may assign S_Q negative and S_F positive.
Figure 3 summarizes the behaviour of the temperature dependence of S at several pressures (10).
At low pressures the TEP becomes more positive below T_H, whereas at elevated pressures (P > 8 kbar) the change of S, accompanying

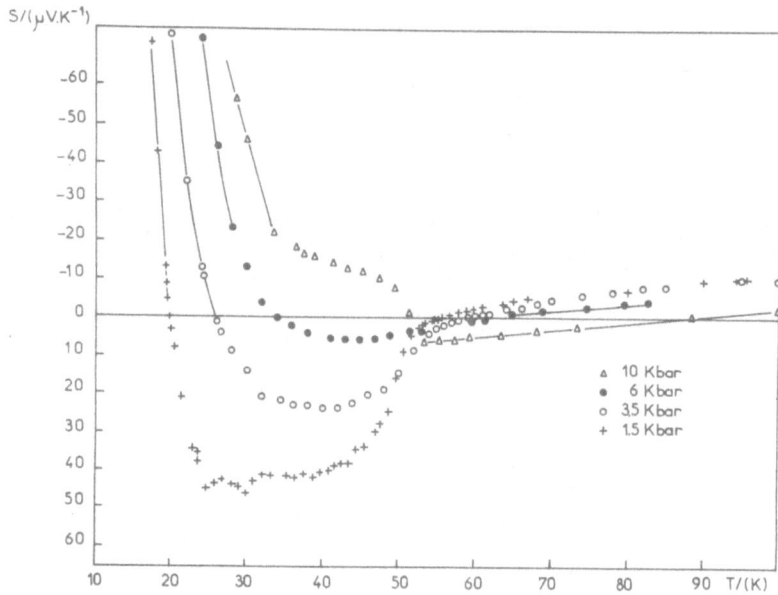

Fig.3 : Absolute thermopower of TTF-TCNQ at different pressures
        in the domain of temperatures where two transitions are
        seen by resistivity.

the transition at T_H, clearly shows a tendency to become <u>more ne-gative</u>. Consequently, following (2),(3) and figure 3 the ambient and low pressure regime can be understood in terms of an energy gap occurring on the TCNQ chain at T_H. At high pressure however the interpretation of figure 3 suggests that the loss of carriers at T_H affects predominantly the TTF chains. The subsequent remo-val, at T_L, of the free carriers left below T_H is accompanied by a sharp increase of S, since, in a semiconducting state according to Boltzmann theory (8)

$$S = - \frac{k_B^2}{|e|} \left(\frac{b-1}{b+1} \frac{\Delta}{k_B T} + \ln \frac{m_h}{m_e} \right) \tag{4}$$

where b is the ratio of electron to hole mobility and m_h and m_e are respectively the effective mass of the holes and of the elec-trons. The sign of S in the semiconducting state is dependent on the impurity content in as much as it is fixed by the position of the Fermi level in the energy gap. It can be inferred from figure 3 that the change of regime occurring at T_H under pressure is con-tinuous, i.e. that the loss of carriers at T_H is continuously trans-fered from the TCNQ chain to the TTF chain, as pressure is increa-sed. As a result of the conductivity and TEP investigations on TTF-TCNQ under pressure, a stand point in which both chains order at T_H is sensible. Thermopower shows conclusively that in the law pressure limit the gaps which open at T_H on the TCNQ chains and on the TTF chains are respectively larger and smaller than kT_H, the converse becoming true above $\sim$ 8 kbar. Such a picture is cor-roborated by recent high accuracy measurements, at ambient pres-sure, of EPR and static spin susceptibility of TTF-TCNQ (11), sho-wing a small but finite drop of the TTF-chain susceptibility oc-curring at T_{II}. 49 K is the temperature below which the transverse "phase sliding" region begins, but no peaking of $\partial \log \rho / \partial T$ has been detected (12). Following a resistivity reinvestigation of the pha-se transitions at 1 bar (13) we believe that 49 K may represent the temperature below which the gap on the TTF chains starts to rise slowly above kT. The phase diagram of TTF-TCNQ derived from the transport properties investigation is displayed on figure 4. This figure shows temperatures such as T_H and T_L detected through the anomalies of the conductivity. More recent data of thermopo-wer and conductivity (10) indicate a very complicated phase dia-gram of TTF-TCNQ in the low pressure regime (P < 5 kbar) below 38K. The salient feature of this diagram is the existence of a temperature domain between T_H and T_L in which the system is nei-ther a conductor since $\partial \rho / \partial T < 0$ nor an insulator since $\sigma > 10^3$ $(\Omega cm)^{-1}$. As already suggested in this section, the 3-D ordered distortion is presumably large on only one of the two chains between T_L and T_H. So far it is still hard to reconcile the "transport properties" phase diagram with the recent high pressure neutron scattering investigations (14). In particular at pressures in excess of 4.6 kbar, the transverse period of the 3-D ordering becomes 2a along the a direction below T_H and no further

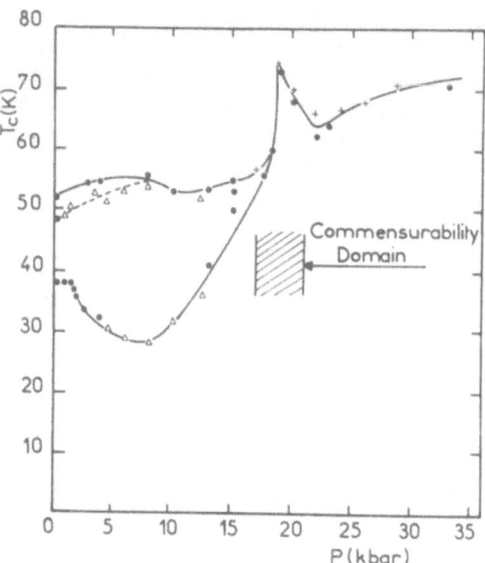

Fig. 4 : Phase diagram of TTF-TCNQ, derived by transport properties
measurements. The transitions at T_H and T_L merge into a
single transition above 17 kbar.

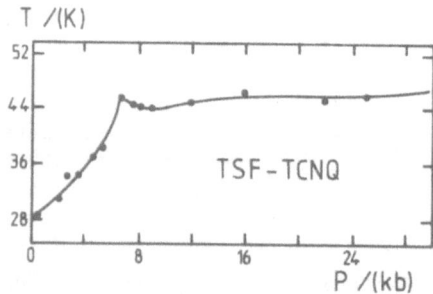

Fig. 5 : Phase diagram of TSF-TCNQ derived by resistivity data,
with indications of commensurability at 7.5 kbar.

change of this ordering has been detected down to 10 K at 4.6kbar. Therefore, neutron scattering fails to show the transition at T_L very clearly observed by conductivity. We have also drawn in figure 4 a tentative dashed transition line starting from 49 K at ambient pressure, rising with pressure and following the maxima of $\partial \log S / \partial T$. Whether this transition line can be related to the recent finding, by neutron scattering data, of a rising 49 K transition under pressure (14) (related to the 3-D ordering of longitudinally polarized distortions) may still be hazardous. It is clear however that clarifying the low pressure diagram requires more extensive work both in transport properties and neutron scattering. A thorough reexamination of the Landau theory of the TTF-TCNQ phase transitions (15) has led to the conclusion that although the transition at T_H may involve primarily the TCNQ chain, a small amplitude distortion must be present on the TTF chains.

II. COMMENSURATE CDW STATE UNDER PRESSURE

Above 15 kbar a sharp peaking of the single phase transition is observed at 19 kbar (12). At the latter pressure, the transition temperature reaches 74 K (16), figure 4. The width of the peaked region in the phase diagram amounts to a domain of $\sim$ 4 kbar in pressure units. In the previous presure domain, the phase transition exhibits a first order character as shown by a one degree hysteresis observed at the transition (12). Comparison effects such a maxima in $\Delta(0)$ and $\Delta(0)/T_c$ are also observed on $\Delta(T)$ around 19 kbar (17). Similarly, on the selenium compound TSF-TCNQ a peaking of T_c has been located at 7.5 kbar (18), figure 5. The features of the TTF-TCNQ phase diagram above 15 kbar have been understood in terms of commensurability between periodic lattice distortion and underlying lattice. Whenever the spanning vector of the Fermi surface at high temperature (the vector 2 k_F for a one dimensional Fermi surface) approaches closely enough a submultiple value of a reciprocal lattice vector, the wavelength of the low temperature ordered state tends to lock on to a multiple of the unit cell distances. The commensurability x 3 ($2k_F = b^*/3$) is suggested by the peaking and the 1st order character of the transition at 19 kbar. The possibility of an enhancement of the charge transfer under pressure is a necessary prerequisite for the observation of a commensurate state at 19 kbar. Such an increase of the charge transfer can reasonably be expected from the broadening of tight binding bands under pressure. The recent observation by neutron scattering of the k_F/b^* value up to 6 kbar (19,14) is in perfect agreement with the 13% increase of k_F/b^* between 1 bar and 19 kbar which is bound to exist if commensurability is achieved at 19 kbar. In this respect, the peaking observed in the TSF-TCNQ phase diagram, figure 5, at 7.5 kbar is in good agreement with the commensurability of TTF-TCNQ at 19 kbar. Under ambient pressure, the charge transfer of TSF-TCNQ (0.63) is already closer to

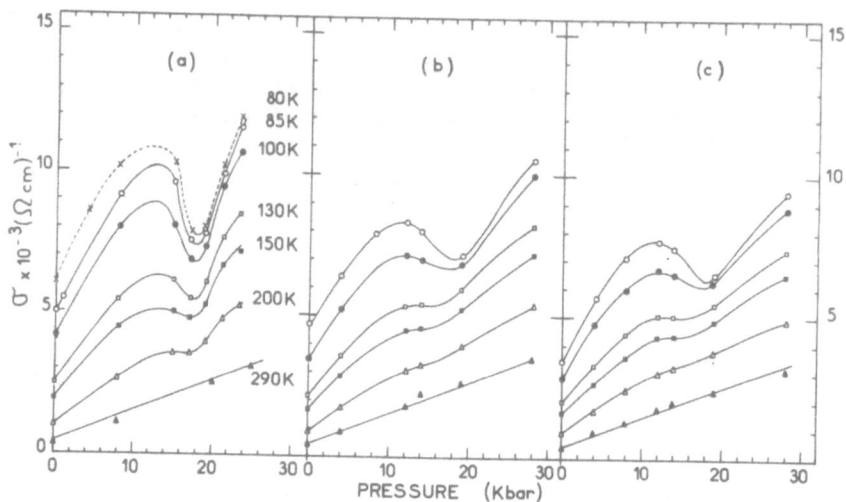

Fig. 6 : Pressure dependence of the longitudinal conductivity of
TTF-TCNQ at constant temperature. The various results
displayed, correspond to samples with CPR = 25, 16 and
11, for (a), (b) and (c) respectively.

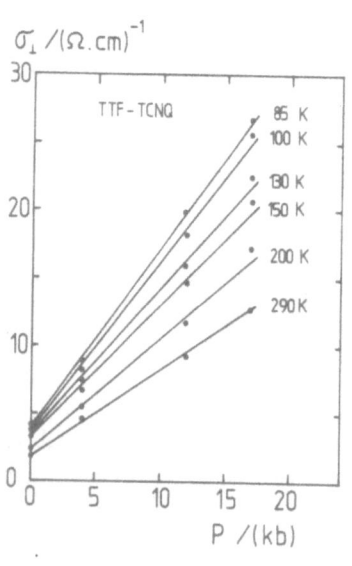

Fig. 7 : Pressure dependence of the
conductivity of TTF-TCNQ
along the a direction. It
is not affected by the commensurability regime.

commensurability (0.66) (10) than that of TTF-TCNQ (0.59) (9).
It is worthwhile mentioning that the pressure dependence of k_F/b^*
is moderate since it only amounts to $\dfrac{\partial \ln k_F/b^*}{\partial \ln b} \simeq -3$. Similar
"weak" volume dependences are also found for all properties dea-
ling with the bare bandwidths (see for example, the plasma edges
(21))but there,much larger volume dependences are observed for
quantities related to the Fermi level properties. This important
     aspect of the physics of 1-D organic conductors will be
discussed more thoroughly in Section V.

III. FLUCTUATION CONDUCTIVITY IN TTF-TCNQ

The interest of 1-D conductors lies mainly in their providing a
new conductivity mechanism to take place at moderately high temperatures (22). This new channel of conduction involves the coherent sliding of all the electrons building up an incommensurate
CDW in a given chain. It is a collective process in as much as it
can give rise to conductivity even in the presence of an energy
gap in the single-particle spectrum. In the 3-D ordered state,
CDW's lose their 1-D character because of the interchain electrostatic interactions. Hence, the free sliding of the CDW's becomes
pinned by infinitesimal amounts of impurities. Above the phase
transition however, when the transverse coherence length remains
smaller than the interchain distance, 1-D fluctuating CDW of finite coherence length can exist in a broad temperature domain as demonstrated by X-ray diffuse scattering. Bardeen and co-workers(23)
suggested that long life time fluctuating CDW of the Fröhlich type
could contribute to the dc longitudinal conductivity. However, the
question concerning the contribution of the CDW's to the metallic
conduction of TTF-TCNQ has not so far received any direct and convincing answer. In fact, this problem has been particularly hot
and controversial for the last 5 years (24). During that time, several authors proposed explanations for the resistivity via single-
particle scattering ; namely, electron-acoustic phonon (25), elec-
tron-intramolecular phonon (26), second order electron-libron (27),
interchain electron-electron (28), electron-spin fluctuations (29).
Admittedly some of the above single-particle mechanisms found their
justification in the unusually large volume dependence of the resistivity observed near ambient temperature (17). However, we shall
see in the rest of the article that most of the arguments put forward in favour of single-particle scattering must be taken with a
grain of salt. Recently, the existence of a significant contribution to the conduction of collective origin has been verified
through the possibility of achieving commensurability under pressure (30,31). As TTF-TCNQ is driven by high pressure towards commensurability there occurs an important drop of conductivity, figure 6. However, the loss of conductivity becomes smaller at high
temperature and practically unobservable above 250 K. High pressure

optical reflectance studies (21) allow to discard the possibility
of drastic (non monotonous) changes in the band structure under
pressure especially at commensurability. Consequently, the data
shown in figure 6 can be discussed in the following two ways :
(i) in the incommensurate case rigid translations of fluctuations
into the CDW (sliding-mode conductivity) are possible since the
energy of a fluctuation is independent of its phase (22). In the
commensurate case, however, the energy of a fluctuation does depend on its phase through a phase dependent cubic term (32) (the
pinning interaction). Therefore, a finite restoring force prevents
translations of the fluctuating CDW so that its phase is pinned to
some equilibrium position thereby suppressing the sliding-mode
contribution to the conductivity (33). Thus, a low order commensurability (x 2 or x 3) is a drastic pinning mechanism for the fluctuating CDW's and it can well explain the lowering of the conductivity in the commensurability domain, as observed in figure 6.
(ii) If the resistivity is dominated by a single-particle like
scattering mechanism, the behaviour around 19 kbar suggests an increase in the scattering rate and/or a decrease in the density of
states at the Fermi level, associated with the commensurability.
In any of the above-mentioned single-particle theories, the resistivity is independent of the phase of the CDW fluctuations but
eventually depends on the amplitude. Measurements of the longitudinal conductivity alone and results such as those of figure 6
are unable to distinguish between the two previous pictures. Fortunately, the transverse conductivity, whenever it is diffusive,
is sensitive to both the density of states at Fermi level and the
smearing of 1-D Fermi surface by the single-particle electron scattering time, $\tau_{//}^{sp}$ (34,35). The diffuse transverse conductivity is
given by the relation

$$\sigma_{\perp} \sim n(E_F)\tau_{\perp}^{-1} \tag{5}$$

where $1/\tau_{\perp}$ is the interchain hopping frequency due to the existence of a finite (but small) interchain transfer integral $t_{\perp}$.
In a Quasi-One-Dimensional conductor the hopping frequency can
be derived with the 1-D Golden rule (34)

$$1/\tau_{\perp} = \frac{2\pi}{\hbar} |t_{\perp}|^2 \, \tau_{//}^{sp} \tag{6}$$

From eqs(5) and (6) $\sigma_{\perp}$ becomes

$$\sigma_{\perp} \sim n(E_F)|t_{\perp}|^2 \, \tau_{//}^{sp} \tag{7}$$

The relation (6) is true in the $\tau_{\perp} > \tau_{//}^{sp}$ limit when the transverse
conductivity is diffusive. This restriction does not constitute a
problem in TTF-TCNQ for which experiments lead to $\tau_{\perp}/\tau_{//} \approx 10^3$ at
room temperature (34). Figure 7, displaying the pressure dependen-
ce of the transverse conductivity shows the non existence of a dip

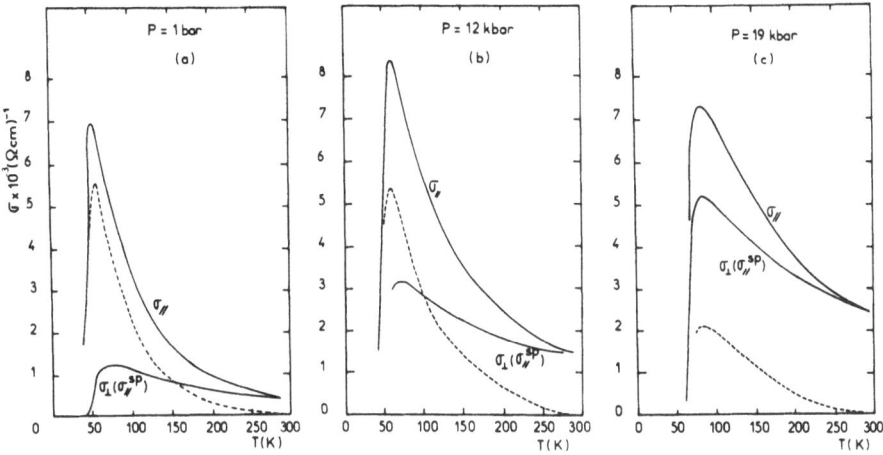

Fig. 8 : Temperature dependence of the absolute value of $\sigma_{//}$ and of the $\sigma_{\perp}$ normalized to $\sigma_{//}$ at room temperature. The dashed-line corresponds to the fluctuating contribution $\sigma_{//}^{F} = \sigma_{//} - \sigma_{//}^{sp}$. Figures (a) and (b) and figure (c) refer to the incommensurate and commensurate situations respectively.

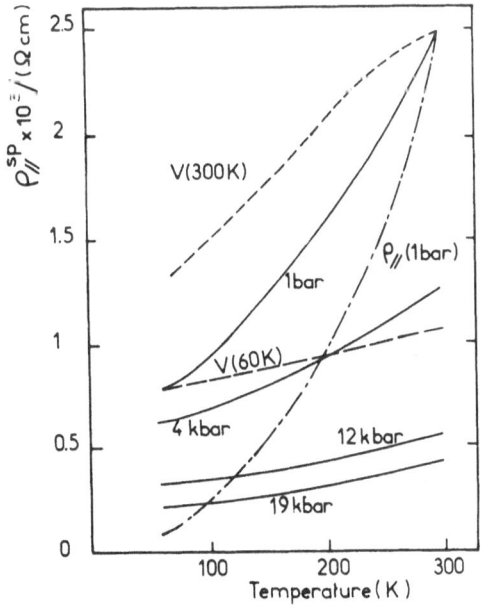

Fig. 9 :
Temperature dependence of $\rho_{//}^{sp} = 1/\sigma_{//}^{sp}$ under constant pressure, continuous line, and under constant volume, dashed line. The ambient pressure behaviour of the total resistivity (after reference 37) is also drawn for comparison.

in the commensurability regime. Therefore figures 6 and 7 and equ(7) lead to the conclusion that neither $n(E_F)$ nor $\tau^{sp}_{//}$ are significantly affected by commensurability. The loss of conductivity at commensurability, figure 6, is thus a purely 1-D effect. It can be attributed to the pinning by a commensurability potential of a fluctuating sliding conductivity. The existence of a conductivity mechanism of fluctuating Fröhlich origin does not prevent the more traditional single-particle contribution from being an important one. Both conduction channels add up to form the total conductivity. A particularly important task for the understanding of the electronic properties of TTF-TCNQ will be the decomposition of the conductivity into two components, namely $\sigma^F_{//}$ and $\sigma^{sp}_{//}$.

IV. DECOMPOSITION OF THE METALLIC CONDUCTIVITY

A salient result of the previous Section is that the longitudinal conductivity must be written

$$\sigma_{//} = \sigma^F_{//} + \sigma^{sp}_{//} \qquad (8)$$

We propose in the present Section a derivation which allows an estimation of both components of (8) at given pressures and temperatures. To accomplish this decomposition two assumptions are required. The first assumption made in the derivation is that $\sigma^F_{//}$ (300 K) is negligibly small compare to $\sigma^{sp}_{//}$(300 K). This assumption is justified by the monotonous increase of $\sigma_{//}$ through the commensurability regime around room temperature. The accuracy of the data of figure 6 shows that $\sigma^F_{//}$(300 K) is smaller than, say, $\frac{\sigma^{sp}_{//}}{10}$(19 kbar, 300 K) thus leading to $\sigma^F_{//}$ (300 K) << 250 $(\Omega cm)^{-1}$.

The second assumption of the derivation is based on the use of equ(7) : in the 1-D regime $(\tau_\perp /\tau^{sp}_{//} >> 1)$, $\sigma^{sp}_{//}$ and $\sigma_\perp$ exhibit the same temperature dependence according to equ(7), i.e. that of $\tau^{sp}_{//}$, since the band structure is only weakly temperature as well as pressure dependent. The model decomposition of the conductivity (31) is illustrated by the experimental situations of figure 8. At every pressure we have plotted the temperature dependence of $\sigma_{//}$, with an initial value $\sigma_{//}$ = 400 $(\Omega cm)^{-1}$ under ambient conditions. We have also displayed the temperature dependence of the transverse conductivity along the a-direction normalized to the value of $\sigma_{//}$ at room temperature.(a) In the incommensurate regime, P = 1 bar and 12 kbar on figure 8a and 8b respectively, the temperature dependences of $\sigma_{//}$ and $\sigma_\perp$ appear very different. The ratio $\sigma_{//}$ (T) /$\sigma_{//}$ (300K) peaks at 20 and 6 at low temperature for 1 bar and 12 kbar respectively, whereas the peaking of $\sigma_\perp(T)/\sigma_\perp$ (300K) does not exceed a factor 2 or 3. Following equ(7), $\sigma^{sp}_{//}$ in (8) is given by the $\sigma_\perp$ curve and the component $\sigma^F_{//}$ in (8) can be derived from the difference at every temperature between the $\sigma_{//}$ and $\sigma_\perp$

curves. This procedure leads back to $\sigma^F_{//}(300\ K) = 0$, as expected
since this was our initial assumption. (b) In the commensurate
pressure regime, the similar decomposition of $\sigma_{//}$, figure 8c, leads
to a much smaller contribution $\sigma^F_{//}$. The non-zero value of $\sigma^F_{//}$ at commensurability may probably result from the thermal excitation of
the sliding mode fluctuating conduction in a situation where the
commensurability pinning is not much larger than kT. The above decomposition leads to a fluctuating collective contribution which
represents quite a significant part of the total conductivity in
incommensurate TTF-TCNQ. For a sample of conductivity peak ratio
(CPR) at 59 K of 17.5 as shown on figure 8a, the fluctuating contribution reaches $\sim 5500\ (\Omega cm)^{-1}$ at 59 K, namely 80% of the total
conductivity. Moreover it appears from figure 8a that $\sigma^F_{//}$ and $\sigma^{sp}_{//}$
are about equal in amplitude when T = 150-200 K. This is also the
temperature regime where $2k_F$ reflections of one dimensional character are first observed when cooling (9). Our analysis brings
a further confirmation of the $2k_F$ CDW's ability to carry a current.
The sensitivity to impurities of the fluctuating conductivity may
be responsible for different CPR in different samples. The pinning
of CDW fluctuations arising from impurities or crystal defects
could also explain diverse data for three different samples on
figure 6 (36). Fluctuating CDW conductivity is probably not restricted to TTF-TCNQ. This is a phenomenon which should arise in
all incommensurate 1-D organic conductors. Actually a preliminary
investigation of the conductivity of TSF-TCNQ reveals a contribution $\sigma^F_{//}$ quite similar to that of TTF-TCNQ (18). A major experimental difference between single-particle and collective contributions to the conductivity is the behaviour under pressure.
On the one hand, the single-particle contribution increases at a
rate $\partial \ln \sigma^{sp}_{//}/\partial P \sim 30\%\ kbar^{-1}$ around room temperature (17,37). According to the data of figure 7, the pressure coefficient of $\sigma^{sp}_{//}$ is
hardly affected by the temperature lowering. On the other hand,
$\sigma^F_{//}$ is only slightly enhanced by high pressure, increasing at a
rate of $\sim 2\%\ kbar^{-1}$ or so at 75 K. Such a moderate volume dependence is also encountered in the behaviour of the phase transition
T_H under pressure (besides the commensurability peaking). It brings
an additional confirmation of the tight relation between the structural properties and the fluctuating CDW contribution to the conductivity. As to the effect of the large volume dependence of the
conductivity on its intrinsic T dependence it has been claimed
that, because of the change of volume provided by the thermal expansion of the lattice, it is important to distinguish between
T-dependence under constant volume and constant pressure (38,17).
At this stage we wish to emphasize that the constant volume reduction of the conductivity must be performed only on the contribution to the total conductivity which is strongly volume dependent:
the single-particle contribution. After the removal of the $\sigma^F_{//}$
contribution at every pressure and temperature, the constant volume reduction of $\sigma^{sp}_{//}$ follows a procedure similar to that used
earlier for the total conductivity (17). In the present situation

the volume dependence of $\sigma_{//}^{sp}$ is given by the volume dependence of $\sigma_\perp$ on figure 7. Temperature dependences of the single-particle resistivity are displayed on figure 9. The important aspect of figure 9 is the quasi linear temperature dependence of $\sigma_{//}^{sp}$ for both V(60 K) and V(300K), and the positive intercept of the constant volume T-dependence with the T = 0 K axis. This latter fact contrasts significantly with the existence of a T = 0 K axis negative intercept previously derived for the total constant volume resistivity (17).

V. THE SINGLE-PARTICLE SCATTERING TIME

From the interpretation of the conductivity in the previous Section we managed to extract the single particle scattering time. We wish to comment in this Section on the importance of τ^{sp} on other physical properties of 1-D organic conductors namely EPR line width and optical reflectance. An important contribution to the spin-lattice relaxation of conduction electron spins in metals is given by the modulation of the spin-orbit coupling by phonons. In the extreme narrowing situation the line width is related to the electron scattering time τ by the Elliott's mechanism of relaxation (39)

$$(\Delta H)_{3D} = \frac{M^2}{\tau} \qquad (9)$$

where M is the matrix element for spin reversal due to spin-orbit coupling. In molecular metals such as TTF-TCNQ or TSF-TCNQ the spin-orbit interaction in equ(9) is dominated by the heavy atoms of the molecules (Sulfur or Selenium). Therefore the observation of very narrow EPR lines at room temperature appears at first look very surprising within the framework of the Elliott's theory with a $\tau \approx 3 \ 10^{-15}$ s derived from optical reflectance (5) or NMR relaxation (34). The remarkably narrow EPR line of most 1-D organic conductors has been attributed to a considerable weakening of the efficiency of the spin-orbit induced relaxation in 1-D conductors (40). For quasi flat Fermi surfaces the reduction factor of the line width becomes $\tau_{//}^{sp}/\tau_\perp$ (35). Therefore, the modification of the Elliott's theory (9) to Quasi-1-D conductors leads to

$$(\Delta H)_{1D} \simeq \frac{M^2}{\tau_\perp} \qquad (10)$$

Taking into account the dependence of the spin orbit interaction on the atomic number Z (of the heteroatom) (41) and the relation between the interchain hopping rate and $\tau_{//}^{sp}$, equ(6), we may rewrite equ(10)

$$\Delta H_{1D} \approx Z^4 \ t_\perp^2 \ \tau_{//}^{sp} \qquad (11)$$

The relation (11) is fairly well obeyed when either P or T are taken

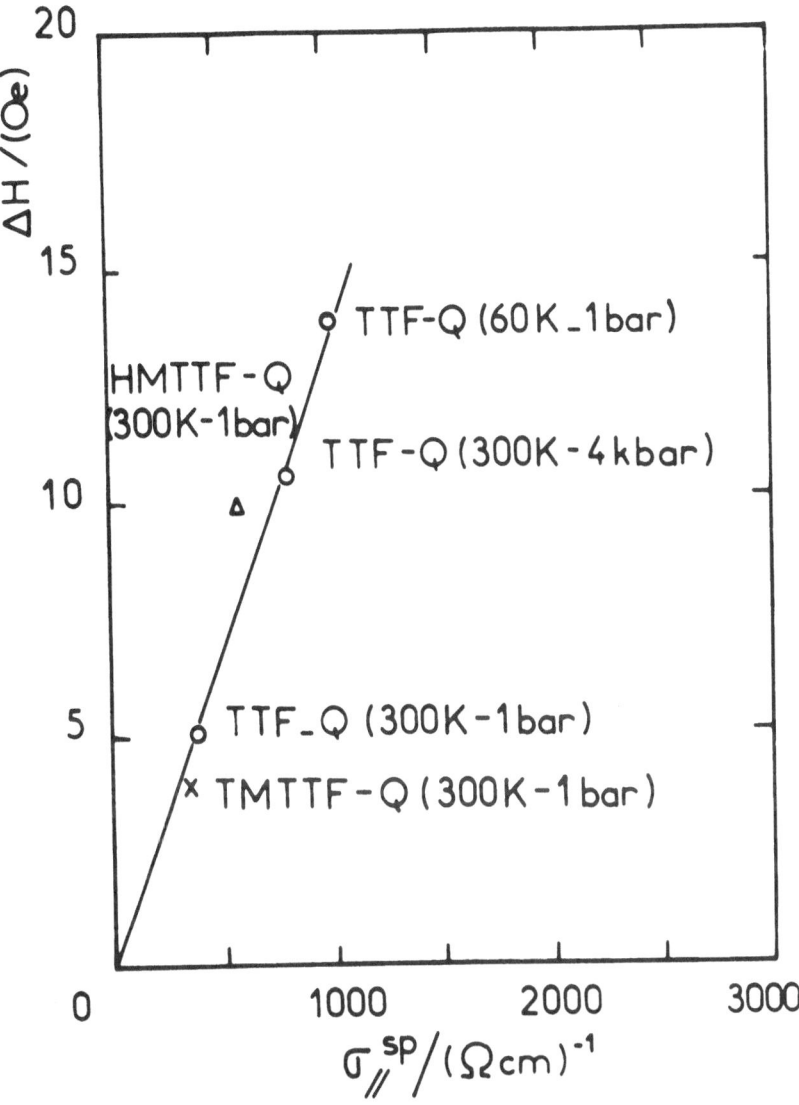

Fig.10 : Proportionnality relation between the EPR line width
         and the single-particle contribution to the conductivi-
         ty in TTF-TCNQ using P or T as parameters

as variable parameters in TTF-TCNQ (42), figure 10. It is also noticed on figure 10 that other sulfur compounds such as HMTTF-TCNQ
(43) and TMTTF-TCNQ exhibit EPR properties in general agreement
with those of TTF-TCNQ. A comparison of EPR line width can be made
between selenium compounds and sulfur compounds following equ(11).
For TSF-TCNQ (40) and TMTSF-TCNQ (44) we derive that way, $t_\perp$ respectively 20% larger and 50% smaller than the $t_\perp$ of TTF-TCNQ in
agreement with the known structure of these compounds. The pressure dependence of the optical reflectance can also be used to derive the pressure dependence of $\tau_{//}^{sp}$. In the frequency range
$1/\tau_{//}^{sp} < \omega < \omega_p$, the reflectance can be approximated by the simple
relation (45)

$$R = 1 - \frac{2}{\omega_p \, \tau_{//}^{sp}} \tag{12}$$

As $\tau_{//}^{sp}$ is increased by pressure the reflectance measured at 0.7eV
tends to increase strongly (21). Taking into account the enhancement of the plasma frequency in equ(12) we derive from the data
of reference (21) an enhancement x 3 for $\tau_{//}^{sp}$. This enhancement is
still smaller than the increase of the dc conductivity at room
temperature   in 36 kbar (17). However, the oversim-
plified description of the optical data in terms of equ(12) does
not allow us to take this discrepancy so seriously. Hence, the
message given by conductivity and optics under pressure is the
very large volume dependence of the single-particle scattering time
$$\frac{\partial \ell n \tau_{//}^{sp}}{\partial \ell n \, b} = -60,$$ about 10 times larger than the volume dependence of
the band parameters reported in Section II. Since the temperature
dependence of $1/\tau_{//}^{sp}$ (at constant volume) is quasi linear, figure
9, there is no reason not to consider first order electron-phonon
scattering as a possible origin for $\tau_{//}^{sp}$. In reference (25) such a
suggestion has been proposed but the volume dependence of $\tau_{//}^{sp}$ was
ascribed to the volume dependence of the lattice modes. The argument was based on the observation of large P dependences of acoustic phonons in other organic crystals. However, we wish to point
out that charged molecular crystals such as TTF-TCNQ are significantly less compressible than uncharged crystals like anthracene
or naphtalene (46). Moreover, the stiffening of the TA(a^*) phonon
branch due to thermal contraction between 300 K and 84 K (equivalent to a 5 kbar pressure) amounts to $\approx$ 20% only (47). Thus, we
feel it is rather hard to attribute the large volume dependence
of $\tau_{//}^{sp}$ (at all temperatures) to the lattice modes only. It has
been suggested many times that electron-electron interactions
play an important role in the electronic properties of organic
conductors (48,). We wish to emphasize here that e-e interactions are known to enhance the el-phonon scattering (this is the
electron-phonon vertex correction). The electron-phonon coupling
constant λ can be derived from the infrared reflectivity studies
(50) using the Hopfield relation

(50) using the Hopfield relation $1/\tau_{//}^{sp} = 2\pi \dfrac{k_B T}{\hbar}\lambda$. A value $\lambda \approx 1.3$ has been found by Heeger and coworkers (50). This value of λ is far too large to agree with the bare electron-phonon coupling constant λ_0 needed for magnetic and structural properties (51) namely $\lambda_0 = 0.12$. The vertex correction of λ_0 which has been estimated within a Parquet theory (51) leads to an enhancement $\Gamma^2(2k_F) \approx 10$ for TTF-TCNQ at room temperature. In the previous theoretical approach (51) the pressure dependence of the vertex correction was attributed mainly to the volume dependence of the bare e-e interactions. Although such a possibility cannot be ruled out we must be aware that the Parquet approximation is probably very crude whenever the e-e interactions are large (52). For TTF-TCNQ in particular what is needed is an improved theoretical treatment which can treat properly the screening and the volume dependence of the electron-electron interactions.

VI. CONCLUSION

This article has tried to present an improved experimental situation of TTF-TCNQ through the most recent investigation performed under pressure. A major achievement is the direct proof of the existence of an important contribution of fluctuating CDW's to the conductivity of the incommensurate metallic state. This collective contribution is a channel of conduction which works in parallel with the more traditional single-particle conductivity. Both channels add to give the measured longitudinal conductivity. Moreover, we have proposed a procedure of decomposition of the longitudinal conductivity into its two components after recognition that (i) the collective contribution is negligible at room temperature and (ii) temperature dependences of both $\sigma_\perp$ and $\sigma_{//}^{sp}$ are similar. Preliminary work performed on other 1-D incommensurate organic conductors shows that fluctuating CDW conductivity is a very general mechanism of conduction. The above mentionned decomposition of the conductivity tells us clearly that all theoretical models based only on single-particle scattering or only on collective conductivity should be handled with some caution. As fas as the single-particle electron scattering is concerned, we have proposed a mechanism based on first order electron-phonon scattering. This model is supported by (i) the experimental temperature dependence of the single-particle resistivity at constant volume, (ii) the large vertex correction of the electron-phonon interaction resulting from the electron-electron interactions which probably lies at the origin of the large volume dependence. Although the understanding of TTF-TCNQ and its derivatives seems to be anchored on rather well established experimental facts now, many challenging problems remain. We may enumerate some of the most obvious : the mapping of the low pressure phase diagram of TTF-TCNQ by transport properties and its relation with the neutron scattering results, the investigation of TTF-TCNQ near commensurability (say $\sim$ 1% off-commensurability) where solitons are likely to be observable, the improvement of the theory treating electron-phonon and electron-electron interaction on the same footing. Finally, one of most exciting aspects of the 1-D research for the near future will be the understanding of the basic physical laws which are behind the experimental observation of a pressure stabilized ground state of a new kind in a whole family of new materials such as TMTSF-DMTCNQ (53,54), $(TMTSF)_2PF_6$ (55) $(TMTSF)_2NO_3$, etc ...

I wish to acknowledge the tight and efficient cooperation with K. Bechgaard in several aspects of this work. I also benefited from the help of A. Andrieux, C. Weyl and P.M. Chaikin. I had many useful discussions with H.J. Schulz, S. Megtert and J.P.Pouget. This work was supported in part by the DGRST Contract n° 78-7-0313.

REFERENCES

(1) R.E. Peierls, Quantum Theory of Solids, Clarendon Press, Oxford, 1955, p.108

(2) F. Denoyer, F. Comès, A.F. Garito and A.J. Heeger, Phys. Rev. Lett., 35, 445, 1975

(3) R. Comès, S.M. Shapiro, G. Shirane, A.F. Garito and A.J. Heeger, Phys. Rev. Lett. 35, 1518, 1975

(4) S. Etemad, Phys. Rev. B, 13, 2254, 1976

(5) A.J. Heeger and A.F. Garito : Low Dimensional Cooperative Phenomena, H.J. Keller editor, Plenum Press, N.Y. 1975

(6) D. Jérome, W. Müller and M. Weger, J. Physique Lett. 35, L-77, 1974
J.R. Cooper, D. Jérome, M. Weger and S. Etemad, J. Physique Lett. 36, L-219, 1975

(7) P.M. Chaikin, J.F. Kwak, T.E. Jones, A.F. Garito and H.J.Heeger, Phys. Rev. Lett. 31, 601, 1973

(8) P.M. Chaikin, R.L. Greene, S. Etemad and E. Engler, Phys.Rev. B, 12, 1627, 1976

(9) R. Comès, Chemistry and Physics of One-Dimensional Metals, H.J. Keller editor, Plenum Press, N.Y. 1976

(10) Data taken from the work of C. Weyl, P.M. Chaikin and D.Jérome, to be published

(11) Y. Tomkiewicz, reported at this meeting, pp. 187.

(12) R.H. Friend, M. Miljak and D. Jérome, Phys. Rev. Lett. 40, 1048, 1978

(13) J.R. Cooper and J. Lukatela, Quasi-One-Dimensional Conductors, Lecture Notes in Physics, 95, Springer Verlag, 1979

(14) S. Megtert and R. Comès, this meeting

(15) E. Abrahams, J. Solyom and F. Woynarovich, Phys. Rev. 16, 5238, 1977

(16) New results obtained at Orsay by A. Andrieux and C. Weyl, to be published

(17) D. Jérome, Molecular Metals, W.E. Hatfield editor, Plenum Press, N.Y. 1979

(18) A. Andrieux and D. Jérome, unpublished results

(19) S. Megtert, R. Comès, C. Vettier, R. Pynn, A.F. Garito, Sol. State Comm. 31, 977, 1979

(20) C. Weyl, E.M. Engler, K. Bechgaard, G. Jehanno, S. Etemad, Sol. State Comm. 19, 925, 1976

(21) B. Welber, P.E. Seiden and P.M. Grant, Phys. Rev. B 18, 2692 1978

(22) H. Fröhlich, Proc. R. Soc. A, 223, 296, 1954

(23) J. Bardeen, Solid State Comm. 13, 357, 1973
D. Allender, J.W. Bray and J. Bardeen, Phys. Rev. B 9, 119, 1974

(24) A.J. Heeger, Chemistry and Physics of One-Dimensional Metals, H.J. Keller editor, Plenum Press, N.Y. 1977

(25) E.M. Conwell, Quasi-One-Dimensional Conductors, Lecture Notes in Physics 95, Springer Verlag, 1979

(26) E.M. Conwell, Phys. Rev. Lett. 39, 777, 1977

(27) H. Gutfreund and M. Weger, Phys. Rev. B 16, 1753, 1977
See also M. Weger, this meeting, pp. 77.

(28) P.E. Seiden and P.M. Grant, Quasi-One-Dimensional Conductors, Lecture Notes in Physics, 95, Springer Verlag, 1979

(29) D. Jérome, J. Physique Lett. 38, L 489, 1977

(30) A. Andrieux, H.J. Schulz, D. Jérome and K. Bechgaard, Phys. Rev. Lett. 43, 227, 1979

(31) A. Andrieux, H.J. Schulz, D. Jérome and K. Bechgaard, J. Physique Lett. 40, L-385, 1979

(32) T. Ohmi and H. Yamamoto, Prog. Theor. Phys. 58, 743, 1978

(33) P.A. Lee, T.M. Rice and P.W. Anderson, Solid State Comm. 14, 703, 1974

(34) G. Soda, D. Jérome, M. Weger, J. Alizon, J. Gallice, H.Robert, J.M. Fabre and L. Giral, J. Physique, 38, 931, 1977

(35) M. Weger, J. Physique, 39, C6, 1456, 1978

(36) H. Fukuyama and P.A. Lee, Phys. Rev. B 17, 535, 1978

(37) R.H. Friend, M. Miljak, D. Jérome, D.L. Decker and D. Debray, J. Physique Lett. 39, L 134, 1978

(38) J.R. Cooper, Phys. Rev. B 19, 2404, 1979

(39) R.J. Elliott, Phys. Rev. B 14, 1, 1963

(40) Y. Tomkiewicz, E.M. Engler and T.D. Schultz, Phys. Rev. Lett. 35, 456, 1975

(41) Y. Yafet, Solid State Physics, 14, 1, 1963

(42) C. Berthier, J.R. Cooper, D. Jérome, G. Soda, C. Weyl, J.M. Fabre and L. Giral, Mol. Cryst. Liq. Cryst. 32, 267, 1976

(43) P. Delhaes, S. Flandrois, J. Amiell, G. Keryer, E. Torreilles, J.M. Fabre, L. Giral, C.S. Jacobsen and K. Bechgaard, J. Physique Lett. 38, L-233, 1977

(44) Y. Tomkiewicz, J.R. Andersen and A.R. Taranko, Phys. Rev. B 17, 1579, 1978

(45) N.F. Mott and Jones, The theory of the properties of Metals and Alloys, Oxford, University Press, 1936

(46) D. Debray, R. Millet, D. Jérome, S. Barisic, L. Giral and
     J.M. Fabre, J. Physique Lett. 38, L 227, 1977
(47) S.M. Shapiro, G. Shirane, A.F. Garito and A.J. Heeger,
     Phys. Rev. B 15, 2413, 1977
(48) J.B. Torrance, Chemistry and Physics of One-Dimensional
     Metals, H.J. Keller editor, Plenum Press, 1977
(49) D. Jérome, Organic Conductors and Semiconductors, Lecture
     Notes in Physics 65, Springer Verlag 1977
(50) A.A. Bright, A.F. Garito and A.J. Heeger, Phys. Rev. B 10,
     1328, 1974
(51) D. Jérome and L. Caron, Quasi-One-Dimensional Conductors,
     Lecture Notes in Physics 95, Springer Verlag 1979
(52) Private Communication from S. Barisic
(53) A. Andrieux, C. Duroure, D. Jérome and K. Bechgaard,
     J. Physique Lett. 40, L-381, 1979
(54) A. Andrieux, P.M. Chaikin, C. Duroure, D. Jérome, C. Weyl,
     K. Bechgaard and J.R. Andersen, J. Physique, December 1979
(55) D. Jérome, A. Mazaud and K. Bechgaard, to be published

SPIN HAMILTONIANS AND THEORETICAL MODELS

Zoltán G. Soos

Department of Chemistry, Princeton University
Princeton, New Jersey   08544   U.S.A.

The enormous variety and range of solid-state magnetic information is conveniently summarized by spin Hamiltonians that typically involve adjustable parameters. Although Heisenberg exchange in magnetic insulators affords an increasingly complete theoretical model for localized moments, spin Hamiltonians and their parameters are generally phenomenological in nature and may not correspond to realistic pictures of the electronic structure. Theoretical models in principle reconcile such accurate, but partial descriptions and provide consistent microscopic parameters. As illustrated with several examples, magnetic resonance methods in low-dimensional solids are particularly effective in probing the dynamics, dimensionality, and location of spins or charge carriers, the internal structure of excitations, and the ground-state charge transfer. The resulting spin Hamiltonians for paramagnetic semiconductors and conductors are unconventional because optical and electric excitations are inextricably linked to magnetic properties.

I. INTRODUCTION

Recent reviews (1-4), conference proceedings (5-8), and previous advanced study institutes (9,10) adequately cover the wealth of static and dynamic (NMR, epr) magnetic phenomena in low-dimensional inorganic insulators (3-5,9) and conductors (2), in π-molecular charge-transfer (CT) and ion-radical salts (1,6, 9,10), and in organic or polymeric conductors (6-10). The magnetic properties of organic conductors and ion-radical salts are also covered in other lectures. My purpose here is to discuss, with the aid of several illustrative examples, the implications of magnetic information for the microscopic

143

L. Alcácer (ed.), The Physics and Chemistry of Low Dimensional Solids, 143–164.

parameters of theoretical models. The focus is on unexpected applications, rather than on broadly-applicable and fairly routine susceptibility, epr, or NMR methods. Systematic studies (1-10) are essential for characterizing various classes of low-dimensional materials, but may not initially be decisive for evaluating microscopic parameters or establishing theoretical models.

The great advantage of magnetic studies is the wide range of available external variables and internal probes. Typical experimental variables include the applied field H_0, the temperature T or pressure P, the orientation Ω of a crystal, the natural or controlled abundance of isotopes, the intrinsic or induced levels of chemical or mechanical defects, etc. Spin Hamiltonians are introduced to present experimental data in compact form, usually with the aid of adjustable parameters. Such Hamiltonians may, or may not, correspond to a sensible physical picture or be consistent with nonmagnetic data. The Drude-Lorentz approach to dielectric properties, where oscillators are introduced to fit an arbitrary spectrum, is an analogous method that also leaves open the microscopic interpretation of the resulting parameters.

Theoretical models, on the other hand, must account for magnetic, optical, and other electronic properties, for example by deriving the form of a successful spin Hamiltonian. The sharp distinction between spin Hamiltonians and theoretical models is unfortunately obscured whenever a reasonably consistent picture has been achieved. The following examples illustrate both spin Hamiltonians that are, and are not, associated with theoretical models. In some cases, additional work may establish a stronger connection. A particularly severe test is to describe temperature dependences or relaxation data with spin Hamiltonians devised originally for static properties. The widely used (4,5) Heisenberg exchange Hamiltonians for magnetic insulators also account (3,11) for dynamic properties in $(CH_3)_4NMnCl_3$ (TMMC) and other ideal systems. They illustrate spin Hamiltonians that have become theoretical models, in contrast to most current spin Hamiltonians for one-dimensional (1D) conductors or semiconductors.

II. ION-RADICAL, CHARGE-TRANSFER, AND SUPERMOLECULAR DIMERS

An antiferromagnetic exchange $2J\vec{s}_1 \cdot \vec{s}_2$ between two s = ½ species produces a singlet (S = 0) ground state and an epr active triplet (S = 1) excited state whose energy 2J can be obtained from the static susceptibility or epr intensity. The spin Hamiltonian for any S = 1 system is

$$\mathcal{H}_T = \mu_B S \cdot g_T \cdot H_0 + D(S_Z^2 - \tfrac{2}{3}) + E(S_X^2 - S_Y^2) + \mathcal{H}_{hyp} \quad (1)$$

where μ_B is the Bohr magneton, the tensor g_T determines the center of the epr spectrum, D and E are the fine structure constants in the principal axis frame X, Y, Z, and $\mathcal{H}_{hyp}$ describes possible interactions with nuclear spins. Any epr spectrum with D,E and X,Y,Z described by $\mathcal{H}_T$ must be a triplet. The magnitudes, possible temperature dependences, and interpretations of the parameters are separate questions that are rather well understood in Cu(II) dimers (12), in ion-radical dimers (1,13,14) $(A^-)_2$ or $(D^+)_2$ based on A = TCNQ (tetracyanoquinodimethane) or D = TMPD (N,N,N',N'tetra-methyl-p-phenylenediamine),and in many other transition-metal complexes, including some with s > ½ when additional terms may occur in Eq. (1). The triplet-exciton splittings (15) in Fig. 1 for the CT complex of M_2P (dimethylphenazine) and TCNQ are the first example of thermally activated D^+A^- dimers.

The observation (13,14) that $\mathcal{H}_T$ fits the epr of several TCNQ salts and of TMPD-ClO₄ suggested most of the properties of triplet spin exciton (1,16) prior to systematic structural, optical, electric, and other studies. The magnitudes of D and E showed the unpaired electrons to be on adjacent ion-radicals; the absence of hyperfine structure proved that exciton motion between dimers was fast compared to $\mathcal{H}_{hyp}$; the resolution of magnetically inequivalent triplet excitons indicated slow motion

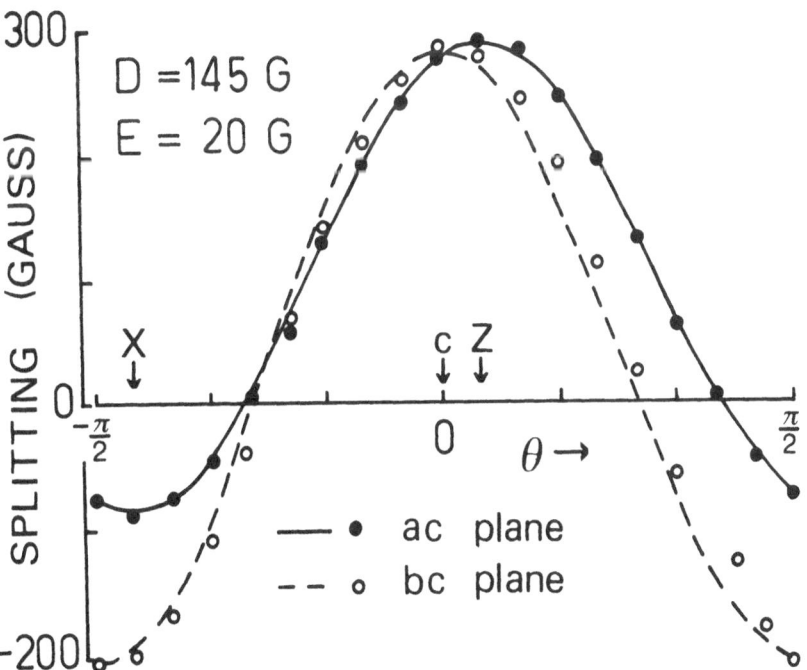

Fig. 1. Fine structure splittings in M_2P-TCNQ. (ref. 15).

between 1D chains of dimers; the g-tensor was consistent with
π-radicals; the epr intensity gave the exchange J and reasonable
fits to a singlet-triplet model. Even the collapse of the fine
structure splittings and subsequent narrowing of the single epr
line with increasing temperature could be understood in terms of
exciton-exciton collisions. The crucial observation of fine
structure splittings and complete averaging of hyperfine interactions, thus permitted a surprisingly complete and even semiquantitative picture of various ion-radical salts, including
complex TCNQ salts, in terms of weakly interacting 1D arrays
of dimers.

The classification (1) of π-molecular solids into alternating
or regular chains on the basis of showing or not showing fine
structure splittings is quicker, but less general, than a complete
structural determination. Triplet excitons have invariably been
observed in alternating chains with several types of π-molecular
overlap along the chain, whereas regular chains with crystallo-
graphically equivalent sites and a single π-overlap have
vanishing D and E. This holds in 1D insulators such as TMMC or
Cu(II) salts, in currently known organic conductors based on
partly-filled ($\gamma < 1$) regular segregated $...A^{-\gamma}A^{-\gamma}A^{-\gamma}...$ or
$...D^{+\gamma}D^{+\gamma}D^{+\gamma}...$ stacks, and in semiconducting CT complexes like
TMPD-TCNQ with mixed regular $...D^{+\gamma}A^{-\gamma}D^{+\gamma}A^{-\gamma}...$ stacks.

The CT complex M_2P-TCNQ crystallizes as an alternating chain,
as presumably do the other triplet-exciton complexes (17) between
TCNQ and substituted phenazines. The fit in Fig. 1 has been
slightly reinterpreted, by taking the principal Z axis to be
tilted by 10° in the ac plane from the chain axis, the crystalline
c axis. This accounts for the observed asymmetry in Fig. 1 on
rotating H_o in the crystallographic ac and bc planes. Direct
computation (15) of the dipole-dipole interaction between the
closer D^+A^- pair in the solid yields D $\sim$ 160G, E $\sim$ 15G, and a
tilt of about 15° in the ac plane. Information about both the
principal axes, X,Y,Z and the magnitudes of D and E in Eq. (1)
naturally provides a more convincing analysis in other systems
as well (18). Whether small deviations between observed and
calculated values of D and E can be related to small changes in
the solid-state spin distribution (19) or to more extended triplets (20) must remain open until the principal axes X,Y,Z are
also treated. Such careful comparisons are needed to go beyond
the very simple, but surprisingly good approximation (1,18) of
identical solid-state and solution spin distributions for
π-radicals.

The similar overlap patterns and reduced D and E for the
complex TCNQ salts $(\phi_3AsCH_3)(TCNQ)_2$, $(\phi_3PCH_3)(TCNQ)_2$, and
$(CH_3CH_2)_3NCH_3(TCNQ)_2$ can be understood (18) in terms of supermolecular A_2^- radicals that form dimers of the type $(A_2^-)_2$. Each

A_2^- dimer has a low-lying excitation involving the out-of-phase combination of the two π-MOs, similar to the $\sigma_g \rightarrow \sigma_u$ excitation of H_2^+. The optical and electric properties of these 1D arrays of supermolecular dimers are consistent (1,21,22) with additional low-lying electronic states, but the parametrization of tetrameric stacks is far from quantitative. The derivation of an alternating Heisenberg exchange Hamiltonian for such an array of supermolecular dimers is also far from satisfactory. The dimer picture $2J\vec{s}_1 \cdot \vec{s}_2$ for $(A_2^-)_2$ chains is consequently still primarily a spin Hamiltonian, even though these complex TCNQ salts were among the first triplet spin exciton systems.

III.   SPIN DYNAMICS IN REGULAR HEISENBERG CHAINS

The width, shape, angular, temperature, and field dependence of the exchange narrowed epr line (11,3) in regular 1D systems like TMMC can all be related to an exchange constant J obtained from the susceptibility and dipole-dipole interactions computed from the crystal structure. Isotropic Heisenberg exchange in these cases clearly accounts for spin relaxation as well. In other cases (23), spin-orbit corrections to isotropic exchange give additional broadenings whose form can usually be anticipated, but whose magnitudes are adjustable. Thus additional parameters can sometimes, but not always, be avoided in discussing spin dynamics of ideal magnetic insulators. Small temperature dependences of J(T) due to thermal expansion or lattice vibrations cannot usually be either demonstrated or excluded.

The fluctuating magnetic fields produced by unpaired electrons often dominate nuclear relaxation. NMR thus provides not only a probe for the local susceptibility (24,25), but for the dynamics of unpaired electronic moments (26-29). Typically, the frequency (27,28) or temperature dependence (29) of the nuclear relaxation is studied under less than ideal circumstances. For example, powders may be required, the hyperfine constants may not be known, and there may be many inequivalent ^1H, ^{13}C or ^{14}N (for NQR) nuclei in the unit cell.

The spin dynamics (30) of α-bis(N-methylsalicyladiminato)Cu(II) illustrate NMR techniques under ideal circumstances. CuNSal is a regular 1D Heisenberg antiferromagnet, with

$$\mathcal{H}_{AF} = 2J\sum_n \vec{s}_n \cdot \vec{s}_{n+1} - g\mu_B H_o \sum_n s_{nz} \qquad (2)$$

and $J/k = 3.04$ K obtained from magnetization, susceptibility, and specific heat measurements. Interchain exchange is small, since 3D ordering occurs (31) at 0.044K. Single-crystal NMR shows resolved NSal protons, one of which is strongly coupled to the nth

Cu(II) spin $\vec{s}_n$ according to

$$\mathcal{H}_{hyp} = a\vec{I}\cdot\vec{s}_n + \vec{I}\cdot A\cdot\vec{s}_n \qquad (3)$$

The isotropic hyperfine constant $a = 5.4 \times 10^{-4}$ cm^{-1} is an order of magnitude larger than the classical electron-nuclear dipolar interaction A. Both a and A can be measured from the <u>position</u> of the proton resonance as a function of Ω and H_0; A can also be computed from the crystal structure. The possibility of measuring coupling constants through the position of the resonance exists for any single crystal with resolved spectra. The possibility of driving the system to a ferromagnetic state, by increasing H_0 past the critical field $H_c = 4J/g\mu_B = 88$kOe for CuNSal, requires a conveniently small J. The T_1^{-1} data (30) in Fig. 2 show the relaxation to be enhanced below the magnon zone edge at H_c and to decrease rapidly thereafter. Orientational considerations indicate the data in Fig. 2 to involve the $a\vec{I}\cdot\vec{s}_n$ term of $\mathcal{H}_{hyp}$, with relaxation due to s_n^+ and s_n^- fluctuations. The theoretical fit contains no adjustable parameters. A similarly complete fit is obtained for other orientations, where the two-magnon s_{nz} part of $\vec{I}\cdot A\cdot\vec{s}_n$ dominates for $H > H_c$.

Single crystals large enough for NMR studies are rarely, if ever, available for the better organic conductors. Single crystal NMR studies on supermolecular dimers in $(\phi_3AsCH_3)(TCNQ)_2$ establish (29) that triplet-exciton motion controls the ^1H relaxation, without resolving TCNQ and ϕ_3AsCH_3 protons. Powder studies offer even more limited opportunities for specifying the coupling constant in $\mathcal{H}_{hyp}$, which in principle require an Eq. (3) for each $I = \frac{1}{2}$ nucleus and contain additional terms for $I > \frac{1}{2}$ nuclei. These quantitative details do not prevent interesting conclusions based on H_0, T, or P dependences in powder NMR studies, which can further be augmented by selective isotopic substitutions or enrichments. Such magnetic studies offer information that usually cannot be obtained otherwise, that is clearly relevant to the spin dynamics of theoretical model of the electronic structure, but that ultimately requires additional adjustable parameters. The obvious challenges are to develop new experimental methods and to draw correct conclusions without necessarily evaluating all the purely magnetic coupling constants.

IV. STRONG CT COMPLEXES WITH MIXED REGULAR STACKS

Strong donors like TMPD or PD (p-phenylenediamine) and strong acceptors like TCNQ or Chloranil are paramagnetic semiconductors that crystallize in mixed regular ...$D^{+\gamma}A^{-\gamma}D^{+\gamma}A^{-\gamma}$ stacks (1). Structural data indicates, by symmetry, a single exchange or transfer integral along the stack, so that the magnetic excitations

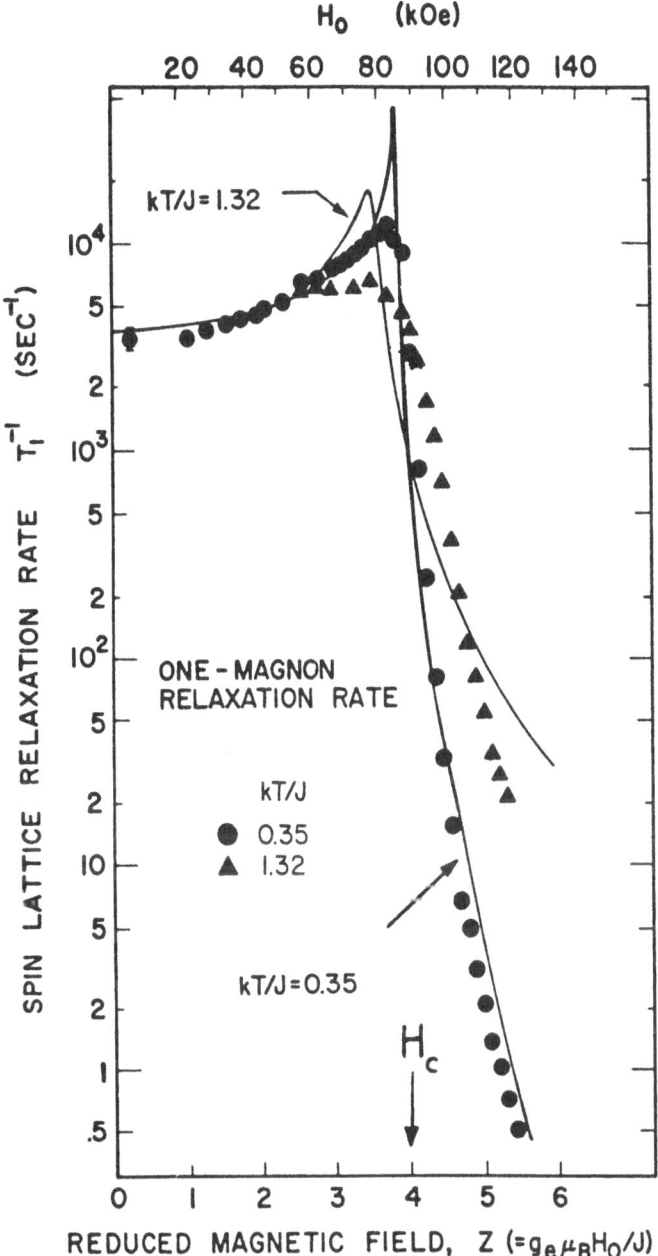

Fig. 2 Spin-lattice relaxation rate of the strongly-coupled NSal
proton as a function of H_O at T = 1.08 K and 4.02 K. The
theoretical curves contain no adjustable parameters. The
crystal orientation insures that only one-magnon (s_n^+ and
s_n^-) terms in Eq. (3) contribute to T_1. The Cu (II) spins
have a ferromagnetic ground state for $H_O > H_c$ = 88 kOe
(ref. 30)

of ionic ($\gamma \sim 1$) complexes are those of a regular Heisenberg chain, Eq. (2). The absence of both fine structure splittings and hyperfine splittings supports a picture (32) of mobile, uncorrelated $s = \frac{1}{2}$ excitations whose spin Hamiltonian is

$$\mathcal{H}_s = - \mu_B \sum_n \vec{s}_n \cdot g \cdot \vec{H}_o + \mathcal{H}_{hyp} \qquad (4)$$

for a g-tensor intermediate between that of D^+ and A^- ion radicals. Single crystal T, P, Ω and H_o studies on PD-Chloranil (33), TMPD-TCNQ (34), and TMPD-Chloranil (35) are described by $\mathcal{H}_s$ for a density $\rho(T) = \exp(-\Delta E_m/kT)$ of uncorrelated $s = \frac{1}{2}$ excitations. The angular dependence of the exchange-narrowed TMPD-Chloranil linewidth follows (35) the secular $\mathcal{H}_{hyp}$ broadening computed from the crystal structure.

$\mathcal{H}_s$ for ionic CT complexes illustrates a successful, but incomplete spin Hamiltonian. The basic problem is that the regular Heisenberg chain has no magnetic gap, whereas the observed (1) $\Delta E_m \sim 0.1$ ev gives an entirely different susceptibility. The exchange parameters required (34,35) by the epr linewidths are some two orders of magnitude smaller than ΔE_m, in contrast to the direct relation in regular Heisenberg chains. The relationship between $\mathcal{H}_s$ and the electronic structure of ionic CT complexes with mixed regular stacks has recently been clarified (36), by showing that $\Delta E_m(\gamma)$ remains finite up to $\gamma_c = 0.68$ for a simple, exactly soluble model of $...D^{+\gamma}A^{-\gamma}D^{+\gamma}A^{-\gamma}...$ stacks. Incomplete ionicity of $\gamma = 0.70 \pm 0.10$ in TMPD-TCNQ was also found (37) in resonance Raman studies of TCNQ vibrations. Given a finite ΔE_m and $\gamma < 1$ in mixed regular stacks, $\mathcal{H}_s$ follows if the unpaired spins of a triplet move independently, with $D = E = 0$.

Incomplete charge transfer clearly indicates that D and A sites must be included in addition to D^+ and A^- ion radicals, thereby going beyond a Heisenberg exchange model. The electron-hole representation (36), excluding D^{+2} or A^{-2} sites, is

$$\mathcal{H}_{eh} = -|t| \sum_{p\sigma} (c^+_{p\sigma} c^+_{p+1\sigma} + c_{p+1\sigma} c_{p\sigma}) - \sum_p (2\delta_o n_p + V_1 n_p n_{p+1}) \quad (5)$$

where $|t|$ is the CT integral $\langle D^+ A^- |\mathcal{H}| DA \rangle$ and the fermion operators $c^+_{p\sigma}$, $c_{p\sigma}$ create, annihilate $A^- \sigma$ sites for even p and $D^+ \sigma$ sites for odd p. The number operators n_p vanish for D or A sites and are unity for D^+ or A^- sites. The solution of $\mathcal{H}_{eh}$ contains linear superpositions of valence-bond (VB) states with specified electronic and spin assignments at each site. The nearest-neighbor Coulomb interaction V_1 is explicitly shown rather than approximated (36) by $2\delta = 2\delta_o + V_1$. Longer-range Coulomb interactions for the 3D crystal can still be added self-consistently to Eq. (5).

The explicit inclusion of V_1 in $\mathcal{H}_{eh}$ is needed for a realistic study of the internal structure of the magnetic excitations, since both adjacent triplet and singlet D^+A^- pairs are stabilized. The experimental result of vanishing D and E in $\mathcal{H}_s$ follows inevitably for $V_1 = 0$ due to statistical considerations. Near the neutral-ionic interface, the strong admixture of neutral and ionic sites may well produce the required uncorrelated $s = \frac{1}{2}$ excitations even for finite V_1. This hypothesis is being tested by explicit VB solutions of $\mathcal{H}_{eh}$ and related models for finite chains.

TMPD-TCNQ illustrates the additional, nonmagnetic information that a theoretical model like $\mathcal{H}_{eh}$ must describe, besides accounting for the epr absorption through $\mathcal{H}_s$. The magnetic gap $\Delta E_m = 0.07$ ev indicates $\gamma \lesssim 0.68$ for $V_1 = 0$ and $\gamma \sim 0.70 - .75$ for finite V_1. The CT excitation at $\Delta E_{CT} \sim 0.95$ ev for these $|t|$, δ_o, and V_1 values is approximately (38) $4|t|$ and is sensitive to the assumed broadening of the dipole-allowed transitions. Smaller values $|t| \sim 0.1$ ev are inadmissible, because fitting ΔE_{CT} produces an ionic lattice with vanishing ΔE_m. The experimentally observed magnetic and optical excitation energies thus impose severe constraints on the microscopic parameters, which in principle are also related to gas-phase data and to approximate molecular calculations.

Direct solutions of $\mathcal{H}_{eh}$ for finite chains produce not only energies but the selection rules and oscillator strengths of optical transitions (38). These numerical methods apply for arbitrary model parameters, including (39) the regime $U \sim 4|t|$ of intermediate correlations in various modified or extended Hubbard models. There is considerable evidence (1,22) that $U \sim 4|t|$ is appropriate for π-molecular solids. Diagrammatic VB methods open the way to rigorous spectral and thermodynamic computations for these models and readily allow for additional near-neighbor interactions. While these theoretical models are surely approximate, direct comparison with both magnetic and other data should allow a proper assessment of their successess and failures. The derivation of spin Hamiltonians like $\mathcal{H}_s$ is virtually assured, while the overall self-consistency of the microscopic parameters is not. The increasingly detailed experimental data for mixed, regular, partly-filled, and alternating π-molecular systems invite systematic applications (40) of approximate, but exactly soluble theoretical models.

V. MODELS FOR RANDOM HEISENBERG CHAINS

Bulaevskii et al (41) pointed out that the low temperature magnetic properties of TCNQ salts with asymmetric donors suggest a random 1D Heisenberg chain

$$\mathcal{H}_R = \sum_n 2J_n \vec{s}_n \cdot \vec{s}_{n+1} - g\mu_B H_o \sum_n s_{nz} \tag{6}$$

and **the** asymmetric donors are in fact structurally disordered in these salts. Detailed subsequent studies (42) of $Q(TCNQ)_2$, where Q = quinolinium, over a wide temperature range (~ 15 mK to 15 K) indicate that the susceptibility $\chi(T)$ goes like $T^{-\alpha}$, the magnetization M as $H_o^{1-\alpha}$, and the magnetic specific heat $C(H_o,T)$ as $T^{1-\alpha}$ for $H_o = 0$ and as $TH^{-\alpha}$ for $g\mu_B H_o \gg kT$. The value of $\alpha \sim 0.80$ is slightly sample dependent in $Q(TCNQ)_2$. There are less complete power-law dependences (41) for the 1:1 TCNQ complex with NMP (N-methylphenazine) and for several other 1:2 salts (43), all with $0.5 < \alpha < 1.0$. These power laws reflect (42) a distribution $f(J)$ of random exchanges in $\mathcal{H}_R$

$$f(J) = \frac{(1-\alpha')}{J_o} \left(\frac{J}{J_o}\right)^{-\alpha'} \qquad 0 \leq J \leq J_o \tag{7}$$

The cutoff J_o insures the normalization of $f(J)$. The important point (41) is that $f(J)$ is finite as $J \to 0$, thereby disconnecting $\mathcal{H}_R$ into noninteracting segments. The relation between α' in $f(J)$ and the observed α is still open (42), although there are arguments (44) for $\alpha' = \alpha$.

These experimental and theoretical results offer interesting applications of successful but incompatible spin Hamiltonians and phenomenological models. The problem was originally bypassed (41) by assuming that $\mathcal{H}_R$ yields noninteracting fermions with a specified density of states $\rho(\epsilon)$. The theoretical problem of relating $\rho(\epsilon)$ to $f(J)$ was subsequently discussed (45) in terms of disordered Hubbard models. The static magnetic properties of fermions with a given $\rho(\epsilon)$ are typical of spin Hamiltonian that summarizes data without necessarily offering a microscopic picture.

Two rather different approximations of $\mathcal{H}_R$ are possible. In the random dimer model (42), every J_{2n+1} is assumed to vanish and the distribution for the remaining J_{2n} of N/2 isolated dimers at sites 2n, 2n+1 is taken to be $f(J)$, with $\alpha' = \alpha$ in Eq. (7). The resulting thermodynamics for noninteracting random dimers yield all the observed power laws. The picture of approximating a random 1D chain by regularly choosing $J_{2n+1} = 0$, is perhaps surprising. The alternative cluster (44) approximation is to disregard any exchange $2J_n < kT$, thereby producing segments whose length is random and temperature dependent. The choice $f(J)$, again with $\alpha' = \alpha$, produces a $T^{1-\alpha}$ power law for $\chi(T)$ due to the interactions between $s = \frac{1}{2}$ odd-length segments. The internal excitations of the segments are neglected entirely, but would be needed for $M(H_o)$ or other properties. The cluster picture of interacting, variable-length, ground-state segments contrasts strongly with that of noninteracting, internally-excited random dimers. These spin Hamiltonians seem incompatible with each other, and neither may be a faithful representation of $\mathcal{H}_R$.

The direct analysis of random segments, up to $N \sim 8$ or 10, by VB methods (46) should clarify the formation of segments defined by $2J_n < kT$, the role of internal excitations, and the relation between $f(J)$ for an infinite chain and random dimers. In particular, the strong peaking of $f(J)$ near $J \to 0$ invites finite-chain calculations, since the frequent occurrence of small J_n disconnects $\mathcal{H}_R$ into small finite segments. The proposed distribution function $f(J)$ in Eq. (7) may have to be revised, especially near $J \sim J_\sim$, which has yet to be related to $\chi(T)$ in the range $20 < T < 400$ K.

The microscopic derivation and parametrization of $\mathcal{H}_R$ in terms of random 1D Hubbard models has been applied (45) principally to the 1:1 compound NMP-TCNQ, for assumed paramagnetic $\gamma = 1$ TCNQ$^{-\gamma}$ stacks and diamagnetic NMP$^+$ cations. The high conductivity and other physical properties (1,22,24,47) of NMP-TCNQ are quite anomalous for half-filled $\gamma = 1$ stacks and suggest partial filling of $\gamma \underset{\sim}{<} 1$. The field dependences of the proton NMR second moments (24) of NMP-TCNQ and DMP-TCNQ, with a deuterated methyl, are shown in Fig. 3. Since M_2 is sensitive to protons near the unpaired electrons, their different behavior demonstrates that $\chi(T)$ is dominated at low T by a few percent of localized spins on the NMP moiety, but NMR does not distinguish (24) between neutral NMP radicals leading to a $\gamma < 1$ TCNQ$^{-\gamma}$ stacks and substitutional NMPH$^+$ cation radicals. The probable contamination of all "NMP-TCNQ" samples by NMPH$^+$ cation radicals has since been demonstrated (48), without ruling out the incomplete charge-transfer suggested by other measurements. Thus NMP-TCNQ is an unfortunate choice for establishing a random 1D Hubbard model and the proposed microscopic parameters (45) for TCNQ$^-$ paramagnetism should probably be reevaluated for a few percent of weakly-interacting, randomly distributed NMP$^+$ or NMPH$^+$ radicals.

The observed structural disorder of TCNQ salts with asymmetric organic donors suggests a random exchange model like $\mathcal{H}_R$, whose solution is still highly approximate and needs further study. A satisfactory theoretical model must provide a plausible derivation of the distribution $f(J)$ in terms of microscopic parameters that are also consistent with nonmagnetic data. TCNQ salts with asymmetric donors in fact show increasingly large differences between the weak interactions needed to fit the thermopower (49) and low-temperature magnetism (41-43) and the far stronger interactions suggested (1,22,8,9) by optical and high-temperature magnetic data. Improved experimental results

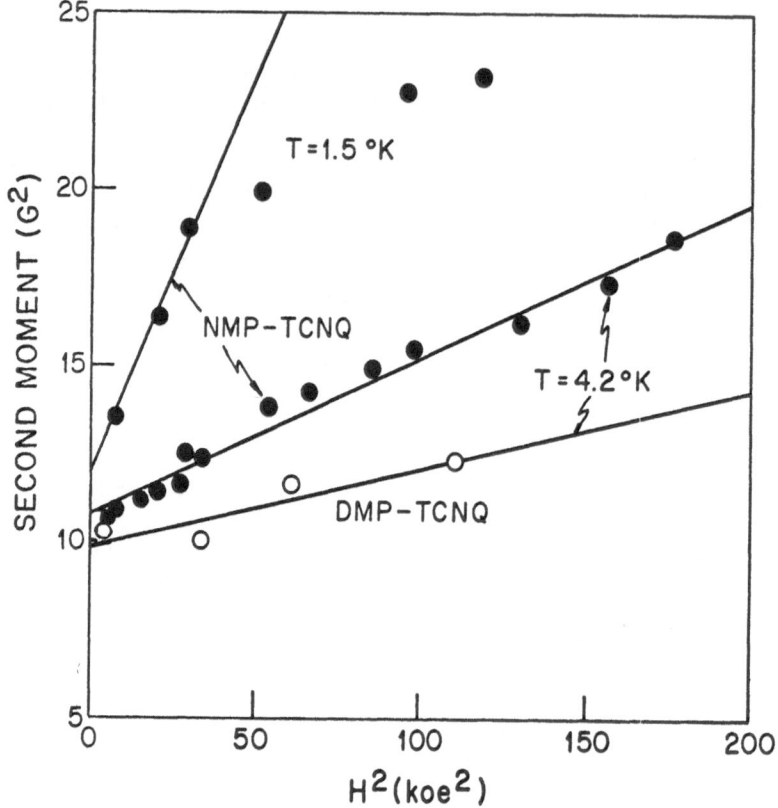

Fig. 3. Field dependence of the second moment of the ^{1}H
NMR in powder samples of NMP-TCNQ and DMP-TCNQ,
with deuterated methyls. (ref. 24).

have tended to sharpen this unphysical dichotomy. There are
several possible resolutions, primarily in terms of electron-
phonon interactions, but such ideas have yet to be incorporated
into well-defined theoretical models.

VI. ATOMIC LIMIT AND SPIN POLARONS

    It has long been recognized (16) that antiferromagnetism in
organic ion-radical salts involves virtual CT excitations around
$\Delta E_{CT} \sim 1$ ev. Such correlation effects (1) are approximated in
Hubbard models with an effective on-site repulsion $U \sim \Delta E_{CT}$ and
a bandwidth $4|t| \sim 0.5$ ev related to the Mulliken CT integral
$|t|$. The qualitative physical properties of ion-radical solids
(1,22,8) can be rationalized by suitable extensions or modifications of Hubbard models. But the thermodynamics and excitation

spectra of such models are poorly understood in the experimentally
relevant regime of intermediate correlations $U \sim 4|t|$, or partial
filling $\gamma = N_e/N$ in regular $\ldots A^{-\gamma}A^{-\gamma}A^{-\gamma}\ldots$ or $D^{+\gamma}D^{+\gamma}D^{+\gamma}\ldots$
stacks. Recent diagrammatic VB methods (36,38,39) yield a com-
plete description for small systems with N_e, $N \lesssim 10$. The special
cases $U \ll 4|t|$ and $U \gg 4|t|$ are, respectively, band theory for
uncorrelated electrons and the atomic limit. A great deal is
known about both for arbitrary γ. In particular, the thermo-
dynamics (50) for 1D Hubbard models with $\gamma < 1$ and $U \gg 4|t|$
amount to decoupled space and spin degrees of freedom. The N_e
spins obey a Curie-law $\chi(T)$, with at most small antiferromagnetic
exchange $J \sim 2t^2\gamma/U$, while the space part gives a metallic band
$-2|t|\cos k$ of spinless fermions filled to $k_F = \gamma\pi$. The partly-
filled atomic limit thus combines high conductivity $\sigma(T)$ and
noninteracting spins. The fact that $\chi(T)$ for π-molecular
conductors does not obey a Curie law immediately rules out the
atomic limit.

Such $\sigma(T)$ and $\chi(T)$ data have recently been reported (51)
for $D^\gamma(I_3^-)_\gamma$ complexes, with $\gamma = 0.36$ and 0.97, of the macrocyclic
donor, $D = ^\gamma 1,4,5,8,9,12,13,16$-octamethyltetrabenzporphinato
nickel (II). Both complexes crystallize in 1D arrays of almost
planar $D^{+\gamma}$ macrocycles separated by I_3^- ions. Similar ligand-
based π-molecular g-tensors are observed in single-crystal epr
for both. The increased $..D^{+\gamma}D^{+\gamma}..$ spacing of 3.78 Å along the
1D chain of the $\gamma = 0.36$ complex, in contrast to spacing of
3.2 - 3.3 Å in ordinary organic conductors, readily rationalizes
the far smaller bandwith $4|t|$ required for a Curie-law $\chi(T)$ in
the atomic limit. These chemically complicated macrocyclic $D^{+\gamma}$
systems are consequently the first realizations (52) of a simple
theoretical model for $U \gg 4|t|$ and $\gamma < 1$. While the precise
values of the microscopic parameters are far beyond current
computations, the atomic limit greatly simplifies the interpretation. The $\chi(T)$ data account for $N_e = \gamma N$ weakly-interacting
localized moments, with $J < 3$ cm^{-1} and 10 cm^{-1}, respectively,
for 1D Heisenberg chains of the $\gamma = 0.36$ and $\gamma = 0.97$ complexes.
The partial filling $\gamma < 1$ is independently known (51) from
stoichiometry and resonance Raman.

The single crystal conductivities of both complexes are
quantitatively described (51,52) by

$$\sigma(T) = AT^{-\alpha} \exp(-\Delta_o/kT) \qquad (8)$$

in the range ~ 150-380 K. The parameters α and Δ_o/k are sample
dependent. A large $\gamma = 0.36$ crystal led (51) to $\alpha = 2.9$,
$\Delta_o/k \approx 900$ K; two additional samples also used for epr gave (52)
$\alpha = 4.1$, $\Delta_o/k = 1300$ K and $\alpha = 3.8$, $\Delta_o/k = 1220$ K; a smaller
$\gamma = 0.97$ sample required (51) $\alpha = 2.7$, $\Delta_o/k = 1160$ K. The
resulting $\sigma(T)$ vs T curves have slightly different maxima. These

differences are tentatively attributed to crystal purity or perfection. The point is that Eq.(8) has been successfully applied (53) to $\sigma(T)$ data for several organic conductors, with similar α, Δ_o/k parameters. But the usual interpretation (53), that $T^{-\alpha}$ is a mobility $\mu(T)$ and $\exp(-\Delta_o/kT)$ is an equilibrium concentration of carriers, is ruled out in these $D^{+\gamma}(I_3^-)\gamma$ crystals, since $\chi(T)$ corresponds to all available, noninteracting D^+ ion radicals. All T dependences in Eq. (8) must therefore be associated with the mobility $\mu(T)$.

The overall mobility of a concentration γ of localized carriers undergoing 1D hopping, with strict exclusion of D^{+2} sites in the atomic limit, obeys (54) an Einstein relation even though individual motions are not diffusive. Thus $\mu(T)$ goes as $T^{-1}\omega_h(T)$, where $\omega_h(T)$ is the nearest-neighbor hopping rate of the localized carriers. The observed $\sigma(T)$ agrees (52), within experimental error, with the predicted (54) $\gamma(1-\gamma)$ concentration dependence. The semiclassical expression (55) for the polaron hopping rate $\omega_h(T)$ has the functional form of Eq. (8), with $\alpha' = \alpha-1 = 0.5$ for a single phonon branch. Since all D^+ ion-radicals are known from $\chi(T)$ to be available for conduction and the atomic limit corresponds to small $|t|$, the observed $\sigma(T)$ supports the occurrence of spin polarons (52) in these $D^{+\gamma}(I_3^-)_\gamma$ systems.

Polaron motion $\omega_h(T)$ provides a relaxation mechanism for the D^+ radicals. Single crystal epr shows small ($\lesssim 3 \times 10^{-3}$) spin delocalization (51) to I_3^-, a specie with several low-energy ($\lesssim 100$ cm^{-1}) vibrations and large spin-orbit coupling. Phonon modulation of the spin orbit coupling leads (56) to a linear T dependence in metals. The epr linewidth $\Gamma(\Omega,T)$ is then proportional to $T\omega_h(T)$, and thus to $T^2\sigma(T)$ for conduction via spin polarons (52). The ratio

$$R(\Omega) = \frac{\Gamma(\Omega,T)}{T^2\sigma(T)} \qquad (9)$$

should be independent of both T and $\omega_h(T)$ when motional modulation of the spin-orbit interaction dominates $\Gamma(\Omega,T)$. $R(\Omega)$ is found (52) to be constant for $T \gtrsim 200$ K in several $\gamma = 0.36$ complexes whose $\sigma(T)$, or $\omega_h(T)$, differ in both magnitude and α, Δ_o parameters. Although the approximate proportionality of $\Gamma(\Omega,T)$ and $T^2\sigma(T)$ is found generally, the precise cancellation of $\omega_h(T)$ requires $\Gamma(\Omega,T)$ and $\sigma(T)$ data for the same sample. In favorable cases such magnetic information thus complements conductivity studies. Small additional (~0.5G) contributions to $\Gamma(\Omega,T)$ below 200 K produce (51) different Ω dependences. Such contributions rationalize the observed (52) positive deviations of $R(\Omega)$ at low T.

The simultaneous realization of $\gamma < 1$ and $U \gg 4|t|$ in these systems immediately allows, through results for the atomic limit, a theoretical model based on spin polarons and a reinterpretation of Eq. (8) entirely in terms of $\mu(T)$. In this case, the proper spin Hamiltonian must still be found, as the different Ω dependences of $\Gamma(\Omega,T)$ for $\gamma = 0.36$ and 0.97 are not understood. The explicit expressions for spin-orbit and other relaxation mechanisms must also be obtained and parametrized in the spin Hamiltonian. While these complicated molecular conductors are not likely to be amenable to detailed modeling, the combination of a Curie-law $\chi(T)$ and single-crystal epr data has interesting implications for any π-molecular conductor whose $\sigma(T)$ is described by Eq. (8).

VII. MODELS WITH ONE VALENCE STATE PER SITE

Some physical properties of 1D π-molecular solids are reminiscent of more familiar atomic, ionic, and molecular solids. These novel open-shell molecular systems are sufficiently diverse to encompass ideas from the band theory of conductors, from exchange in magnetic insulators, from the optical properties of closed-shell molecular solids, and from electrostatic binding in inorganic salts. Their novelty lies in the fact that none of the earlier theoretical models suffices, although elements of each are retained. For example, the almost van der Waals spacing of π-radicals suggests narrow bands, with small $|t|$, while the semiconducting nature of $\frac{1}{2}$-filled regular stacks demonstrates the need for including electron-electron correlations in any theory of conductors. Low-lying optical and conducting states point to generalizations of magnetic insulators, while the lowest CT transitions differ from those of typical molecular solids. The delocalized charge and spin distributions of π-radicals and the occurrence of partial charge transfer complicate the analysis of the Madelung energies of ionic solids. Intramolecular vibrations and librations open new possibilities for electron-phonon interactions.

These similarities and differences with other types of solids probably account for the strikingly different pictures proposed by early workers. Thus conductors and 1D instabilities were discussed in terms of band models that completely ignored the molecular nature of the solids. Conversely, their molecular spectroscopies were related to nonconducting solids, thus minimizing the complications of additional low-energy CT transitions in complex salts. Magnetic properties of ion-radical salts at low temperature were described by exchange Hamiltonians that explicitly exclude charge motion. Such partial studies inevitably led to successful spin or phenomelogical Hamiltonians with quite different microscopic parameters. A closer examination reveals such determinations to hinge on the model adopted, the experimental data selected, and on the reasonability of various coupling constants.

The initial challenge of modeling a new class of solids is to highlight key features, since truly quantitatively descriptions hardly require a model but merely the solution of a many-electron Coulomb Hamiltonian. For example, the structural and physical properties of all organic ion-radical solids are consistent with weakly interacting molecular building blocks (1). Their widely different physical properties are classified in Fig. 4 according to the nature of the 1D stacks, the degree of electron filling, and the π-overlap with neighbors. Frenkel (triplet) spin excitons occur in alternating stacks, be they segregated or mixed, $\frac{1}{2}$-filled or complex. Organic conductors involve complex ($\gamma < 1$) regular segregated stacks and complex salts have additional low-lying CT excitons.

The underlying theoretical model is that D^+ or A^- ion radicals have a single valence state, the π-MO of the unpaired electron, that is largely transferable from one solid to another. Grossly different physical properties reflect differences in the degree of filling, the overlap patterns, and the type of neighbors. A single valence state per site leads (1) in general to a fermion site representation (57) with 0, 1, or 2 electrons in the π-MO. Conventional band theory follows when electron-electron interactions are neglected or treated self-consistently; a VB description results when correlations are explicitly included. The restriction to a single valence state per site will in my opinion suffice for solid-state properties, while higher (eg. $\pi \rightarrow \pi*$) molecular excited states involve molecular exciton theory.

The restriction to one valence state per site occurs in many other contexts, starting with the H_2 covalent bond constructed from two 1s AOs. Huckel and other theories of π-bonding are based on a single $2p_z$ AO at each C or N; their application (58) to polyenes as the Pariser-Parr-Pople (59) model is an earlier and more general form of Hubbard models. Theories of simple metals like alkalis start with a single s-state AO, while simple inorganic semiconductors can be rationalized in terms of one or more hybrid AOs per site. The resulting model parameters like $|t|$ or U vary greatly, as do the approximations suitable for treating molecules, metals, or π-radical solids. Nevertheless, these theories all share a common denominator of one valence state per site.

The importance of correlations in narrow-band organic solids requires theoretical methods, such as VB diagrams (36,38,39), that explicitly describe these contributions. The current limitations to finite systems is tolerable for 1D problems, but quite

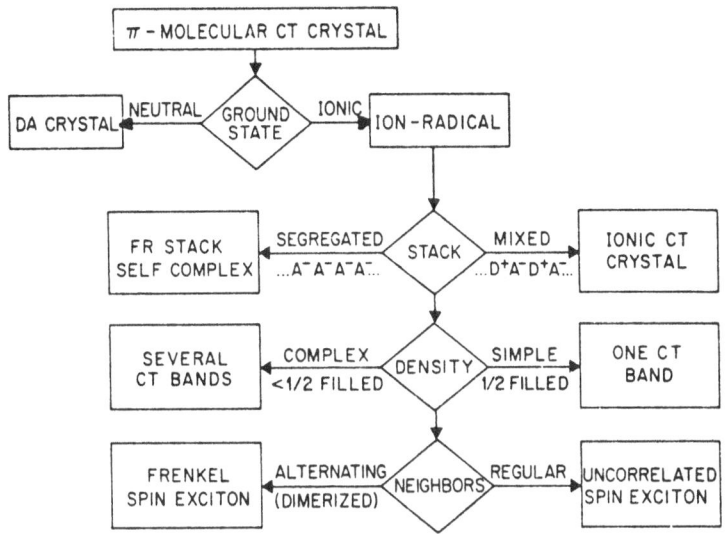

Fig. 4. Schematic classification of open-shell π-molecular CT
        crystals with emphasis on the type of neighbors, the
        degree of filling and the overlap patterns.

restrictive for extended 2D or 3D structures. Approximate methods
must clearly be found, once the exact role of correlations in the
regime $U \sim 4|t|$ is established in small systems. For example,
highly excited configurations could be omitted, thereby reducing
the configuration-interaction problem. More general reductions
are possible in other cases: the exclusion of D^{+2} or A^{-2} sites
in mixed stacks leads to three, rather than four, possibilities
per sites; the restriction to one electron per site in magnetic
insulators leads to two states, α and β, for each $\vec{s}_n$. These
different limits of one valence state per site are realized by
at least some π-radical solids.

It is important to remember that the valence state can be
defined (1,57) in principle to include all intrasite (molecular)
contributions as well as many averaged crystal contributions.
The resulting theoretical model is phenomelogical. Its parameters require quantum chemical computations, including a proper
assessment of vibronic contributions. Direct computations of
J, t, U, V_1, and other parameters typically involve geometrical
assumptions, rely on parametrized theoretical method, and yield
rough estimates. Such adjustable parameters reduce theoretical
models to spin Hamiltonians on the molecular level, and may
occasionally blur any distinctions between them.

Models with one valence state per site highlight the role
correlations in partly-filled narrow bands. Their extension to
electron-phonon or exciton-phonon coupling must certainly be
considered in systems with phase transitions, just as other
extensions arise in disordered systems. Magnetic properties
(1-10) are important for characterizing phase transitions, but
such spin Hamiltonians have been omitted. Spin Hamiltonians
bearing on charge transport have also been neglected, except for
spin polarons in the atomic limit. The inevitable theoretical
choice is to concentrate on correlations, or electron-phonon
coupling, or disorder, while awaiting adequate simultaneous
treatments. Thus models for electron-phonon coupling tend to
ignore correlations, and vice versa. In my opinion, correlations
are generally important in open-shell π-molecular solids and
suffice for many properties. Electron-phonon interactions or
disorder are critical in some systems, where they must be added
to correlation effects. In any case, many new spin Hamiltonians
can be anticipated to summarize additional solid-state magnetic
data, thereby showing the way to more satisfactory theoretical
models for the electronic structure, which in turn may occasionally
permit direct quantum mechanical comparisons. Both spin Hamiltonians and theoretical models will continue to find many
applications.

It is a pleasure to acknowledge the financial support of
this work through NSF-CHE-27418.

REFERENCES

1. Z.G. Soos and D.J. Klein, in "Molecular Association," Vol. 1
   (ed. R. Foster, Academic, N.Y. 1975) pp. 1-109; Z.G. Soos,
   Ann. Rev. Phys. Chem. $\underline{25}$, 121 (1974).

2. H.R. Zeller, Adv. Sol. State Phys. $\underline{13}$, 31 (1973).

3. P. M. Richards, in "Local Properties at Phase Transitions"
   (Editrice, Bologna, 1975) p. 539; D.W. Hone and P.M. Richards,
   Ann. Rev. Mater. Sci. $\underline{4}$, 337 (1974).

4. L. S. DeJongh and A.R. Miedema, Adv. Phys. $\underline{23}$, 1 (1974)

5. L.V. Interrante, ed."Extended Interactions Between Metal Ions
   in Transition Metal Complexes"(American Chemical Society,
   Washington, D.C. 1974) No. 5.

6. J.S. Miller and A.J. Epstein, eds. "Synthesis and Properties
   of Low-Dimensional Materials," Ann. N.Y. Acad. Sci. $\underline{313}$ (1978).

7. L. Pál, G. Grüner, A. Jánossy, and J. Sólyom, eds.
   "Organic Conductors and Semiconductors" (Springer-Berlag,
   Berlin, 1977).

8. W.E. Hatfield, ed. "Molecular Metals" (Plenum, N.Y. 1979)

9. H.J. Keller, "Low-Dimensional Cooperative Phenomena" (NATO
   ASI Series B7, Plenum, N.Y. 1975).

10. H.J. Keller, ed. "Chemistry and Physics of One-Dimensional
    Metals" (NATO ASI Series B25, Plenum, N.Y. 1977).

11. T.T.P. Cheung, Z.G. Soos, R.E. Dietz, and F.R. Merritt,
    Phys. Rev. B17, 1266 (1978).

12. B. Morosin, R.C. Hughes, and Z.G. Soos, Acta Cryst. B31,
    762 (1975); G.F. Kokoszka and G. Gordon, Trans. Metal Chem.
    5, 181 (1969).

13. D.B. Chesnut and W.D. Phillips, J. Chem. Phys. 35,1002 (1961).

14. D.D. Thomas, H.J. Keller, and H.M. McConnell, J. Chem. Phys.
    39, 2321 (1963).

15. D. Nöthe, W. Moroni, H.J. Keller, Z.G. Soos, and S. Mazumdar,
    Solid State Commun. 26, 713 (1978).

16. P.L. Nordio, Z.G. Soos, and H.M. McConnell, Ann. Rev. Phys.
    Chem. 17, 237 (1966).

17. D. Nöthe and H.J. Keller, unpublished work; I. Fujita and
    Y. Matsunaga, preprint (1979).

18. A.J. Silverstein and Z.G. Soos, Chem. Phys. Letters 39,525(1976)

19. S. Flandrois and J. Boissonade, Chem. Phys. Letters 58,596(1978)

20. T. Hibma, G.A. Sawatzky, and J. Kommandeur, Chem. Phys. Letters
    23, 21 (1973); R. Lynden-Bell and H.M. McConnell, J. Chem.
    Phys. 37, 794 (1962).

21. A. Brau, P. Bruesch, J.P. Farges, W. Hinz, and D. Kuse, Phys.
    Stat. Sol (b) 62, 615 (1974).

22. J.B. Torrance, Acct. Chem. Res. 12, 79 (1979); J.B. Torrance,
    B.A. Scott, and F.B. Kaufman, Solid State Commun. 17,1369(1975)

23. Z.G. Soos, K.T. McGregor, T.T.P. Cheung, and A.J. Silverstein,
    Phys. Rev. B16, 3036 (1977); K.T. McGregor and Z.G. Soos,
    J. Chem. Phys. 64, 2506 (1974).

24. M.A. Butler, F. Wudl, and Z.G. Soos, Phys. Rev. B12,4708(1975).

25. E.F. Rybaczewski, L.S. Smith, A.F. Garito, A.J. Heeger and
    B. Silbernagel, Phys. Rev. B14, 2746 (1976).

26. M.A. Butler, L.R. Walker, and Z.G. Soos, J. Chem. Phys. 64,
    3592 (1976).

27. F. Devreux and M. Nechtschein, Mol. Cryst. Liq. Cryst. 32,
    251 (1976); J. Avalos, F. Devreux, M. Guglielmi, and
    M. Nechtschein, Mol. Phys. 36, 669 (1978).

28. G. Soda, D. Jerome, M. Weger, J. Alizon, J. Gallice,
    H. Robert, J.M. Fabre, and L. Giral, J. Physique 38,931(1977);
    J. Alizon, G. Berthet, J.P. Blanc, J. Gallice, H. Robert,
    J.M. Fabre, and L. Giral, Phys. Stat. Sol. (b) 85,603(1978)

29. G.M. Semeniuk and D.B. Chesnut, Chem. Phys. Letters 25,251
    (1974); D.B. Chesnut and S.R. Bondeson, J. Chem. Phys. 68,
    5383 (1978).

30. L.J. Azevedo, A. Narath. P.M. Richards, and Z.G. Soos, Phys.
    Rev. B (submitted); Phys. Rev. Letters (submitted).

31. L.J. Azevedo, W.G. Clark, and E.O. McLean, Proc. XIVth Int.
    Conf. on Low Temp. Phys. LT14, Helsinki, Finland,p. 369
    (1975); L.J. Azevedo, Ph.D. Thesis UCLA, 1975 (unpublished).

32. Z.G. Soos, J. Chem. Phys. 46, 4284 (1967).

33. R.C. Hughes and Z.G. Soos, J. Chem. Phys. 48,1066 (1968).

34. B.M. Hoffman and R.C. Hughes, J. Chem. Phys. 52, 4011 (1970);
    Solid State Commun. 7, 895 (1969).

35. T.Z. Huang, R.P. Taylor, and Z.G. Soos, Phys. Rev. Letters
    28, 1054 (1972).

36. Z.G. Soos and S. Mazumdar, Phys. Rev. B18, 1991 (1978);
    Chem. Phys. Letters 56, 515 (1978).

37. R.P. Van Duyne, private communication; Z.G. Soos, S.Mazumdar,
    and T.T.P. Cheung, Mol. Cryst. Liq. Cryst. 52, 397 (1979).

38. S.R. Bondeson and Z.G. Soos, Chem. Phys. (in press).

39. S. Mazumdar and Z.G. Soos, Synthetic Metals (in press).

40. S.R. Bondeson and Z.G. Soos, in Extended Linear Chain Compounds,
    Vol. 1 (ed. J.S. Miller, Plenum, N.Y.) in preparation.

41. L.N. Bulaevskii, A.V. Zvarykina, Yu. S. Karimov,
    B.B. Lyobovskii, and I.F. Shchegolev, Sov. Phys. - JETP
    $\underline{35}$, 384 (1972); I.F. Shchegolev, Phys. Stat. Sol. $\underline{12}$, 9(1972)

42. W.G. Clark and L.C. Tippie, Phys. Rev. B (in press);
    L.J. Azevedo and W.G. Clark, Phys. Rev. $\underline{B16}$, 3252(1977);
    W.G. Clark, L.C. Tippie, G. Frossati, and H. Godrin,
    J. de Phys. $\underline{39}$, C6-365 (1978)

43. G. Mihály, K. Holczer, G. Grüner, and L.D. Kunstelj, Solid
    State Commun. $\underline{19}$, 1091 (1976); G. Mihály, K. Holczer,
    A. Jánossy, G. Grüner, and M. Milják, in ref. 7, p.553(1977).

44. G. Theodorou, Phys. Rev. $\underline{B16}$, 2264 (1977).

45. G. Theodorou and M.H. Cohen, Phys.Rev. Lett. $\underline{37}$, 1014 (1976);
    G. Theodorou, Phys. Rev. $\underline{B16}$, 2273, 2254 (1977).

46. Z.G. Soos, unpublished results.

47. J.F. Kwak, G. Beni, and P.M. Chaikin, Phys. Rev. $\underline{B13}$,641 (1976).

48. H.J. Keller, D. Nöthe, W. Moroni, and Z.G. Soos, Chem. Comm.
    $\underline{1978}$, 331; D.J. Sandman, Mol. Cryst. Liq. Cryst. $\underline{50}$, 235(1979).

49. J.F. Kwak and P.M. Chaikin, Phys. Rev. $\underline{B13}$, 652 (1976).

50. G. Beni, T. Holstein, and P. Pincus, Phys. Rev. $\underline{B8}$ 312(1973);
    D.J. Klein, ibid $\underline{B8}$, 3452 (1973).

51. T.E. Phillips, R.P. Scaringe, B.M. Hoffman, and J.A. Ibers,
    J. Amer. Chem. Soc. (in press).

52. T.E. Phillips, B.M. Hoffman, and Z.G. Soos, Solid State
    Commun. (submitted).

53. A.J. Epstein, E.M. Conwell, and J.S. Miller, Ann. N.Y. Acad.
    Sci. $\underline{313}$, 183 (1978).

54. P.M. Richards, Phys. Rev. $\underline{B16}$, 1393 (1977).

55. M. Trlifaj, Czech. Journ.Phys. $\underline{6}$, 533 (1956);T. Holstein,
    Ann. Phys. (N.Y.) $\underline{8}$, 343 (1959); D. Emin, in "Electronic
    and Structural Properties of Amorphous Semiconductors",
    (eds. P.G. LeCombre and J. Mort, Academic Press, N.Y. 1973)
    p. 261-327.

56. Y. Yafet, Sol. State Phys. $\underline{14}$ (ed. F. Seitz and D. Turnball,
    Academic, 1963) p. 1-98.

57. D.J. Klein and Z.G. Soos, Mol. Phys. <u>20</u>, 1013 (1971).

58. K. Schulten, I. Ohmine, and M. Karplus, J. Chem. Phys. <u>64</u>, 4422 (1976); F.A. Matsen, Acct. Chem. Res. <u>11</u> 387 (1978).

59. R. Pariser and R.G. Parr, J. Chem. Phys. <u>21</u>, 446, 767 (1953); J.A. Pople, Trans. Far. Soc. <u>42</u>, 1375 (1953).

MOLECULAR VIBRATION STUDIES OF QUASI-ONE-DIMENSIONAL ORGANIC
CHARGE-TRANSFER COMPOUNDS

Renato Bozio and Cesare Pecile

Institute of Physical Chemistry, The University,
2, Via Loredan, 35100 PADOVA, Italy

1. INTRODUCTION

By nature, investigations of very complex systems such as
the quasi 1-D organic charge transfer (CT) compounds imply the
use of a multidisciplinary approach. Thus, to be succesful, experimental studies have to combine a wide variety of physical
techniques. In recent years vibrational spectroscopy has entered
the field and proved itself a valuable and informative tool.
Sufficient data have been collected on the intramolecular vibrational behaviour of large electron donor or acceptor (EDA) molecules and their radical ions to allow a full exploitation of vibrational spectra in the study of CT solids. The interest in this
type of investigations is further justified by the fact that the
interaction between electrons and intramolecular vibrations has
been shown to play a relevant role in determining the physical
properties of 1-D organic conductors and their temperature dependence (1-4).

In the present lecture vibrational studies aiming at one or
more of the following goals will be discussed: (i) use the intramolecular vibrational frequencies to check whether, in the time
scale of Raman or infrared measurements, the electronic charge
along the 1-D columns appears as localized or delocalized and,
in the latter case, try to evaluate the degree of charge transfer;
(ii) provide the basic information on the intramolecular vibrational behaviour for an a priori evaluation of the electron-molecular vibration (e-mv) coupling (1); (iii) obtain a thorough
interpretation of the vibronic effects typical of infrared spectra of organic semiconductors making possible the experimental
measurement of the e-mv coupling constants; (iv) use the tempera-

L. Alcácer (ed.), The Physics and Chemistry of Low Dimensional Solids, 165–186.
Copyright © 1980 by D. Reidel Publishing Company.

ture dependence of the intensity of the vibronic infrared absorptions to monitor the distortions of the linear chains in those
systems undergoing the typical Peierls transition (5).

2. VIBRATIONAL ANALYSIS OF NEUTRAL AND IONIZED MOLECULES AND THEIR APPLICATIONS

The experimental techniques and the interpretative tools for
a sound vibrational analysis of large polyatomic molecules are
well settled (6, 7). We limit ourselves to mention the methods
used and some peculiar problems encountered with the corresponding radical ions. Our aim is to convey the idea that the collection of basic data on the vibrational behaviour of new molecules
is the rate determining step in the vibrational study of organic
conductors.

A thorough vibrational assignment of the neutral molecules
must rely on as much as possible direct experimental information
on the symmetry species of the vibrational modes. This information comes from Raman depolarization ratios in solution and intensity ratios from polarized infrared and Raman spectra of oriented crystalline samples through the use of the "oriented gas model" (6b) whose validity, generally verified, must be checked
for each system.

Some features of the vibrational study of the corresponding
radical ions must be pointed out. (i) The fact that the effective
molecular configuration is retained upon ionization allows one
to establish a pairwise correspondence between vibrational modes
of the neutral and ionized molecules. This is advantageous because, on the one hand, it facilitates the assignment of the radical ion by resorting to the correlation with the neutral molecule, on the other hand, it makes meaningful the notion of ionization frequency shifts, that is, the changes in vibrational frequencies in going from neutral to fully ionized molecules. (ii)
Since the open-shell molecular ions under discussion are characterized by strong electronic absorptions in the visible region,
the Raman spectra obtained by standard cw lasers display a resonance or preresonance character (8). Consequently, they are dominated by or almost limited to features related with totally symmetric fundamentals and combination tones. This fact, advantageous
for extracting informations on excited electronic states and for
studying very dilute samples, is a drawback for the purpose of
a complete vibrational assignment because it hides the experimental identification of non-totally symmetric Raman-active modes.
(iii) When radicals partecipate to a strong intermolecular CT
interaction (9, 10) new vibronic absorptions appear in the infrared spectra. As we shall see afterwards, this is due to a mechanism of intensity borrowing from the CT electronic transition.

The vibronic features must be clearly identified to avoid misassignment of the infrared active modes of the non-interacting
radical ion.

The information on the vibrational benaviour of a molecular
structure gained from infrared and Raman spectra must be extended to the description of the normal mode displacements. This is
done by performing a normal coordinate analysis (NCA) (6a) through
a modified valence force field (MVFF). Such analysis allows also
a scrutiny of the assignments particularly of those left tentative by the experimental data. A further function of a NCA peculiar to parallel vibrational studies of neutral molecules and
their ions is the bridging between observed ionization frequency
shifts and bond order changes upon ionization.

It is well known that finding a MVFF for a polyatomic molecule is an under-determined problem (6a, 11). In fact the same
set of observed frequencies can be reproduced by different MVFF's.
If the aim is only the rationalization of the vibrational assignment, the preferential choice of one force field in respect to
another is a marginal problem provided that one rejects force
constant values physically unacceptable in the light of empirical
knowledge. However, different MVFF's reproducing the same set of
observed frequencies may give substantially different descriptions of the normal modes. If one needs the latter information
as an input data for calculation of physical parameters influenced by molecular vibrations the choice of the MVFF is more critical. Some guidelines are recommendable to cope with this problem. (i) A set of force constants well determined on similar
molecules (possibly by the "overlay" technique (7d, 11)) must
be transferred to the molecule under study to produce a zeroth
order frequency calculation. (ii) The transferred MVFF must be
tested by comparing the calculated frequencies with those safely
assigned by experiment. (iii) The number of variations performed
on the initial MVFF in order to obtain an acceptable overall fit
must be kept to a minimum. In this way the degree of reliability
of the calculated normal mode displacements is largely determined
by that of the transferred force constants. The only further test
is the capability of the calculated eigenvectors to correctly pre
dict physical properties influenced by molecular vibrations.

The knowledge of the intramolecular vibrational behaviour
of strong electron donor or acceptor molecules gained by applying the above mentioned criteria and methods to the experimental
study and NCA is summarized and exemplified in Table I for the
case of the well famous TCNQ (12-14) and TTF (15, 16) structures.
For the sake of conciseness, only the a_g, b_{1u} and b_{3u} fundamental
modes are included being the most involved in the applications
to be discussed later on. Furthermore (see below) the a_g modes
are the only ones allowed by symmetry to couple with electrons

TABLE I. OBSERVED AND CALCULATED FUNDAMENTAL FREQUENCIES (ν/cm^{-1}) OF TCNQ AND TTF NEUTRAL MOLECULES AND RADICAL IONS

sym. species	TCNQ⁰ (a)			TCNQ⁻ (b)		Δν_obs	sym. species	TTF⁰ (c)			TTF⁺ (c)		Δν_obs
	obs.	calc.	pot. energy distr. %	obs.	calc.			obs.	calc.	pot. energy distr. %	obs.	calc.	
a_g ν1	3048	3052	$K_6(99)$	-	3052		a_g ν1	3083	3076	$K_5(99)$	-	3075	
ν2	2229	2230	$K_5(87)$	2206	2186	-23	ν2	1555	1556	$K_1(42),K_2(46)$	1505	1508	-50
ν3	1602	1602	$K_1(46),K_2(22),K_3(25),H_3(20)$	1615	1589	+13	ν3	1518	1514	$K_1(46),K_2(46)$	1420	1442	-98
ν4	1454	1454	$K_1(30),K_3(59)$	1391	1403	-63	ν4	1094	1089	$H_5(89)$	1078	1071	-16
ν5	1207	1207	$H_3(81)$	1196	1206	-11	ν5	735	735	$K_4(97)$	758	754	+23
ν6	948	932	$K_2(45),K_4(21)$	978	954	+30	ν6	474	464	$K_3(68),H_2(20)$	501	494	+27
ν7	711	726	$K_2(32),K_4(17)$	725	742	+14	ν7	244	232	$H_1(23),H_2(31)$	265	244	+21
ν8	602	600	$K_4(23),H_6(43),H_7(18)$	613	607	+11	b_{1u} ν13	3108	3076	$K_5(99)$	3079	3075	-29
ν9	334	338	$H_1(18),H_6(18)$	337	337		ν14	1530	1537	$K_2(91)$	147?	1457	-52
ν10	144	126	$H_6(28),H_7(49)$	148	126		ν15	1090	1090	$H_5(89)$	1072	1073	-18
b_{1u} ν18	3065	3051	$K_6(99)$	-	3052		ν16	781	779	$K_3(65),H_1(17)$	836	828	+55
ν19	2228	2230	$K_5(87)$	2181	2186	-47	ν17	734	735	$K_4(97)$	751	754	+17
ν20	1545	1542	$K_2(17),K_3(66),H_3(18)$	1504	1514	-41	ν18	427	431	$K_3(27),H_2(44)$	460	463	+33
ν21	1405	1399	$H_3(69)$	1361	1386	-44	b_{3u} ν34	639			705		+66
ν22	998	993	$K_4(16),H_2(34)$	1008	1003	+10	ν35	247			-		
ν23	962	958	$K_2(58),H_3(16)$	987	981	+25	ν36	110			-		
ν24	600	600	$K_4(26),H_6(40),H_7(17)$	-	607								
ν25	549	524	$K_3(20),K_4(23),H_6(16)$	541	524								
ν26	146	160	$H_6(37),H_7(50)$	-	160								
b_{3u} ν50	859	-		836		-23							
ν51	-			585									
ν52	475			483									
ν53	220			225									
ν54	103			-									

(a) Observed frequencies from Ref. (12) whereas calculated ones are slightly different due to the correction of a minor error in the definition of a symmetry coordinate. Note that the value of the interaction force constant F_6 was misprinted in Table 7 of Ref. (12). The correct value is 0.145 mdyn/rad. Only contributions to the potential energy distribution >15% are included. Coordinates involved: K_1 and K_3, ring and external C=C stretchings; K_2 and K_4, ring and external C-C stretchings; K_5, CN stretching; H_1 and H_2 ring CCC bendings; H_3, CCH bending; H_4, C-C=C ring-external bending; H_5 to H_7, CCC and CCN external bendings. (b) From Ref. (14). (c) From Ref. (16). Coordinates involved: K_1 and K_2, ethylenic and ring C=C stretchings; K_3 and K_4, C-S stretchings; K_5, CH stretching; H_1, SCS bending; H_2, CSC bending; H_3, ring SCC bending; H_4, external CCS bending; H_5, CCH bending.

The fundamental frequencies of the neutral and ionized molecules are paired according to the correspondence of their approximate normal mode description in terms of potential energy distribution (PED). In fact, in going from neutral molecules to radical ions the changes in the PED's are not such that the qualitative description is substantially modified. The observed frequency differences ($\Delta\nu_{obs}$) are inclusive of the ionization shifts as well as of all the effects due to the variation in the environmental interactions (solvent or crystal field effects). As a consequence the small $\Delta\nu_{obs}$ can give only qualitative indications about the trend of the ionization shifts. Conversely, the greatest ones are predominantly due to the ionization process. A simple rationalization of the ionization shifts has been performed by using the force fields available for the neutral molecules (12, 16) and modifying only some diagonal force constants. Whenever possible these variations were argued from empirical correlation diagrams bond lengths or bond orders versus force constants. Despite an extremely limited number of force constant variations the observed trend and size of frequency shifts are substantially reproduced.

This result succesfully correlates the frequency shifts with the bond order changes upon ionization through the NCA. The information on the shape of the normal vibrations drawn from the NCA can be set in a form suitable for utilization as input data for calculation of other physical parameters, namely, in terms of nuclear cartesian displacements for each normal mode. A vectorial representation of the totally symmetric (a_g) modes of TCNQ and TTF is given elsewhere (15, 17). However, even from the very rough description in terms of PED's (Table I) it is apparent that most of the modes consist of complex nuclear displacements which cannot be described by the common notion of group vibrations. In view of the applications to be discussed afterwards it is worth noting that the totally symmetric modes displaying the greatest ionization shift in both TCNQ and TTF (a_g, ν_4 and ν_3 respectively) consist of the simultaneous stretching of the ring and ethylenic C=C bonds.

For few key EDA structures like, e.g., the whole family of chlorinated-parabenzoquinones (7) including the tetrachloro derivative (chloranil) and tetrabromo-parabenzoquinone (bromanil) (18) the vibrational knowledge has achieved a degree of completeness comparable or even better than for TCNQ and TTF. For other interesting EDA molecules such as tetramethyl-paraphenylenediamine (TMPD) (19, 20), dichloro-dicyano-parabenzoquinone (DDQ), and tetracyanobenzene (TCNB) (21) only incomplete or preliminary vibrational data are available and the needed further vibrational work is currently in progress in our laboratory.

2.1 Electron charge distribution along the stacks

The ionization frequency shifts can be used to study in the peculiar time scale of vibrational spectroscopy: $10^{-13} - 10^{-14}$ sec) the electronic charge distribution in those systems in which the average charge per molecule is or is suspected to be less than unity. In fact a localized electron charge distribution would imply the presence in the vibrational spectra of the frequencies of both the neutral and ionized molecules, whereas a delocalized one would be characterized by single frequencies with intermediate values. In the latter case one can try to evaluate the degree of charge transfer since, at least in principle, the vibrational frequencies should be related in a simple direct way to the charge density on a molecular site.

The presence of localized charges previously suggested for $Cs_2(TCNQ)_3$ (22) and $Cu(TCNQ)_2$ (23) by X-ray or photoelectron data has been confirmed by the observation of the frequencies of both neutral TCNQ and its monoanion in their infrared and Raman spectra (13, 24). Analogous results have been reported for $(TTF)_3(CuCl_2)_5$ in which the Raman frequencies of TTF cation and dication are detectable (25) and obtained for $(TTF)_2Ni(dithiola-$

te)$_2$ (26) whose infrared spectrum displays the frequencies of
neutral TTF, $(TTF^+)_2$ dimer and Ni(dithiolate)$_2$ anion. The above
results are consistent with the low conductivity of the materials.
A rather surprising result has been recently reported (27) for
the highly conducting materials based on the electron donor tetrathiotetracene (TTT) and its selenium analogue TSeT. The resonance Raman spectra of TTT(TCNQ)$_2$, (TTT)$_2$I$_3$, (TSeT)$_2$Cl, (TSeT)I$_{0.75}$
and (TSeT)I$_{2.2}$ show lines which have been assigned either to neutral TTT, TSeT and TCNQ or to their fully charged ions.

A delocalized charge distribution is more likely in the
case of highly conducting materials. The possibility of obtaining an estimate of the degree of charge transfer (ρ) from molecular vibrational frequencies is particularly relevant in those
cases in which this information cannot be simply deduced from
the stoichiometry of the compound. This is the case, e.g., of
the family of TTF-TCNQ and derivatives, of NMP-TCNQ and of the
newly sinthetized (28) class of compounds containing the para-
benzoquinone structures as acceptors. The knowledge of ρ is a
crucial step for the physical understanding of the organic conductors since it directly determines the band filling, i.e. the
Fermi energy, and allows the study of the relation between band
filling and physical properties (29). Other experimental techniques have so far been used to deal with the problem of determining ρ, namely, photoelectron spectroscopy (30), diffuse X-ray
(31) and neutron (32) scattering and NMR methods (33). However
it has been shown (30) that photoelectron data must be considered
with much care because of the many factors affecting the photoelectron lineshapes. In addition, surface effects can cause the
electronic structure to differ substantially from that of the
bulk material. On the other hand, X-ray (32) and neutron (33)
scattering, which in recent years gave a tremendous impetus to
the understanding of 1-D organic conductors, are very subtle
techniques and the preparation of suitable samples may be in
some case a difficult problem.

In view of its simplicity, vibrational spectroscopy seems
a very promising technique whose exploitation is now made possible by the accumulation of sufficient basic vibrational data.
However, since we are dealing with complex vibrational spectra
typical of systems containing large polyatomic molecules some
caution is needed. One can establish the following useful criteria of conduct when using vibrational frequencies for the determination of ρ. (i) Raman spectra should be preferred (34) in
respect to infrared ones being simpler and more intelligible since
they are dominated by the resonantly enhanced totally symmetric
modes. On the other hand the infrared spectra are complicated by
the overlapping of different phenomena like low lying electronic
absorptions typical of highly conducting materials (35), vibronic absorptions due to e-mv interactions (36) and interference

effects (37). (ii) The diagnostic frequencies must be carefully choosen among those which display the highest ionization shift in order to minimize errors in the estimated ρ due to the environmental effects on the intramolecular frequencies. (iii) The linear dependence of vibrational frequencies vs. charge density, which is commonly assumed, must be tested by investigating systems for which the average charge per molecule is directly determined by stoichiometry. The results of first attempts to use Raman frequencies for the determination of ρ are summarized in Table II. Many of the data, except those concerning TMTTF systems, are taken from literature (38-40) but are reinterpreted in the light of our determination of ionization frequency shifts. In the upper part of the table are indicated the frequencies choosen as diagnostic for TTF, TMTTF and TCNQ neutral molecules and fully charged ions. The frequencies for the former are from the corresponding molecular crystals (12, 15, 41) and those for the latter are from dimerized free radical salts (13, 16, 41). This choice seems preferable if one assumes that similar intermolecular effects are operating in the systems for which the value of ρ has to be evaluated. The middle part shows the results of checks on the linear dependence of Raman frequencies vs. charge density. The values of ρ deduced from Raman data are very close to the average charge per molecule directly indicated by the stoichiometry of the compounds. Concerning the highly conducting systems, the result for TTF-TCNQ is, within the approximation of the Raman method, in good agreement with the value of

TABLE II - DEGREE OF CHARGE TRANSFER (ρ) AS DETERMINED FROM RAMAN FREQUENCIES (cm^{-1})

	TTF a_g, ν_3	$(\rho)^a$	TMTTF[b] a_g, ν_2	$(\rho)^a$	TCNQ a_g, ν_4	$(\rho)^c$	Ref.
TTF°	1518	(0)					(16)
TTF⁺	1420	(1)					(16)
TCNQ°					1454	(0)	(12)
TCNQ⁻					1394	(1)	(14)
TMTTF°			1537	(0)			(41)
TMTTF⁺			1418	(1)			(41)
(TTF)I₀.₇₁	1452	(0.67)					(38)
(TTF)Br₀.₇₆	1448	(0.71)					(38)
(TTF)Cl₀.₈₀	1441	(0.79)					(38)
TEA(TCNQ)₂					1421	(0.5)	(40)
Q(TCNQ)₂					1425	(0.5)	(40)
(TMTTF)₂BF₄			1476	(0.51)			(42)
TTF-TCNQ	1456	(0.63)			1420	(0.6)	(38,39)
TMTTF CA			1473	(0.54)			(42)
TMTTF BA			1476	(0.51)			(42)
TMPD TCNQ					1407	(0.8)	(40)

[a] Estimated accuracy ±0.05; [b] Vibrational treatment made considering methyl groups as point masses. [c] Estimated accuracy ±0.1

ρ previously determined by diffuse X-ray scattering (31). The
results (42) obtained for the new materials TMTTF-CA and TMTTF-BA
confirm the presence of partial charge transfer already indicated
by optical measurements (28) and represent the first determination of the ρ value for these compounds. Finally, the interesting
case of the donor-acceptor CT crystal TMPD-TCNQ containing mixed
stacks is included. The preliminary indication (40) from Raman
data of an incomplete charge transfer contrasts with the early
and commonly accepted suggestion (43) of a sharp neutral/fully
ionic separation in donor-acceptor crystals. The uncertainty in
the ρ values comes from the limited resolution used in recording spectra of low intensity and from the crystal field effects
on the intramolecular frequencies. The accuracy of the data reported in the table can be empirically estimated in ± 0.05 for
ρ values deduced from TTF and TMTTF frequencies and ± 0.1 for
those deduced from TCNQ.

2.2 A priori calculation of the e-mv coupling constants

A second important application of the basic vibrational
knowledge gained by spectroscopic studies at the molecular level
is the a priori evaluation of fundamental parameters describing
the interaction between electrons and intramolecular vibrations
(1). It seems appropriate to spend few words about the general
problem.

A significant step forward in the understanding of the
quasi 1-D organic conductors was the experimental verification
of the existence of a charge density wave (CDW) ground state
in some highly conducting compounds like TTF-TCNQ (35b) and
mixed valence TTF-halides (44). This finding directly implies
that the electron-phonon interaction has a vital role in determining the physical properties of these compounds. On the other
hand, the molecular nature of the organic CT solids brought out
the idea that the coupling with the numerous intramolecular vibrational modes of the large constituent molecules may dominate
the overall electron-phonon interaction (2). The interaction
between electrons and intramolecular vibrations accounts for the
fact that the energy ε of the molecular orbital (MO) which acco-
modates the radical electron, is a function of the instantaneous
nuclear geometry of the molecule. As a consequence, for small
nuclear displacements, the MO energy may be expanded in a power
series of the dimensionless normal coordinates $\{Q_\alpha\}$ truncated
after the linear term according to

$$\varepsilon = \varepsilon_o + \sum_\alpha g_\alpha Q_\alpha \qquad (1)$$

where ε_o denotes the equilibrium MO energy and

$$g_\alpha = (\partial \varepsilon / \partial Q_\alpha)_o \qquad (2)$$

referred to as the linear e-mv coupling constants are the fundamental parameters describing the e-mv interaction. Simple symmetry arguments (1) show that, for non-degenerate radical electron MO's, the only non-zero coupling constants are those for
the totally symmetric intramolecular modes. For instance they
amount to 10 for TCNQ (17, 45) and 7 for TTF (17).

An a priori evaluation of the parameters g_α can be obtained
by using an approximate calculation of the electronic structure
and of the Cartesian displacements for each normal mode. The latter are directly available from the normal coordinate analysis,
the former can be conveniently calculated by a CNDO/S method (46)
parametrized to reproduce the photoelectron and electronic absorption spectra. The procedure for the determination of the linear
coupling constants consists of three steps: (i) calculate the atomic positions in the molecule according to the values of the Cartesian displacements given for each normal mode; (ii) calculate
the electronic structure for this geometrical structure; (iii)
extract the coupling constants from the slopes of the electronic
orbital energy versus the magnitude of the normal mode displacement. The results for TCNQ (17, 45, 47) and TTF (17) are at present the only available in the literature and have been recently
reviewed (1). They succeeded in showing that the overall e-mv interaction is indeed relevant. A measure of the strength of this
interaction can be given in terms of the small polaron binding
energy calculated for the radical electron MO which is 0.15 eV
for TCNQ and 0.21 eV for TTF, values (17) which cannot be neglected in a realistic attempt of interpretation of the physical pro
perties of 1-D organic systems.

3. VIBRONIC INFRARED ABSORPTIONS OF DIMERIZED ION-RADICAL SYSTEMS

Up to this point we have considered the vibrational behaviour of isolated molecular systems. The study of the effects on
the vibrational spectra of the CT intermolecular interactions (9)
gives further informations pertinent to 1-D organic conductors
and open other fruitful ways to applications.

The observation of unusual spectroscopic effects in the infrared spectra of donor-acceptor CT complexes (9) has been reported long time ago and related with the interaction between
electronic states and intramolecular vibrations. The first example, that is the strong intensity enhancement of some infrared
forbidden bands of benzene and iodine in the infrared spectrum
of the binary molecular complex they form in solution, was reported and discussed in the context of the CT theory (9) by Ferguson
and Matsen (48) two decades ago. Subsequently a rather more for-

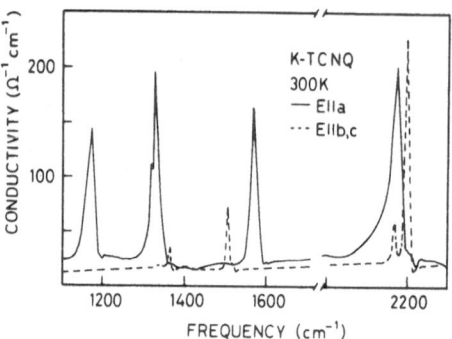

FIG. 1 - Real part of the infrared conductivity of KTCNQ (Ref.51)

mal theory of the infrared intensity effects in donor-acceptor
complexes was reported by W.B. Person (49).

Unusual spectroscopic effects attributable to electron-molecular vibration coupling are observed also for CT crystals (9)
and free radical salts (50) containing segregated stacks of only
one kind of molecule. Fig. 1 shows the infrared conductivity of
KTCNQ at room temperature in the spectral region 1100-2250 cm^{-1}
obtained by Kramers-Krönig transformation of polarized reflectivity data (51). The spectrum is dominated by strong bands polarized along the a crystal axis that is the stacking axis of the
planar TCNQ radicals. These bands display an out-of-plane polarization and lie in a spectral region in which only in-plane molecular vibrations are expected. In addition, their frequency is
nearly coincident with that of the totally symmetric modes of
TCNQ radical which are normally infrared inactive. Both these
facts strongly suggest that the unusual absorptions are due to
a vibronic activation in infrared of the totally symmetric intramolecular modes.

A detailed interpretation of this spectroscopic effect and
its correlation with the analogous effect observed in donor acceptor complexes like the benzene-I_2 system take advantage of
the consideration of the actual crystal structure of KTCNQ (52).
The latter consists of distorted stacks of TCNQ anions with alternating interplanar distances resembling a structure in which
the TCNQ anion dimer is the basic unit. Such a radical-radical
π-dimer may be viewed as a CT complex in which each radical ion
acts as both the donor and the acceptor (10, 53). Alternatively,
adopting a viewpoint more familiar to physicists and considering
the half occupied molecular orbitals alone, a radical-radical
dimer may be considered a materialization of the Hubbard dimer
(54). Starting from these two somewhat different viewpoints but
adopting equivalent physical arguments it is possible to work out
theoretical models accounting for the anomalous infrared spectra

of dimerized ion radical systems. M.J. Rice (55) has recently
reported a theory for the vibronic infrared absorptions of a
Hubbard dimer. Here we prefer to adopt the more chemically orient
ed language of the Mulliken's CT theory (9) in order to enphasize
the analogies between infrared vibronic absorptions of radical-
radical self-dimers and of the donor-acceptor complexes. This
can be done (56) by adapting to the case of the radical-radical
dimer the vibronic model originally proposed by Person (49) for
a donor-acceptor dimer. This model shows how one vibrational
mode of the dimer may gain infrared intensity through the modulation of the mixing coefficients of the CT configurations with
the no-bond one in the electronic ground state of the dimer.
For the vibrational modes of a symmetric dimer corresponding to
the out-of-phase coupling of the totally symmetric vibrations
of the two component radicals, the part of the dipole moment derivative accounting for the intensity gain referred before (the
Person's "delocalization moment" (49) is given by

$$(\vec{\mu}_{del})_\alpha \simeq - \frac{4t}{U^2} (g_\alpha^- + g_\alpha^=)\vec{\mu}_{CT} \qquad (3)$$

In this expression t is the electron transfer integral, $U = U_o - V$
is the effective Coulomb interaction where U_o is the intramolecular Coulomb repulsion energy between two electrons on the same
molecule and V is the Coulomb repulsion energy between adjacent
radicals. $\vec{\mu}_{CT}$ is the transition dipole moment of the electronic
CT absorption, g_α^- and $g_\alpha^=$ are the e-mv coupling constants with
the highest molecular orbital half occupied in the anion and
fully occupied in the dianion respectively. If one considers
the in-phase coupling of the same vibration of the two radicals
one obtains $\vec{\mu}_{del} = 0$. Summarizing, the vibrational modes of the
dimer corresponding to the out-of-phase coupling of totally symmetric vibrations of the component molecules may borrow infrared
intensity from the CT transition. Pictorially, the mechanism may
be viewed as an electron oscillation back and forth from one radical to the other induced by the molecular vibrations. Hence
they should appear in the infrared spectrum of dimerized free
radical systems as strong vibronic absorptions polarized perpen-
dicularly to the molecular planes. Note that equation 3 implies
the possibility of an experimental evaluation of $(g_\alpha^- + g_\alpha^=)$. In
fact, $|(\vec{\mu}_{del})_\alpha|$ can be measured from the integrated absorption
intensity of the vibronic infrared bands whereas the parameters
t, U and $|\vec{\mu}_{CT}|$ can all be determined from a study of the CT electronic absorption.

      The validity of the dimeric model for the interpretation
of the vibronic infrared bands has been substantiated by studying
the effect on the infrared spectra of the dimerization of radicals
in solution (14) and of the typical monomer-dimer phase transitions in (1:1) free radical salts (57). The infrared spectra of
solutions of LiTCNQ in DMSO (14) of increasing concentration

show new absorptions whose intensity increases with the concentration of the dimer. They display frequencies coincident with
those of the main bands polarized parallel to the dimerized
stacks in the KTCNQ reflectance spectrum (Fig. 1) and are therefore of similar origin. This finding demonstrates that the appearance of new infrared bands due to a vibronic intensity enhancement of the totally symmetric modes caused by their interaction with electrons is an intrinsic property of the radical-radical π-dimer. The sensitivity of the infrared bands of vibronic
origin to a dimerization process in the solid state can be checked by studying the effect on the infrared spectra of the well
known monomer-dimer phase transitions of (1:1) alkaline salts of
TCNQ (52, 58). The plots of the relative peak intensity of the
vibronic bands against temperature (57) show abrupt changes at
temperatures coincident with the transition temperatures as determined by direct X-ray structural studies (52, 58) or by magnetic susceptibility measurements (59). The exploitation of this
result introduces a fruitful use of infrared spectroscopy for
investigating simple segregated free radical salts. When a monomer-dimer phase transition has been characterized by other physical measurements the temperature dependence of the infrared
spectra allows one to single out the vibronic features. When the
latter are already known from studies of similar systems they can
be used for investigating the phase transitions in new compounds.

3.1 Experimental evaluation of the e-mv coupling constants

The identification for dimerized TCNQ free radical systems
of infrared absorptions of vibronic origin and their unequivocal
interpretation in terms of a simple dimeric model suggest that
experimental estimates of the linear e-mv coupling constants
could be obtained through a quantitative analysis of these spectroscopic effects. It is to be noted that infrared absorptions
of vibronic origin have been observed (60) and theoretically
analysed (61) also for some intermediate conductivity compounds
like TEA(TCNQ)$_2$. Actually, the theoretical analysis carried out
in terms of phase oscillations of intramolecularly stabilized
charge density waves components (62) (see below) provided the
first experimental estimate of the e-mv coupling constants for
TCNQ which however suffered for the many necessary assumptions
about the electronic structure. Conversely, the study of the
vibronic features in dimerized systems enables the e-mv coupling
constants to be experimentally determined indipendently of the
details of the actual nature of the electronic states in the
crystal. With this aim, M.J. Rice et al. (63) have carried out
a detailed analysis of the polarized reflectance spectra of KTCNQ
(64). They made use of the results of a theoretical model somewhat different from both those cited above (55, 56). The starting point was the Hamiltonian of a dimerized free radical chain
including a term accounting for the e-mv interaction but with

TABLE III - EXPERIMENTAL AND CALCULATED COUPLING CONSTANTS (g_α)
OF TCNQ ANION

	K TCNQ			MEM(TCNQ)$_2$	TCNQ$^-$
a_g modes $\underline{a}$ (cm^{-1})	vibronic $\underline{b}$ bands (cm^{-1})	g_α(exp) $\underline{c}$ (meV)		g_α(exp) $\underline{d}$ (meV)	g_α(calc) $\underline{e}$ (meV)
ν_1 (3052)	---	---		6	0
ν_2 2216	2185	73		43	54
ν_3 1610	1579	83		67	110
ν_4 1397	1346	71		62	61
ν_5 1207	1184	41		37	37
ν_6 984	983	42		10	39
ν_7 730	721	21		24	26
ν_8 622	621	0		6	1
ν_9 348	330	22		22	31
ν_{10} (148)	--	N.A.		9	11

$\underline{a}$ Chi, C.K., and Nixon, E.R.: 1975, Spectrochim. Acta 31A, pp. 1739-1747.
Parenthesized values are from Ref. (13): that for ν_1 is from calcula-
tion; that for ν_{10} is from the Raman spectrum of Rb(TCNQ). $\underline{b}$ This work.
The frequency values are from transmission measurements and substantial-
ly agree with those from reflectivity reported in Refs. (50) and (51).
The frequencies 822 and 477 cm^{-1} of Ref. (51) must be rejected being
due to a misassignment of out-of-plane infrared active modes. $\underline{c}$ From
Ref. (63, 66). $\underline{d}$ From Ref. (65). $\underline{e}$ From Ref. (67).

an arbitrary crystal interaction term which, however, was intended to include implicity both the electron transfer interaction
inside and between dimers and the on-site Coulomb interaction.
By using the linear response theory an expression was obtained
for the frequency dependent conductivity $\sigma(\omega)$ in a frequency
region including both molecular vibrations and CT electronic
transitions. However, contrary to the isolated Hubbard dimer (55)
and due to the arbitrariness of the crystal interaction term,
the $\sigma(\omega)$ associated with the CT transitions could not be written
down explicitly. None the less, by fitting the experimental conductivity data, values for the e-mv coupling constants (g_α)
could be obtained (63, 66). They are shown in column third of
Table III which also reports the experimental frequencies of the
corresponding Raman-active totally symmetric modes and those of
the vibronically activated infrared bands (columns first and
second respectively). Column fourth gives the new experimental
values of g_α reported at this NATO A.S.I. by C.S. Jacobsen (65)
and obtained from the reflectivity data of MEM(TCNQ)$_2$. For the
analysis of the data a microscopic model was used which in some
way differ from that used for KTCNQ (63). (i) The dimeric unit
in MEM(TCNQ)$_2$ contains only one electron instead of two and thus
it was not necessary to allow for the on-site Coulomb repulsion U.
(ii) The theoretical model takes into account a slight inequiva-
lence of the two monomers in the dimer. The agreement between

the two sets of experimental values of g_α is satisfactory with
the only exception of the constant g_6. Its value of 42 meV measured for KTCNQ (63) cannot be accepted being the result of the
misassignment of the infrared-active out-of-plane mode at 848
cm^{-1} to a vibronically activated band deriving from the totally
symmetric mode ν_6 at 984 cm^{-1}. Owing to this fact and considering the high quality of the experimental data and the use of a
simple but suitably adapted theoretical model in analysing them
we believe that the new g_α values deduced from MEM(TCNQ)$_2$ are
the best experimental values so far available. The last column
of Table III gives the a priori calculated values of the coupling
constants of TCNQ anion (67). The agreement with the experimental ones is semiquantitative and must be considered satisfactory
bearing in mind the unavoidable approximations introduced in the
calculation particularly concerning the normal mode descriptions.

The described approach to the evaluation of e-mv coupling
constants appear to have a great potential which has only just
begun to be exploited. In fact the spectroscopic evidence of e-mv
interaction is not limited to TCNQ. Wide and exciting perspecti-
ves are opened by the preliminary results we have obtained on
other key electron donor or acceptor molecules. A study of the
influence of the dimerization on the infrared spectrum of TTF$^+$
has been carried out (16) according to the lines outlined before
for TCNQ$^-$. It has shown that some infrared absorptions of the
dimer are attributable to vibronic intensity enhancement of totally symmetric modes of the constituent radicals, namely, those
dominated by C=C and C-S stretchings (a$_g$ modes ν_3, ν_5 and ν_6,
see Table I). Corresponding bands have been observed (16) in

TABLE IV - VIBRONIC INFRARED BANDS (cm^{-1}) AND CALCULATED E-MV
           COUPLING CONSTANTS (meV) OF TTF

	(TTF)Br$_{1.0}$ [a]		g_α(calc) [b]
	a$_g$ modes	vibronic bands	
ν_1	(3075)	--	12
ν_2	1505	not detected	44
ν_3	1420	1370 vs, br	117
ν_4	1073	overlapped	21
ν_5	758	751 mw	45
ν_6	501	487 ms	78
ν_7	264	overlapped	5

[a] From Ref. (16). Parenthesized frequency value from calculation.

[b] From Ref. (17).

the infrared spectrum of crystalline (1:1) (TTF)Br whose structure (68) contains almost isolated $(TTF^+)_2$ dimers. Work is in
progress to get a quantitative analysis of the vibronic bands
from which experimental values of the e-mv coupling constants
can be extracted. However, some meaningful comparison between
experimental data and calculated coupling constants is possible.
Table IV shows that the strongest vibronic bands correspond to
the totally symmetric modes for which the highest values of the
e-mv coupling constants have been calculated (17).

For the case of chloranil (69), bromanil (70) and TMPD (56)
radical ions, the vibronic absorptions in infrared have been
identified by studying the monomer-dimer phase transitions of
the corresponding simple segregated free radical salts.

4. PHASE PHONON ABSORPTIONS AND THE METAL-INSULATOR TRANSITIONS
   IN 1-D ORGANIC CONDUCTORS

In the high conductivity quasi 1-D organic compounds the
study of the temperature dependence of the infrared spectra can
give experimental evidence of the participation of intramolecular vibrations to the formation of CDW's and to the related phenomena such as the typical metal-insulator transition. In the
first stage of such an investigation it is convenient to avoid
the complexity of having two kinds of conducting linear chains
like in TTF-TCNQ. The crystal structure of the highly conducting
mixed valence TTF halides and pseudohalides $(TTF)Br_{0.7}$ (71),
$(TTF)I_{0.71}$ (72), $(TTF)SCN_{0.58}$ (44) and $(TTF)SeCN_{0.58}$ (44) consists of segregated eclipsed stacks of TTF molecules with the
counter anions occupying the channels between the stacks. Recently direct X-ray evidences of a CDW ground state have been reported by Bell's researchers (44) for the single linear chain
compounds $(TTF)SCN_{0.58}$ and $TTF(SeCN)_{0.58}$. Furthermore, the similarity of the physical behaviour of the whole family of mixed
valence TTF halides and pseudohalides allows one to consider
all of them as CDW systems (73).

When electrons interact with both inter- and intra-molecular phonons the resulting CDW's consist of different components
one for each interacting vibrational mode (3). We have shown
before that in TTF systems the e-mv interaction is indeed relevant. In the fluctuation regime the motion of the CDW's is described by the time variation of their phase. For temperatures
below the metal-insulator transition temperatures various pinning
mechanisms contribute to fix the phase of the CDW's (74). However, in the pinning regime, the dynamics of the CDW's can be
described in terms of harmonic oscillations of their phase and
amplitude about the equilibrium values. The new collective modes
corresponding to the phase oscillations of the CDW components,

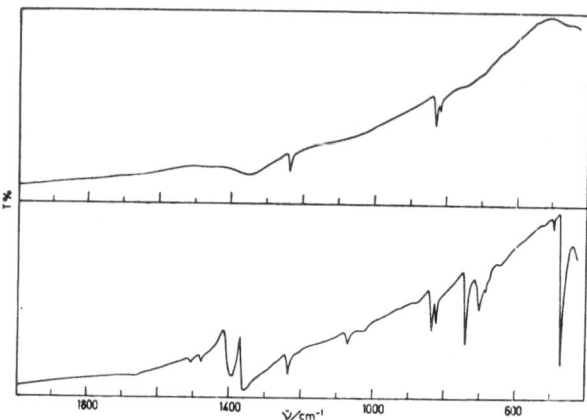

FIG. 2 - Powder IR spectra of (TTF)Br$_{0.76}$: top 298K; bottom 20K

known as <u>phase phonons</u>, are the subject of a detailed microscopic theory developed by M.J. Rice (62). The main results of this
theory valid for frequencies below the single particle gap 2Δ
and of relevance for the subsequent discussion of the infrared
spectra are that the phase phonons originated by molecular vibrations occur at frequencies slightly below those of the totally symmetric molecular modes (about -10%) (62, 75). Such phase
oscillations involve a bodily displacement of the appropriate
component of the condensed electronic charge and are therefore
optically active along the chain direction. The collective modes
of frequency ω > 2Δ become damped <u>via</u> electron-hole pair excitation and hence they should give rise to indentations in the
continuous electronic absorption resembling the antiresonance
effect described by Fano (37, 76).

        Fig. 2 shows the powder absorption spectrum of (TTF)Br$_{0.76}$
at room temperature (upper part) which is representative of all
the family of mixed valence compounds. The broad absorption extending from about 600 cm^{-1} up to 2000 cm^{-1} is attributable to
the continuum of electronic transitions. The narrow bands at
820, 833 and 1238 cm^{-1} are attributable to the infrared active
vibrational modes of the TTF molecule (16). The corresponding
spectrum at 20 K (lower part of Fig. 2) shows, superimposed to
the broad electronic absorption, additional strong bands at 470,
705 and 742 cm^{-1} and a complicated feature with two antiresonance dips in the region 1350-1450 cm^{-1}. Note that the frequency
location of three of the additional features of the low temperature spectrum, namely 470, 742 cm^{-1} and the region 1350-1450 cm^{-1}
nearly coincides with that of the vibronic absorptions of the
dimeric (TTF$^+$)$_2$ species and of (1:1) (TTF)Br deriving from the
coupling of electrons with the totally symmetric ν_6, ν_5 and ν_3
modes of TTF (Table IV). The remaining additional absorption at

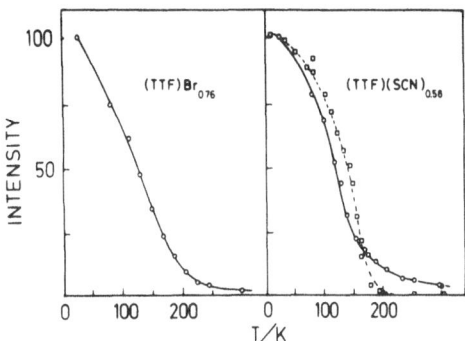

FIG. 3 - Peak intensity of the 470 cm^{-1} band vs. T. Dashed curve:
X-ray superlattice intensity vs. T (Ref. 44)

705 cm^{-1} must be assigned, by its frequency, to the infrared
active out-of-plane C-H bending mode of TTF^{+} (16) and its strong
temperature dependence is likely attributable to an effect of
narrowing with lowering temperature.

Fig. 3 shows the plot of the intensity against temperature
of the absorptions near 470 cm^{-1} observed for $(TTF)Br_{0.76}$ and
$(TTF)SCN_{0.58}$. The intensity is expressed as peak absorbance normalized to its value at 20 K. Similar curves have been obtained
also for $(TTF)I_{0.71}$ and $(TTF)SeCN_{0.58}$ (56). The curves display
"knees" at temperatures nearly coincident with those reported
for analogues "knees" in the magnetic susceptibility or ESR
intensity (77) and for singularities in the derivative of the
electrical conductivity (78) against temperature. These changes
of magnetic and electrical properties were attributed to the occurrence of a metal-insulator transition. But, what is even more
important is that for the case of $(TTF)SCN_{0.58}$ the curve infrared intensity vs. temperature is in fair agreement with that
reported by Bell group (44) and obtained by plotting the normalized superlattice intensity observed in the X-ray study (broken
curve in Fig. 3). Note that the latter quantity is proportional
to the square of the order parameter Δ of the CDW and that the
smoother shape of the infrared curve around the transition point
is probably due to the fact that our work was performed on powder and hence highly defective samples.

All the above findings indicate that the strong additional
absorptions at 742 and 470 cm^{-1} which appear <u>only</u> in the semiconducting low temperature phase are attributable to the excitation of phase phonons (62) mainly related to the ν_5 and ν_6
totally symmetric modes of the TTF molecules. The antiresonance
dips observed in the spectral region 1350-1450 cm^{-1} are attributable to the damping of the phase phonon mode originated by
the ν_3 totally symmetric mode. The fact that indentations in

the continuous electronic absorptions were observed only for
this mode agrees with the estimates of the electronic energy
gap 2 Δ obtained by electrical conductivity (78) and magnetic
susceptibility (77) measurements which indicate a value in the
range 1200-1600 cm^{-1}.

Summarizing, the presence of strong electron-molecular vibra
tion interaction in conducting quasi one-dimensional TTF systems
clearly manifests itself in the infrared spectra as the appearance of unusual spectroscopic effects. They find a satisfactory
explanation in terms of phase phonon absorptions and antiresonance
effects, and their magnitude correlates well with the a priori
calculated values of the e-mv constants (Table IV). The temperature dependence of these effects appears closely related to the
evolution of the order parameter Δ associated with a Peierls
transition and this shows that infrared spectroscopy can be successfully used to monitor the metal-insulator transitions of
conducting TTF systems.

5. ENDING REMARKS

In this lecture we have tried to present the main ideas and
results which came out from the relatively recent impact of vibra
tional spectroscopy with the field of low dimensional solids. The
accumulation of sufficient basic data on intramolecular vibrations has made vibrational spectroscopy: (i) an advantageous tool
to solve specific problems like the determination of the degree
of charge transfer and the monitoring of linear chain distortions;
(ii) an unique source of information on the interaction between
conduction electrons and molecular vibrations.

We have not attempted a comprehensive coverage of the literature and have limited ourselves to a discussion of organic CT
systems. We barely mention that important results (79) have been
obtained by applying vibrational spectroscopy to the study of
other systems like, e.g., conducting polymers, metal chain compounds, etc. and other problems like lattice modes and electron-
lattice phonon interactions.

Acknowledgments. Much of the research reviewed is the result
of the long term work of the Padua group to which the association
of A. Girlando and I. Zanon was and currently is essential. The
technical assistance of F. Marzola, M. Zanetti and C. Ricotta has
been much appreciated. We also acknowledge the incitement received by many workers in the field of 1-D systems and particularly
we thank M.J. Rice, J.B. Torrance, A.N. Bloch, J. Kommandeur and
Z.G. Soos for their interest in our work and fruitful discussions.
The cooperation of G. Pasetto in preparing the "camera-ready"
typescript is gratefully acknowledged.

REFERENCES

1. Duke,C.B., in "Synthesis and Properties of Low-Dimensional Materials" edited by Miller,J.S.,and Epstein,A.J.: 1978, Ann. New York Acad. Sci. 313, pp. 166-178.
2. Rice,M.J., Duke,C.B., and Lipari,N.O.: 1975, Solid State Commun. 17, pp. 1089-1093.
3. Rice,M.J.: 1978, Solid State Commun. 25, pp. 1083-1086.
4. Conwell,E.M.: this volume.
5. Peierls,R.E.: "Quantum theory of Solids", 1955, Oxford University Press, London, p. 108.
6. (a) Wilson,E.B., Decius,J.C.,and Cross,P.C.: "Molecular Vibrations", 1955, McGraw Hill, New York; Califano,S.: "Vibrational States", 1976, Wiley, New York; (b) Turrell,G.: "Infrared and Raman Spectra of Crystals", 1972, Academic Press, New York; (c) Decius,J.C.,and Hexter,R.M.: "Molecular Vibrations in Crystals", 1977, McGraw Hill, New York.
7. For typical examples of vibrational studies of polyatomic molecules see: (a) Girlando,A., and Pecile,C.: 1973, J. Chem. Soc., Faraday II 69, pp. 1291-1303; (b) 1974, Spectrochim. Acta 31A, pp. 1187-1200; (c) 1978, Spectrochim. Acta 34A, pp. 453-463; (d) 1979, J. Mol. Spectry. 77, in press.
8. Johnson,B.B., and Peticolas,W.L.: 1976, Ann.Rev. Phys. Chem. 27, pp. 465-491.
9. Mulliken,R.S., and Person,W.B.: "Molecular Complexes", 1969, Wiley, New York; Yarwood,J., ed., "Spectroscopy and Structure of Molecular Complexes", 1973, Plenum Press, New York.
10. Soos,Z.G., and Klein,D.J., in "Molecular Association", edited by Foster,R.: 1975, Academic Press, Vol. 1, pp. 1-109.
11. Duncan,J.L., in "Molecular Spectroscopy", edited by Barrow, R.F., Long,D.A, and Millen,D.J.: 1975, Specialist Periodical Reports, The Chemical Society, London, Vol. 3, pp. 104-162; Zerbi,G., in "Vibrational Spectroscopy - Modern Trends", edited by Barnes,A.J., and Orville-Thomas,W.J.: 1977, Elsevier, Amsterdam, pp. 261-284.
12. Girlando,A., and Pecile,C.: 1973, Spectrochim. Acta 29A, pp. 1859-1878. See also: Takenaka,T.: 1971, Spectrochim. Acta 27A, pp. 1735-1752; Kaplunov,M.G., Panova,T.P., Yagubskii,E.B., and Borod'ko,Yu.G.: 1972, J. Struct. Chem. 13, pp. 411-417.
13. Bozio,R., Girlando,A., and Pecile,C.: 1975, J. Chem. Soc. Faraday II, 71, pp. 1237-1254.
14. Bozio,R., Zanon,I., Girlando,A., and Pecile,C.: 1978, J. Chem. Soc. Faraday II, 74, pp. 235-248.
15. Bozio,R., Girlando,A., and Pecile,C.: 1977, Chem. Phys. Letters, 52, pp. 503-508; See also: Berlinsky,A.J., Hoyano,Y., and Weiler,L.: 1977, Chem. Phys. Letters 45, pp. 419-421; Temkin,H., Fitchen,D.B. and Wudl,F.: 1977, Solid State Commun. 24, pp. 87-92.
16. Bozio,R., Zanon,I., Girlando,A., and Pecile,C.: 1979, J. Chem. Phys., in press.

17. Lipari,N.O., Rice,M.J., Duke,C.B., Bozio,R., Girlando,A., and
    Pecile,C.: 1977, Int. J. Quantum Chem. Symposium, 11, pp. 583-
    594; 1978, Ibid. Symposium 12, pp. 545.
18. Girlando,A., Zanon,I., Bozio,R., and Pecile,C.: 1978, J. Chem.
    Phys. 68, pp. 22-31.
19. Sabitskii,A.V., and Kuznetsov,M.: 1971, J. Struct. Chem. 12,
    pp. 391-396.
20. Jeanmaire,D.L., and Van Duyne,R.P.: 1975, J. Electroanal. Chem.
    66, pp. 235-247.
21. Takenaka,T., Umemura,J., Tadokoro,S., Oka,S., and Kobayashi,T.:
    1978, Bull. Inst. Chem. Res. Kyoto Univ. 56, pp. 176-191.
22. Fritchie,C.J., and Arthur,P.: 1966, Acta Cryst. 21, pp. 139-
    145; Borod'ko,Yu.G., Kaplunov,M.G., Moravskaya,T.M., and Pokho-
    dnya,K.I.: 1977, Phys. Stat. Sol. 83, pp. K141-K144.
23. Sano,M., Ohta,T.,and Akamatu,H.: 1968, Bull. Chem. Soc. Japan
    41, pp. 2204-2206; Ikemoto,I., Thomas,J.M. and Kuroda,H.:
    1973, Bull. Chem. Soc. Japan 46, pp. 2237-2238.
24. Girlando,A., Bozio,R., and Pecile,C.: 1974, Chem. Phys. Let-
    ters 25, pp. 409-412.
25. Siedle,A.R., Candela,G.A., Finnegan,T.F., Van Duyne,R.P.,
    Cape,T., Kokoszka,G.F., and Woyciesjes,P.M.: 1978, J. Chem.
    Soc., Chem. Comm., pp. 69-70.
26. Bozio,R., Pecile,C., Interrante,L.V. et al., unpublished
    results.
27. Kachapina,L.M., Kaplunov,M.G., Yagubskii,E.B., and Borod'ko,
    Yu.G.: 1978, Chem. Phys. Letters 58, pp. 394-398.
28. Torrance,J.B., Mayerle,J.J., Lee,V.Y. and Bechgaard,K.: 1979,
    J. Am. Chem. Soc., in press.
29. Miller,J.S. and Epstein,A.J., in "Molecular Metals", edited
    by Hatfield,W.E.: 1979, Plenum Press, New York, pp. 35-41.
30. Ritsko,J.J., Epstein,A.J., Salaneck,W.R., and Sandman,D.J.:
    1978, Phys. Rev. B17, pp. 1506-1509, and references therein.
31. Denoyer,F., Comès,R., Garito,A.F., and Heeger,A.J.: 1975, Phys.
    Rev. Letters 35, pp. 445-448; Kagoshima,S., Anzai,H., Kaji-
    mura,K., and Ishiguro,T.: 1975, J. Phys. Soc. Japan 39, pp.
    1143-1144; Flandrois,S., and Chasseau,D.: 1977, Acta Cryst.
    B33, pp. 2744-2750.
32. Comès,R., Shapiro,S.M., Shirane,G., Garito,A.F., and Heeger,
    A.J.: 1975, Phys. Rev. Letters 35, pp. 1518-1521.
33. Butler,M.A., Wudl,F., and Soos,Z.G.: 1975, Phys. Rev. B12, pp.
    4708-4719; Ehrenfreund,E., and Garito,A.F.: 1976, Solid State
    Commun. 19, pp. 815-816.
34. It is worth noting, however, that at this NATO A.S.I., Aaron
    Bloch has reported on the finding of a linear relation bet-
    ween infrared-active C≡N stretching frequencies of TCNQ and
    ρ in numerous organic CT compounds.
35. (a) Torrance,J.B., in "Chemistry and Physics of One-Dimensio-
    nal Metals", edited by Keller,H.J.: 1977, Plenum Press, New
    York, pp. 137-166; (b) Heeger,A.J., ibid., pp. 87-135.

36. Rice,M.J.: 1976, Phys. Rev. Letters 37, pp. 36-39.
37. Kràl,K.: 1977, Chem. Phys. 23, pp. 237-242; Gor'kov,L.P., and Rashba,E.I.: 1978, Solid State Commun. 27, pp. 1211-1217; Bloch,A.N.: this volume.
38. Kuzmany,H., and Kundu,B.: 1979, Lecture Notes in Physics 95, pp. 259-265.
39. Temkin,H., and Fitchen,D.B. in "Lattice Dynamics", edited by Balkanski,M.: 1978, Flammarion, Paris, pp. 587-590.
40. Van Duyne,R.P. in Soos,Z.G., Mazumdar,S., and Cheung,T.T.P.: 1979, Mol. Cryst. Liq. Cryst. 52, pp. 93-102; Soos,Z.G.: this volume, pp. 143.
41. Bozio,R., Meneghetti,M., Zanon,I., and Pecile,C.: unpublished results.
42. Torrance,J.B., Bozio,R., Pecile,C. et al.: to be published.
43. McConnell,H.M., Hoffman,B.M. and Metzger,R.M.: 1965, Proc. Natl. Acad. Sci. U.S. 53, pp. 46-50; Nordio,P.L., Soos,Z.G., and McConnell,H.M.: 1965, Ann. Rev. Phys. Chem. 17, pp.236-260.
44. Thomas,G.A., Moncton,D.E., Wudl,F., Kaplan,M., and Lee,P.A.: 1978, Phys. Rev. Letters 41, pp. 486-490; Thomas,G.A.: this volume, pp. 31.
45. Lipari,N.O., Duke,C.B., Bozio,R., Girlando,A., Pecile,C., and Padva,A.: 1976, Chem. Phys. Letters, 44, pp. 236-240.
46. Lipari,N.O., and Duke,C.B.: 1975, J. Chem. Phys. 63, pp. 1748-1757 and pp. 1768-1774.
47. A non-quantum-chemical evaluation of the e-mv coupling constants on the basis of structural data has been reported by Kaplunov,M.G.: 1978, Sov. Phys. Solid State 20, pp. 881-883.
48. Ferguson,E.E., and Matsen,F.A.: 1958, J. Chem. Phys. 29, pp. 105-107.
49. Friedrich,H.B., and Person,W.B.: 1966, J. Chem. Phys. 44, pp. 2161-2170.
50. The first reported examples are those of the alkali TCNE salts and KTCNQ. Hinkel,J.J., and Devlin,J.P.: 1973, J. Chem. Phys. 58, pp. 4750-4756, and references therein; Anderson,G.R., and Devlin,J.P.: 1975, J. Phys. Chem. 79, pp. 1100-1102.
51. Tanner,D.B., Jacobsen,C.S., Bright,A.A., and Heeger,A.J.: 1977, Phys. Rev. B16, pp. 3283-3290.
52. Konno,M., Ishii,T., and Saito,Y.: 1977, Acta Cryst. B33, pp. 763-770.
53. Hausser,K.H., and Murrell,J.N.: 1957, J. Chem. Phys. 27, pp. 500-504.
54. Harris,A.B., and Lange,R.V.: 1967, Phys. Rev. 157, pp.295-314.
55. Rice,M.J.: 1979, Solid State Commun. 31, pp. 93-98; see also: Kràl ,K.: 1976, Czech. J. Phys. B26, pp. 660-669; 1977, Czech. J. Phys. B27, pp. 200-210.
56. Bozio,R., and Pecile,C.: to be published.
57. Bozio,R., and Pecile,C.: 1977, J. Chem. Phys. 67, pp. 3864-3868.
58. Terauchi,H.: 1978, Phys. Rev. B17, pp. 2446-2452, and references therein; Kobayashi,H.: 1978, Acta Cryst. B34, pp. 2818-2825.

59. Vegter,J.G.,and Kommandeur,J.: 1975, Mol. Cryst. Liq. Cryst. 30, pp. 11-49.

60. Kaplunov,M.G., Panova,T.P., and Borod'ko,Yu.G.: 1972, Phys. Stat. Sol. A13, pp. K67-K69; Brau,A., Brüesch,P., Farges,J.P., Hinz,W., and Kuse,D.: 1974, Phys. Stat. Sol. B62, pp. 615-623; Farges,J.P.: this volume, pp. 223.

61. Rice,M.J., Pietronero,L.,and Brüesch,P.: 1977, Solid State Commun. 21, pp. 757-760.

62. Rice,M.J.: 1976, Phys. Rev. Letters 37, pp. 36-39.

63. Rice,M.J., Lipari,N.O. and Strässler,S.: 1977, Phys. Rev. Letters 39, pp. 1359-1362.

64. These spectra were analyzed also in terms of CDW's phase oscillations in Ref. (51). However, we confine the attention on the more appropriate analysis in terms of the dimeric charge oscillation model of Ref. (63).

65. Rice,M.J., Jacobsen,C.S. and Yartsev,V.M.: poster session of this NATO A.S.I. and to be published.

66. Rice,M.J.: 1979, Lecture Notes in Physics 95, pp. 230-243.

67. Bozio,R., Girlando,A., Ragazzon,D., Zanon,I., and Pecile,C.: to be published. The coupling constants for the radical molecular orbital of TCNQ anion have been recalculated by using the improved set of force constants reported in Ref. (14).

68. Scott,B.A., La Placa,S.J., Torrance,J.B., Silverman,B.D., and Welber,B.: 1977, J. Am. Chem. Soc. 99, pp. 6631-6639.

69. Bozio,R., Girlando,A.,and Pecile,C.: 1977, Chem. Phys. 21, pp. 257-263.

70. Girlando,A., Zanon,I., Bozio,R., and Pecile,C.: to be published.

71. La Placa,S.J., Corfield,P.W.R., Thomas,R.,and Scott,B.A.: 1975, Solid State Commun. 17, pp. 635-638.

72. Johnson,C.K., and Watson,C.R.: 1976, J. Chem. Phys. 64, pp. 2271-2286.

73. Tomkiewicz,Y.,and Taranko,A.R.: 1978, Phys. Rev. B18, pp. 733-741.

74. Lee,P.A., Rice,T.M.,and Anderson,P.W.: 1974, Solid State Commun. 14, pp. 703-709.

75. Horovitz,B., Gutfreund,H., Weger,M.: 1978, Phys. Rev. B17, pp. 2796-2799.

76. Fano,U.: 1961, Phys. Rev. 124, pp. 1866-1878.

77. Wudl,F., Schafer,D.E., Walsh,W.M., Rupp,L.W., Disalvo,F.J., Waszczak,J.V., Kaplan,M.L.,and Thomas,G.A.: 1977, J. Chem. Phys. 66, pp. 377-385; Sugano,T., and Kuroda,H.: 1977, Chem. Phys. Letters 47, pp. 92-95.

78. Warmack,R.J., Callcott,T.A.,and Watson,C.R.: 1975, Phys. Rev. B12, pp. 3336-3338; Somoano,R.B., Gupta,A., Hadek,V., Novotny, M., Jones,M., Datta,T., Deck,R.,and Hermann,A.M.: 1977, Phys. Rev. B15, pp. 595-601.

79. A comprehensive review of infrared and Raman properties of 1-D materials is in preparation: Bozio,R. and Pecile,C.: "Extended Linear Chain Compounds" edited by J.S. Miller, Plenum Press.

EPR OF ORGANIC CONDUCTORS

Yaffa Tomkiewicz

IBM T. J. Watson Research Center
Yorktown Heights,
New York 10598, U.S.A.

The contribution of EPR measurements of organic metals to the understanding of these compounds is described. The particular examples discussed are: 1. The respective roles of the different stacks in the different phase transitions of TTF-TCNQ. 2. The EPR linewidth magnitude as a measure of the deviation of the band structure from one dimensionality.

In general, analysis of spin resonance spectra yields valuable information about the magnetic properties of the investigated material. The purpose of this talk is to show that some of the evaluated parameters provide unique answers to some of the outstanding questions related to the understanding of this fascinating class of organic materials. The magnetic parameters which will be discussed in this paper are the g-value and the linewidth. Let us start with the g-value: In organic conductors in general, there are two kinds of stacks - one built of donor molecules, while the other consists of acceptor units. The physical basis for the presence of the two kinds of stacks is the requirement for charge neutrality. One of the reasons for the existence of multitude of phase transitions in these materials[1] is that different stacks have different tendencies to undergo a Peierls distortion, and as a consequence, they order at different temperatures. A very important step in understanding this multitude of phase transitions is to identify which stack is undergoing a transition at a particular temperature. This information could be obtained in principle by measuring the magnetic susceptibility of each kind of stack as a function of temperature, and determining which stack's susceptibility opens a gap at the transition temperature of interest. The underlying assumption is that a gap in the single particle excitation spectrum will give rise also to a gap in the magnetic excitation spectrum. Such a measurement would be feasible if each kind of stack gives rise to a resolved EPR line. Then double integration of each line[2] will yield the respective spin susceptibility of that particular stack. However, since the difference between the Larmor frequencies of the donor and acceptor stacks is small in comparison to the spin hopping rate between them, their EPR absorptions merge into one line,

L. Alcácer (ed.), The Physics and Chemistry of Low Dimensional Solids, 187–195.

eliminating the possibility of this direct measurement. In some special cases though, the individual stacks' susceptibility can be obtained indirectly by a method which we refer to as the *g decomposition technique*.

What is this technique all about? If the interstack spin hopping rate is much faster than the difference between the imaginary Larmor frequencies of the respective stacks, having characteristic g tensors g_Q and g_F, then the g-value of the single absorption line will be:

$$g_{obs} = \alpha_Q g_Q + \alpha_F g_F \tag{1}$$

where α_Q and α_F are the fractional spin susceptibilities

$$\alpha_Q = \frac{\chi_Q}{\chi_Q + \chi_F} \tag{2}$$

$$\alpha_F = \frac{\chi_F}{\chi_Q + \chi_F} \tag{3}$$

and χ_Q and χ_F are the separate-stack spin susceptibilities. This reflects the fact that the g tensors of the individual stacks are weighted in proportion to the time a spin excitation spends on each stack and this, in turn, is proportional to the density of spin excitations and, therefore, to the spin susceptibility of each stack. Therefore, for a compound for which equation (1) is applicable, the *ratio* of the partial spin susceptibilities is directly measurable at any temperature at which g_Q and g_F are known. If in addition one knows the total spin susceptibility $\chi_T = \chi_Q + \chi_F$, then one can decompose χ_T into its donor and acceptor contributions. Let us now consider how a given compound can be tested experimentally, i.e., whether the extreme narrowing approximation is valid for it, namely, if equation (1) can be used. There are three necessary conditions for the applicability of the decomposition technique, which can be checked experimentally.1. Only one EPR absorption line is observed and it should include *all* the spin susceptibility. 2. This single line should have a Lorentzian shape. 3. The measured linewidth should be frequency independent. However, even if a compound does meet the necessary conditions for applicability, the decomposition results will not be useful unless an additional set of conditions is met. These conditions are the following. 1. g_Q and g_F must be known over the whole range of temperatures in which the decomposition takes place. 2. g_Q and g_F should be significantly different. The reason for this requirement is that the error in the determination of the donor and acceptor fractional susceptibilities, as shown in equation (4), is inversely proportional to $g_F - g_Q$.

$$\Delta\alpha_Q = -\Delta\alpha_F = \frac{\alpha_F \Delta g_F + \alpha_Q \Delta g_Q - \Delta g_{obs}}{g_F - g_Q} \tag{4}$$

Since both g_F and g_Q are tensors, the difference between them depends on the orientation of the magnetic field relative to the crystallographic axis. Namely, even if in a certain orientation the above mentioned condition is met, it does not mean that the decomposition results would be meaningful in any arbitrary orientation of the field. The error bar should be determined for every orientation.

TTF-TCNQ is the example for which the decomposition technique applicability, usefulness and results will be demonstrated. As was shown for this compound[3], all the spin-susceptibility appears in the single EPR absorption line, with a Lorentzian shape[4] and frequency independent width as checked[5] at 10MHz, 10GHz and 36GHz. Thus, all the necessary conditions for applicability are met in TTF-TCNQ. Now about the conditions for the usefulness of the decomposition technique: The actual tensors of g_Q and g_F in TTF-TCNQ were measured at 4 K by Walsh et.al.[6] At this temperature, two separate absorption lines originating from the TTF stacks and the TCNQ stacks were resolved. The principal g-values of the TTF stack were in a good agreement with those evaluated[7] for TTF-$Br_{0.76}$, a compound in which all the spin excitation reside on the TTF stack, while the values obtained for TCNQ agreed with the g-values of NMP-TCNQ. Resolved absorption lines of either donor or acceptor, and in some cases of both stacks, were observed also in HMTTF-TCNQ[8] and HMTTF-$TCNQF_4$[9]. These lines, originating most probably from crystalline imperfections, are very useful for determination of g_Q and g_F. Good agreement exists between the molecular g-values as obtained for solutions and the values derived from the low temperature measurements for these compounds.

In the particular case of TTF-TCNQ, the g-values of the single EPR absorption line agree within experimental accuracy with the corresponding molecular values of TTF over the temperature range[10] $(20 < T \leq 35)$K. Therefore, one can conclude for that particular temperature regime that all the intrinsic magnetic susceptibility is localized on the TTF stacks and that the g-value is temperature independent. However, no similar experimental evidence exists with regard to the TCNQ stacks and even with regard to the TTF stacks. The temperature independence of the g-values is confirmed only for temperatures lower than 35 K. Therefore, let us assume that indeed the donor and acceptor stacks g-values are temperature independent, use the 4 K results as obtained by Walsh et.al., and evaluate α_Q as a function of the angle between the magnetic field and the a axis in the ac* plane, θ. In Figure 1, g_Q, g_F and g_{obs} at 293 K, 50 K and 45 K are shown as a function of θ. It is clearly seen that all of them, except g_Q, have an appreciable angular dependence. Thus, if g_F and g_Q, as measured at low temperatures, are incorrect values at higher temperatures, one would expect an α_Q depending on θ. Even if strange cancellation of angular dependence can occur at any particular temperature, since g_{obs} is strongly temperature dependent, it is highly impossible for the same cancellation to occur at a different temperature. α_Q as a function of θ is shown in Figure 2 for 293 K, 50 K and 45 K. It should be pointed out that since g_{obs} is practically temperature independent[11] between 293 K and 80 K, α_Q as evaluated for 293 K, actually represents α_Q over a much broader temperature range. As is clearly seen from this figure, α_Q is independent of θ, as expected in the regime of of the validity of equation (1) when, the proper tensors of g_Q and g_F are used.

Figure 2 shows the error in α_Q using equation (4). Clearly the error bar has an angular dependence with a minimum in an orientation in which $g_F - g_Q$ is largest - $H_o \mid \mid c^*$.

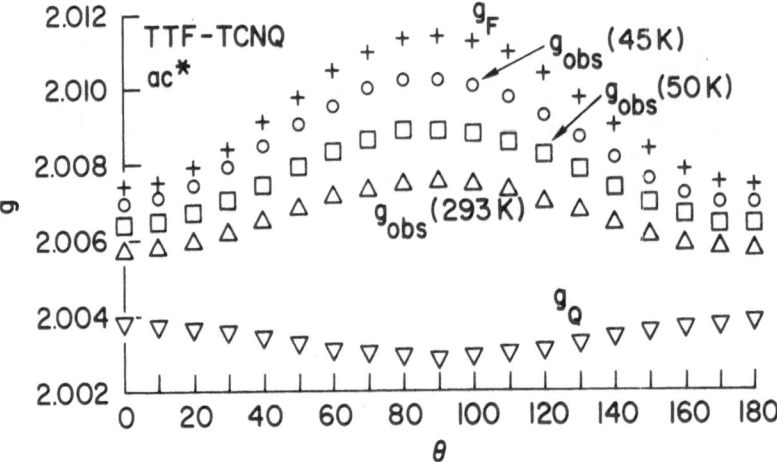

Fig. 1.   g as a function of the angle between the magnetic field and the a
crystallographic axis, θ, in the ac* plane of TTF-TCNQ.

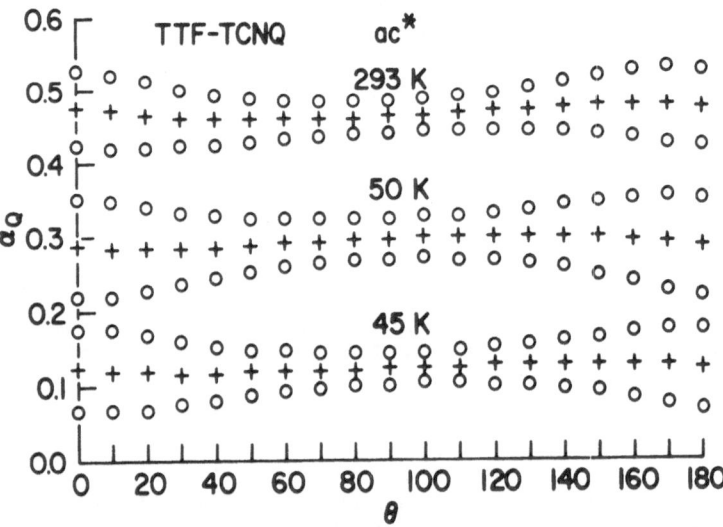

Fig. 2. α_Q, marked as +, as a function of θ in the ac* crystallographic plane
for 293 K, 50 K and 45 K. Each pair of vertical dots corresponds to
the respective error bar in the measurement.

In this orientation, α_Q was measured as a function of temperature and combined with the high resolution static susceptibility data of Herman et.al.,[12] yielded the susceptibility decomposition shown in Figure 3. χ_Q and χ_F are shown in a temperature range in which TTF-TCNQ is known[1] to undergo three phase transitions at 52.6 K in., 48 K and 38K. The conclusions of the decomposition are the following: 1. The largest effect of the metal to insulator transition at 52.6 K is on the TCNQ stacks, therefore, these stacks drive the corresponding Peierls instability. The implications of this result on stabilizing the metallic state in organic conductors are discussed in detail in reference 13. 2. The 48 K transition in which a three dimensional order is expected[14] to develop on the TTF stacks does not seem to have an effect on the magnetic susceptibility of these stacks. 3. Significant effect, though, is found at 38 K, where the transverse periodicity of the superlattice jumps discontinuously from 3.8a to 4a.

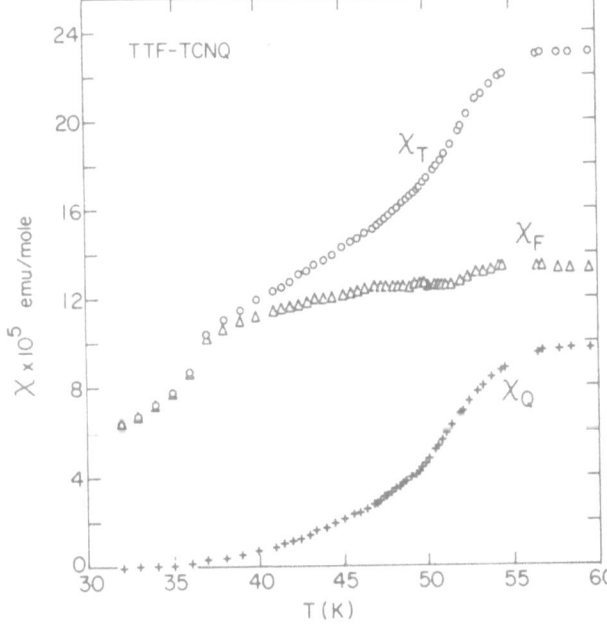

Fig. 3.   The total susceptibility χ_T and and the susceptibilites of the donor and acceptor stacks respectively, χ_F, and χ_Q, as a function of temperature.

The above described susceptibility decomposition of TTF-TCNQ is only one example in which the *g decomposition technique* was utilized. Other examples are TMTTF-TCNQ[15], TMTSeF-DMTCNQ[13], HMTTF-TCNQ[8] and HMTTF-TCNQF4[9]. In all these cases, the decomposition contributed significantly to the understanding of the physics of these materials.

Now about the linewidth: In isotropic metals, the dominant relaxation process was shown to be[16] the spin-lattice relaxation caused by the scattering of conduction electrons by acoustical phonons. Thus, both the resistivity and the spin resonance linewidth are determined by the electron-phonon scattering rate. The proportionality constant between the linewidth and the electron-phonon scattering rate is the square of the spin orbit coupling. Let us check whether this mechanism is also the dominant relaxation process in quasi 1D metals of the TTF-TCNQ family. This can be done for the isostructural[17] family $(TSeF)_x(TTF)_{1-x}$ (TCNQ), $0 \leq x \leq 1$, in the following way: a characteristic feature of the spin-phonon relaxation mechanism is that, other things being equal, the linewidth should vary as $(\Delta g)^2$, where Δg is the deviation of the g-value from the free-electron g-value of 2.0023, because Δg is a measure of the spin-orbit coupling strength. Since the spin-orbit coupling of TSeF is significantly larger than that of TTF, a continuous variation of the average g value in the isostructural family can be achieved by doping TTF-TCNQ with varying amounts of TSeF-TCNQ. As the concentration of TSeF-TCNQ in TTF-TCNQ is increased, one would expect an increasing linewidth, which is indeed observed (Fig. 4). This increase of linewidth with doping, however, could also be due to the growth of another relaxation mechanism different from that which dominates the relaxation in the pure compound. To check if this is indeed the case, we measured the angular dependence of the linewidth in the pure and doped compounds, and found that in spite of the big variation in the magnitude of the linewidth, the angular dependences obtained look strikingly similar, indicating a similar relaxation mechanism over the whole range of solid solutions.

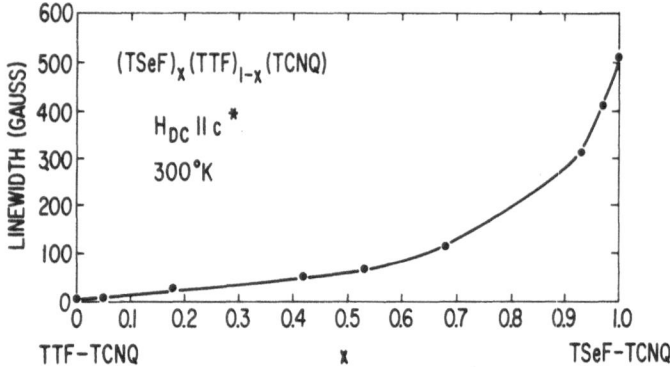

Fig. 4.   Linewidth of $(TSeF)_x(TTF)_{1-x}(TCNQ)$ $0 \leq x \leq 1$ as a function of the concentration, x, of TSeF in the donor stack.

    To see if the single spin-relaxation mechanism is indeed the spin-phonon interaction via spin-orbit coupling, we have normalized the measured linewidth to $(\Delta g_F)^2$, where Δg_F is the g shift measured for the donor stack. The results are shown in Fig. 5. Two concentration ranges emerge, which we discuss separately:
(i) $0 \leq x \leq 0.68$. In this range of x, the strong dependence of the linewidth on dopant concentration is appreciably reduced by the $(\Delta g)^2$ normalization. For example, comparison of the unnormalized linewidths for x = 0 and x = 0.68 shows a dramatic increase by a factor of 25, in going to the latter concentration, while a

comparison of the normalized linewidths shows an increase by only a factor of 3. The behavior in this concentration regime is thus consistent with the origin of the linewidth being the spin-phonon interaction. The slow increase of the normalized linewidth with x observed in this concentration regime can be the residue of the phenomenon observed in the second concentration regime.

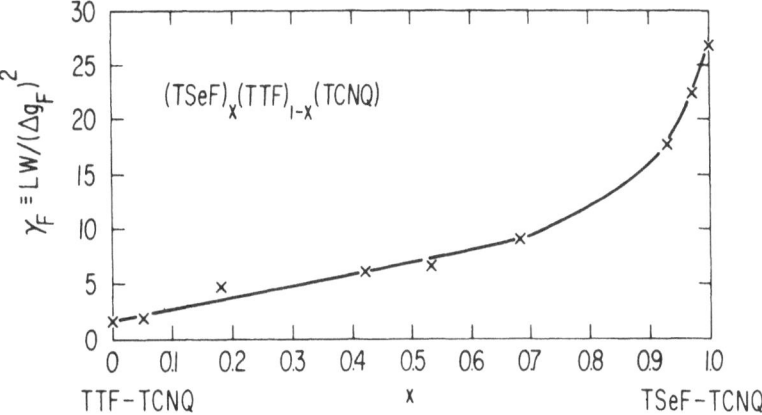

Fig. 5.   Linewidth of $(TSeF)_x(TTF)_{1-x}(TCNQ)$ $0 \leq x \leq 1$ at 300 K normalized to $(\Delta g_F)^2$, where Δg_F is the deviation of the donor stack g-value from the free-electron value, as a function of x.

(ii) $0.68 \leq x \leq 1$. In this range of x, there is still a strong dependence on x, even for the normalized linewidth. For example, the normalized linewidth is 16 times larger at $x = 1$ than at $x = 0$. In order to explain this difference, we have to further our understanding[18] of electron-phonon scattering in one dimension.

For systems with flat energy surfaces, there exist two possible kinds of scatterings in terms of their effect on the longitudinal component of the electron wave vector: (a) Scattering in which the change in the longitudinal component of the wave vector, $k_{||}$, is small. We shall refer to this scattering as "forward" scattering despite the fact that Δk is not necessarily scattered in exactly the forward direction since $k_{\perp}$ does not have to be conserved. (b) Scattering in which the change in $k_{||}$ is large ($\sim 2k_F$). This scattering we shall refer to as "backward" scattering. Both the "forward" and "backward" scatterings are sharply reduced from their values in three-dimensional systems, but the reduction of their contribution to the linewidth is much greater than to the resistivity. The longitudinal component of the phonon wave vector is confined, in the case of one-dimensional forward scattering, to small values because of energy conservation. This limitation significantly reduces the probability of a spin flip in a forward-scattering process since the spin-flip matrix element is proportional to $\Delta k_{||}$. The contribution of backward scattering with spin-flip to the linewidth is also greatly reduced in electronically one-dimensional systems, because the matrix element for strictly backward spin-flip scattering is zero, the respective states between which the scattering occurs being related by time-reversal symmetry. Thus, for a given scattering rate as measured by resistivity, and for a given spin-orbit coupling strength, the rate of spin-flip scattering is much smaller in electronically 1D systems than in 3D systems. Actually the magnitude of the normalized linewidth can be used as a measure of the deviation of a given band structure from one-dimensionality. For example, the difference between the normalized linewidths of TTF-TCNQ and TSeF-TCNQ was attributed[18] to a smaller anisotropy in the band structure of the latter compound. The lowering of[19] the metal to insulator transition of TSeF-TCNQ in comparison to TTF-TCNQ was interpreted as a direct consequence of this smaller anisotropy.

In conclusion, we have shown that the following information can be derived from EPR of organic metals: 1. The g-values combined with the susceptibility lead to an understanding of the roles of the different stacks in the phase transitions. 2. The normalized linewidth magnitude can be used as a measure of the deviation of a system from one dimensionality.

Acknowledgments

The technical assistance of A. R. Taranko is appreciated and particular thanks to Ms. E. Marino for her help in getting the manuscript ready on time. My collaborators in the various stages of this work, in particular, T. D. Schultz are acknowledged.

REFERENCES

(1)   See for example, Kagoshima, S., Anzai, H., Kajimura, K., and Ishiguro, T.,
        1975, J. Phys. Soc. Jpn. 39, pp. 1143;   Ellenson, W. D., Shapiro, S. M.,
        Shirane, G, and Garito, A. F., 1977, Phys. Rev B 16, pp. 3244.

(2)   The reason for double integration is that the EPR absorption is measured in
        the derivative mode.

(3)   Gulley, J. E., and Weiher, J. F., 1975, Phys. Rev. Lett. 34, pp. 1061.

(4)   Tomkiewicz, Y., Taranko, A. R., and Torrance, J. B., 1977, Phys. Rev. B
        15, pp. 1017.

(5)   See for discussion on the frequency independence of the linewidth. Tom-
        kiewicz, Y, Taranko, A. R., and Torrance, J. B., 1976, Phys. Rev. Lett 36,
        pp. 751.

(6)   Walsh, W. M., Jr.,   Rupp, L. W., Jr., Schaefer, D. E., ad Thomas, G. A.,
        1976,Bull. Am. Phys. Soc. 19, pp. 296.

(7)   Tomkiewicz, Y., and Taranko A. R., 1978, Phys. Rev. B 18, pp. 733.

(8)   Tomkiewicz, Y., Taranko, A. R., and Schumaker, R., 1977, Phys. Rev. B 16,
        pp. 1380.

(9)   Tomkiewicz, Y., Torrance, J. B., Bechgaard, K., and Mayerle, J. J., 1979,
        Bull. Am. Phys. Soc 24, pp. 232.

(10)   The temperature corresponding to the lower boundary is strongly sample
        dependent.

(11)   Tomkiewicz, Y., Scott, B. A., Tao, L. J. and Title, R. S., 1974, Phys. Rev.
        Lett. 32, pp. 1363.

(12)   Herman, R. M., Salamon, M. B., De Pasquali, G., and Stucky, G., 1976,
        Solid State Commun. 19, pp. 137.

(13)   Tomkiewicz, Y., Anderson, J. R., and Taranko, A. R., 1978, Phys. Rev. B
        17, pp. 1579.

(14)   Bak, P., and Emery, V. J., 1976, Phys. Rev. Lett. 36, pp. 978.

(15)   Tomkiewicz, Y., Taranko, A. R., and Green, D. C., 1976, Solid State
        Commun. 20, pp. 767.

(16)   See for summary Y. Yafet in Solid State Physics, Vol. 14, edited by Ehren-
        reich, H., Seitz, F., and Turnbull, D., (Academic Press New York 1965).

(17)   Engler, E. M., Scott, B. A., Etemad, S., Penney, T., and Patel, V. V., 1977,
        J. Am. Chem. Soc. 99, pp. 5909; Etemad, S., Engler, E. M., Schultz, T. D.,
        Penney, T., and Scott, B. A., 1978, Phys. Rev. B 17, pp. 513.

(18)   Tomkiewicz, Y., Engler, E. M., and Schultz, T. D., 1975, Phys. Rev. Lett.
        35, pp. 456.

(19)   Etemad, S., 1976, Phys. Rev B 13, pp. 2254.

ESR IN ALKALI TCNQ-SALTS

Jan Kommandeur

Laboratory for Physical Chemistry
University of Groningen
Nijenborgh 16, 9747 AG GRONINGEN
The Netherlands

"Om der wille van de smeer
Likt de kat de kandeleer"

Joost van den Vondel

L. Alcácer (ed.), The Physics and Chemistry of Low Dimensional Solids, 197–212.
Copyright © 1980 by D. Reidel Publishing Company.

Introduction

Electron Spin Resonance is a useful tool for the study of TCNQ-salts. It allows to determine pure spin magnetism, without the need for corrections, the fine structure of the observed lines can say something about the correlation of the spins, and line-widths may give information on the dynamics of the species observed. It is particularly useful in one-dimensional systems. Because the technique is slow compared to the usual electronic motions in three dimensions, usually only three-dimensional averages are obtained. In a one-dimensional system, i.e. where electronic (or spin) motion in the other directions is severely limited, rather detailed information can be gleaned. In this lecture I purport to show, how we used ESR for the determi-nation of spin susceptibilities, for an assessment of the Hubbard U and for the determination of the anisotropy in the alkali-TCNQ-salts

1. Spin-susceptibilities

ESR is usually obtained as a derivative signal, i.e. through the modulation detection technique a derivative with respect to field is taken of the absorption signal. To estimate the total absorption, which is proportional to the spin susceptibility a double integration is then required. Double integrations are notorious , since baseline problems lead to divergencies and cut-off problems. We therefore decided to at least eliminate one integration by recording the ESR signal directly. Since, however, such a signal is very weak one must record it repeatedly to obtain reasonable signal to noise. Or, rather, since one only wants the integral, one should obtain the integral a large number of times. This is exactly what we have done.

The field is swept by a trapezoidal sweep, which encompasses the whole ESR signal, which is detected by the crystal, fed to a voltage to frequency (V-F) converter and then to a reversible (up-down) counter. When the field is changing in the flanks of the trapezoid modulation the V-F converter is switched to the positive channel, when the field is constant it goes to the negative channel. After several hundred runs a number (on the counter) is obtained, which is proportional to the spin susceptibility. As usual the signal to noise ratio of the integral (the integration is performed by the VFC in combination with the counter) is improved over a single integration by the factor $\sqrt{N}$, where N is the number of times the signal has been swept. Obviously a small portion of the signal, depending on the lineshape, will always fall outside the sweepwidth, thus limiting the accuracy of the method. It appears to be the best for any signal to obtain the integral for a number of sweep-widths, plot these as a function of the reciprocal sweep and then extrapolate to zero. With that procedure a reproducibility

of better than 1% for one sample situation can be obtained.
To obtain the spin susceptibility the internal conditions of the
cavity must be probed, which is conveniently made possible by a
small ruby crystal glued to the inside of the cavity in such an
orientation that it's signal occurs at 1500 Gs and 5000 Gs, well
away from the TCNQ signal at 3000 Gs. After every sample measurement, a ruby measurement is performed by just changing the
magnetic field. Since this doesn't change conditions in the
cavity at all, a good measure of the cavity situation is
obtained. Using the ruby correction a reproducibility of about
1% is obtained for samples under different conditions, such as
with or without dewar equipment, with or without liquid N_2 or
He, etc. Finally the system has to be calibrated. To this
purpose the sample is replaced by a weighed crystal of $Cu\ SO_4$.
5 H_2O of which the spin susceptibility can be calculated from
the weight. Altogether, the determinations yield an accuracy
in absolute susceptibility of about 3%.

   We used the determination of spin susceptibility as a
function of temperature for the observation of the by now well
known $2k_f$ transition in the alkali-TCNQ's, as is shown in
fig. 1 (1).

   Roughly speaking first order, second order and "continuous"
phase transitions were found. These phase-transitions are now
recognized as (Spin)-Peierls transitions. The increase of the
susceptibility with temperature arises from the combination of
the electron transfer t and the on-site Coulomb interaction U,
leading to the well-known phenomenon of anti-ferromagnetic
exchange with $J = -4t^2/U$. Another material studied more recently
is $MEM(TCNQ)_2$. Fig. 2 shows the behaviour of the susceptibility
around 19K and around 340K. The major change at low and the small
($\approx$ 7%) change at high temperature are clearly observed. (2)

2. Triplet Excitons and the Hubbard U

   Very soon after the discovery of the TCNQ-salts it was
found, that a fine structure was present in the ESR signal. This
structure derives from a spin-spin dipolar interaction. Two
spins must be bound to obtain this effect. For TCNQ-salts there
are two explanations, which represent the limiting cases of what
is probably the real situation. We consider a half-filled band,
i.e. one electron per site of the crystal.

   i) Limit of U >> 4t , the on-site interaction is large
compared to the transfer integral. All electrons are localized,
the important interaction is anti-ferromagnetic exchange, the
ground state can be visualized as a chain of alternating spins.
We obtain the first excitation of this system by turning over
one spin. Since it takes the same energy to turn over any other
spin there will be N such excitations (where N is the number
of spins in a chain) and the excitations will broaden into a
band.

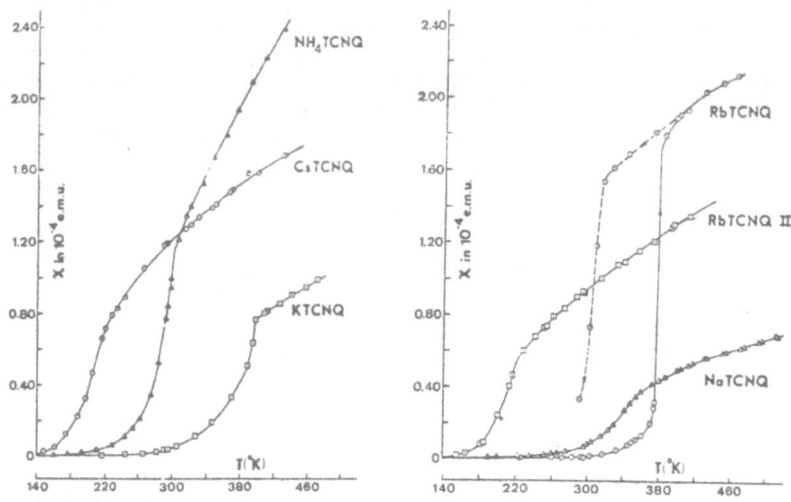

Fig. 1.   Experimental behavior of the spin susceptibility
           with temperature for a number of crystalline M⁺TCNQ⁻
           salts. A Curie type 'impurity' contribution (table
           V.III) has been subtracted from each curve.

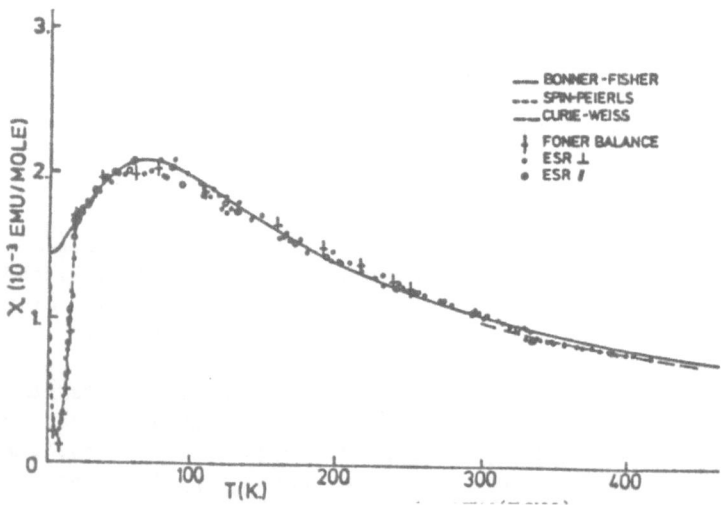

Fig.2. Spin Susceptibility in MEM(TCNQ)₂

If the system is regular, i.e. all spins have the same
interaction with their neighbours, then the lowest energy
excitation will be the one, where the spin flip of π is
obtained by turning all spins over the angle Π/N. If we must
apply the energy $2J$ to locally flip one spin in this
environment, the energy for the first excitation in a regular
band will be of order $2J/N$ and since N is very large that energy
goes to zero. The band of spin wave excitations in a regular
system starts at the ground state. We cannot identify a pair of
spins forming a triplet, and thus it does not seem proper to
speak about triplet states in such a system (3). A different
situation obtains when the chain is dimerized, i.e. when a
spin has an interaction with its right-hand neighbour, different
from the interaction with its left-hand neighbour, say J' and J''.
Again the first excitation is turning over one spin, now
costing an energy $J' + J''$. There are, of course, again N such
excitations, but, except for spin direction (m_s!), turning over
the other spin of the pair to which the first one belonged,
leads to the same state. There are thus $\frac{1}{2}N$ such pair excitations,
the pair state with parallel spins (a triplet !) is thus $\frac{1}{2}N$-
fold degenerate and these states will thus form a band, the
triplet exciton band, with a width determined by J'', if that is
the smallest exchange energy. The limit of dimerization of course
is the chain of isolated dimers, where $J'' = 0$, and a gap J'
exists between the singlet and triplet states. In the monomeric
chain J' equals J'', and the gap vanishes. The dimerized
system thus has a triplet exciton band separated from the ground
state by a gap of order $J' - J''$. In this system triplet
excitons can be recognized, they will have the typical spin-
dipolar splitting. Conversely, if a fine-structure on the ESR
signal is found, the system cannot be a regular monomeric chain.
In the monomeric as well as in the dimerized state there are
also excitations of an electron to a site, where an electron is
already present. Such an excitation will be at an energy close
to the electron-electron repulsion energy U and will contribute
to conductivity through both the electron excited and the hole
left behind. The general order of these states for a monomeric
and a dimeric chain (in this case calculated for a chain of
four) is given in figure 3.

ii) Limit $4t >> U$, the transfer integral is bigger than the
electron-electron interaction.

The regular chain is now simple. It is a metal with a half
filled band. The electron-electron interaction mixes the one-
electron k-states. Two-electron states with a lot of double
occupancy will go up, leaving low double occupancy at the
bottom of the energy spectrum. Since high double occupancy can
only occur for a pair of electrons with opposite spin a non-zero
value of U enhances the susceptibility.

Without U in the dimerized system, i.e. where the transfer
integrals between left and right-hand neighbours differ ($t' \neq t''$)

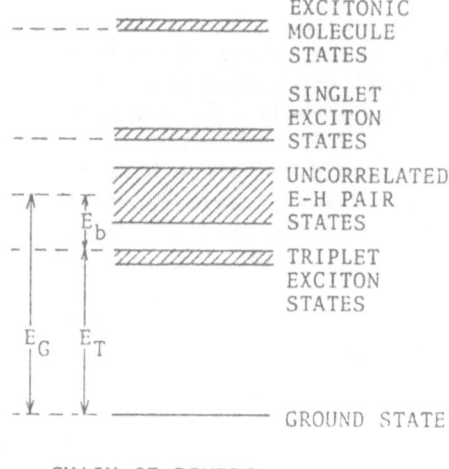

Fig. 3   Energy level scheme for a system of weakly
         interacting dimers.

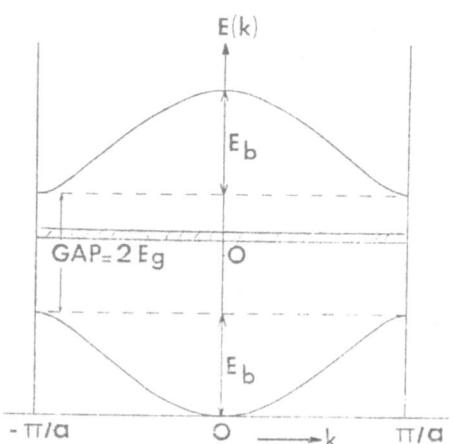

Fig. 4   Band system with triplet Excitons for
         alterning linear chain.

we obtain two bands, one basically deriving from the bonding
orbital at -t' and one from the anti-bonding orbital at +t'. The
widths of the bands are given by t'', the resulting gap is
2(t' - t''). Electron-electron repulsion is of course the same
as electron-hole attraction. Therefore, if we excite an alectron,
it may lower it's energy by remaning bound to the hole it left
behind, again forming an exciton, and if the spins are parallel,
a triplet exciton. These states will be at energies lower than
those needed for the creation of an uncorrelated electron-hole
pair. Usually one situates them within the forbidden gap,
although this amounts two mixing one-electron and two-electron
diagrams. Fig. 4   gives the ordering of the states for this
limit.

In perturbation theory limits i) and ii) can be calculated
and expressions can be given for the various energies (4).
Algebraically they look quite different, and one hopes that in
the "difficult" region, where U≈4t they may go into one another.

It will be clear that the activation energy for conductivity
can be measured. With our spin integration technique we measured
the temperature dependence of the magnetic susceptibility from
which one obtains the energy gap for the triplet excitons. For
the values of U, t' and t'' we then have two equations, i.e. we
are one short. In RbTCNQ the intermolecular distances of the
TCNQ-ions are not very different, but their overlaps are. We
took for the ratio t'/t'' the ratio of the geometrical overlaps
μ'/μ'', known from the crystal structure (5), thus giving
ourselves a third equation. Analysis then yielded U≈1.1 eV,
t'≈0.35, and t''≈0.10 eV, putting us right in the "difficult"
region. It came as somewhat of a surprise, however, that it
didn't matter, whether we used limit i) or limit ii) for our
calculation, numerically they led to the same result.

This was, we think, the first almost unambiguous determination of U in a half-filled band of TCNQ-ions.

The influence of the transfer integral t expresses itself
in a modification of the dipolar interaction. Because of the
transfer the triplet exciton is not completely localized on a
single dimer. The neighboring dimers will obtain some of the
unpaired spin. Therefore, the average distance of the dipoles is
increased, which leads to a reduction of D, the dipolar splitting
constant, which goes as $<1/r^3>$. Analysis shows that about 10% of
the triplet spins is on the neighboring dimers (6)

3. The anisotropy of a TCNQ-chain

The degeneracy of the triplet excitons extends along the
chain, as well as to other chains. They will move along as well
as at right angles to these chains. How often do they hop on,
and how often do they hop at right angles to the chain direction ?
ESR can give the answer, because it is such a slow technique that
it is sensitive to the dynamics.

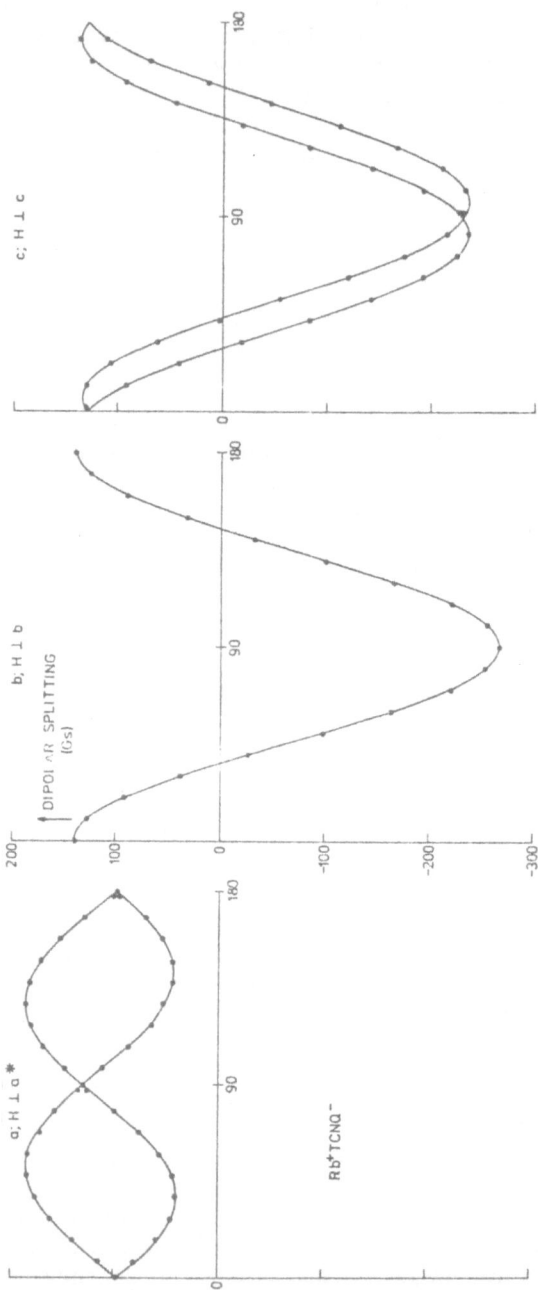

Fig.5. Angular dependence of spin-spin dipolar splitting of triplet excitons in Rb.TCNQ.

As we pointed out before the fine structure splitting is
due to the spin-spin dipolar interaction. Because the spins
adhere to electrons spread over the two molecules of the pair,
this interaction is tensorial in nature, i.e. it depends on the
angle that the spin direction (in the high-field condition the
direction of the magnetic field $\vec{B}$) makes with the planes of the
paired molecules. Fig. 5   gives an example of the angular
dependence of the splitting as measured for the excitons in
RbTCNQ.

In this compound, however, there are two sets of chains,
differing by the crystallographic orientations of the molecular
planes. Therefore, in a general direction, one finds two ESR
signals with different splittings, because the spin directions
make different angles with the molecular planes in the two sets
of chains. If an exciton jumps from one chain to the other it
starts contributing to the other ESR signal with a different
splitting. The spin "feels" its neighbour in a different way.
The process amounts to a jumping between two inequivalent sites,
and as is well known from NMR this leads to line broadening.
How much is the line broadened ? This is easy to determine
because, when the magnetic field is in a position symmetrical
with respect to the two chains, the sites become equivalent and
the broadening contribution should disappear. Figure 6   shows
this effect as observed very clearly in RbTCNQ. The line has a
constant line width for almost all orientations, but when we
get within 3° of symmetrical orientation it first broadens
and then narrows severely at the symmetrical orientation. As
is clear from figure 6,   , the "fast exchange" width (at the
symmetrical minimum) and the "slow exchange" width (far away from
the symmetrical situation) can be determined quite accurately.
For both limits there are very well established expressions
containing the transverse jumping frequency ν_t , the energy
difference between the two sites $\hbar\omega$ and the extra linewidth
$$\Delta H = (g\beta\hbar)^{-1}(\hbar\omega)^2/8\nu_t$$
It will be clear that from our experiments the lateral
jumping frequency ν_t can be calculated. By doing these measurements as a function of temperature it turned out that the jumping
was activated: $\nu_t = \nu_t^{\,o}$ exp $-$ $^{E_L/KT}$ with $E_L = 0.08$ eV. A plot
of ln ν_t vs $\frac{1}{T}$ is given in fig. 7.   This means that the
exciton is   incoherent, the exchange bandwidth is narrowed by a
polaron type interaction.

Sofar for the transverse jumping rate ν_t. The longitudinal
jumping rate ν_1 can also be obtained. The excitons can be
thought of as trains on a single track traveling in both directions. Therefore they must collide. In the collision of two
species with spin $S = 1$ one can form $S = 0$, $S = 1$ and $S = 2$ states,
but important is, that the original quantum numbers m_S
(± 1 or 0) of the triplets are lost. All that is required after
collision is the conservation of total quantum number. They can
for instance come in as $S = 1$, $m_S = +1$, and $S = 1$, $m_S = -1$ and

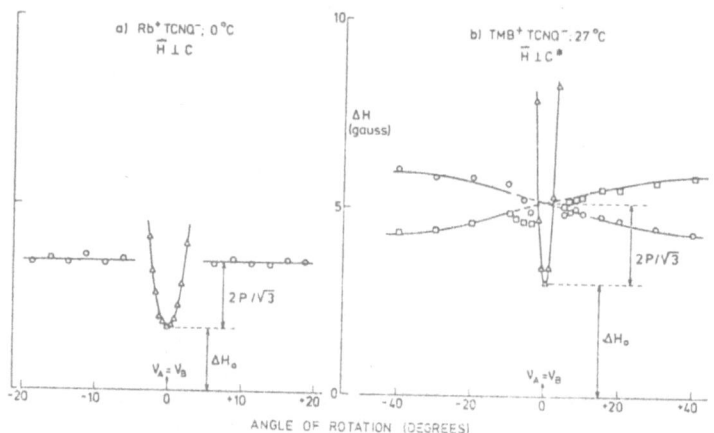

ANGLE OF ROTATION (DEGREES)

Fig.6.   The width of the lines due to excitons at trans-
          lationally inequivalent sites as a function of the
          orientation of the crystal with respect to H. At
          angle zero the lines coincide and the line width
          is considerably reduced. Note that the width of the
          lines in Rb$^+$TCNQ$^-$ is isotropic, whereas the width
          of the lines in TMB$^+$TCNQ$^-$ (squares and circles) is
          anisotropic. P=hν_t/gβ_0 is the jumping rate in gauss.

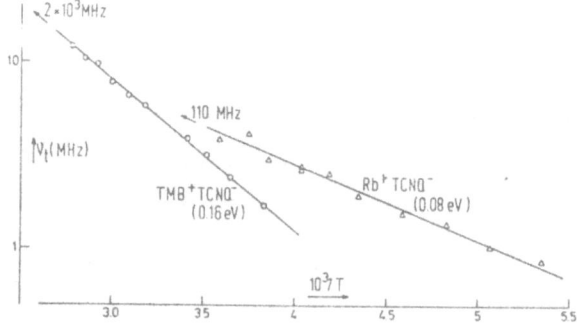

Fig.7.   The transverse jumping rate of the excitons in
          Rb$^+$TCNQ$^-$ and TMB$^+$TCNQ$^-$ as a function of temperature.

leave as S = 1, m_S = 0 and S = 1, m_S = 0. Collisions limit the
life-times of the states the excitons are in. Therefore, if there
are many collisions per unit time, the pairing of the electrons
is lost, they start to "feel" the average field of all other
electrons, which averages out to zero. Therefore the spin-spin
dipolar interaction is washed out, the signal averages from two
separate split lines to one single line in the center. The process
is quite similar to the increasingly rapid rotation of two
protons in NMR and can again be handled with the theory developed
for that problem. The frequency of the collisions is given by
the number density of the excitons ρ times their hopping rate ν_1,
longitudinally along the chain. In the limit of fast collisions we
have for the resulting linewidth : $\Delta H = \beta(\hbar\omega)^2(\rho\nu_1)^{-1}$, where β
is a constant and $\hbar\omega$ the fine structure splitting
     In the limit of slow collisions only the widths and positions
of the seperate components are affected and we have: $\Delta H = A\rho\nu_1$,
where A is another constant.
          Both the number density and the longitudinal
transfer rate can be temperature activated, we expect for the
temperature dependence
$$\Delta H = A\ \rho_0\nu_1^{\ 0}\ \exp - E_T/kT\ \exp^{-E_L/kT},$$
where E_T and E_L are the temperature dependencies of the triplet
number density and the transfer rate, respectively.
     Fig. 8   gives a plot of $\ln(\rho\nu_L)$ vs. $\frac{1}{T}$ for RbTCNQ. Table I
compares the data with those for the transverse hopping rate.
We conclude that in RbTCNQ the anisotropy of exciton motion is
$\nu_1^{\ 0}/\ \nu_t^{\ 0} \underset{\sim}{} 10^5$.
     To check ourselves we also studied the excitons in
trimethyl-benzidylazolium - TCNQ: TMB(TCNQ), where the TCNQ ions
are arranged in pairs that are stacked side by side. In this
particular compound we can   study the sideways chain hopping
rate as compared to the interchain transfer. Since the overlap
in the chain is now very unfavorable one expects a much smaller
anisotropy. We found $\nu_1^0/\nu_t^0 \underset{\sim}{} 10$, as shown in table I, which
gives a lot of confidence for our earlier measurement.
     It might be questioned, whether the nuclear hyperfine inter-
action would lead to a line-width, as well. The hopping rates
found, however, are so fast, that the electron spin "sees" the
average of many proton spins. At the temperatures, where the
triplet spins are present this average is zero, therefore the
hyperfine interaction is averaged out.
     In TMB TCNQ a small remnant can still be found, since the
hopping rates are lower there then in RbTCNQ. For this compound
an estimate of the hopping rate could be obtained from the
remanent hyperfine width (7).

4. The excitonic polaron and the Peierls instability

     As pointed out above, the excitonic motions ν_1 and ν_t are
thermally activated, i.e. a lattice distortion must accompany

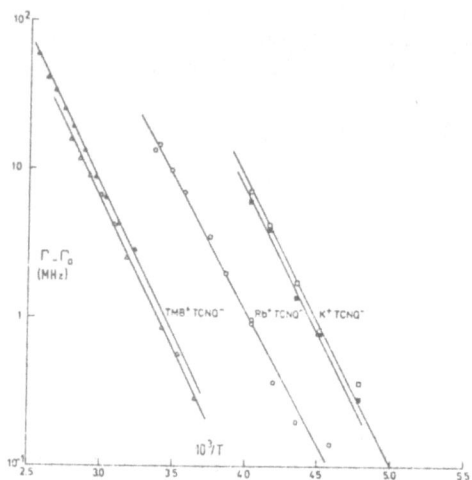

Fig. 8.   The difference between the actual and low tempe-
           rature limiting values of the linewidth as a
           function of 1/T for Rb^+TCNQ^-, K^+TCNQ^- A and B and
           TMB^+TCNQ^-.

Table I.   Exciton motion in Rb^+TCNQ^- and TMB^+TCNQ^- (Jum-
                 ping rates are in sec^{-1}, activation energies
                 in eV)

	inter-chain jumping $\nu_t = \nu_{to}\exp(-E_L/kT)$			intra-chain jumping $\nu_j = \nu_{jo}\exp(-E_L^1/kT)$		
	ν_{to}	E_L	$\nu_t(200^\circ K)$	E_L^1	$\nu_j(200^\circ K)$	A
Rb^+TCNQ^-	1.1×10^8	0.08	1.1×10^6	0.10	$>10^{11}$	<0.01
TMB^+TCNQ^-	2×10^9	0.15	3.3×10^5	0.20	$\sim10^7$	$\sim2/9$

The jumping rate along the chains is at least five orders of
magnitude larger than the transverse jumping rate. In TMB^+TCNQ^-
the transverse and intra-chain motion differ only one or two
orders of magnitude. The difference in the anisotropy of
the motion of the excitons in the two salts is clearly a consequence of the different structures of the chains of $TCNQ^-$
ions.

the exciton. It can therefore be called an excitonic polaron.
This polaron effect is a prelude to the (spin) Peierls transition, happening at much higher temperature, where the
dimerized chain turns into a regular one (1). After all, the
dimerization is due to the exchange interaction which through
its strong dependence on distance pulls two ions together.
Of course, this exchange interaction is only operative in the
singlet state. When the pair is excited to the triplet, the
exchange is lost, the ions are no more bound, and one can except
the pair to "open up". The bonding energy of the pair is then
the 0.08 eV found for the activation energy.

A very rough estimate for the Peierls transition temperature
T_c may be obtained from this number. Assuming that the transition
goes from an almost diamagnetic dimerized lattice to an almost
totally paramagnetic regular lattice, thus putting the spin
entropy gain at Nkln2, while using for the energy to destroy
$\frac{1}{2}N$ dimers $\frac{1}{2}NE_T$, we have with $\Delta H = T_c \Delta S$,

$T_c \approx \frac{\Delta H}{\Delta S} = \frac{0,04}{K\ln2} \approx$ 700 K, which overestimates the real transition

temperature of 376 K by about a factor of two, at least showing
that the order of magnitude of the numbers is correct. It should
be noted that this estimate of T_c completely neglects the
difference in phonon entropy between dimerized and regular
lattice, which would lower the estimate of T_c considerably.

5. The "Impurity" Signal

In all TCNQ-salts showing a spin-spin dipolar splitting
a so-called "impurity signal" has been found. It is a central
line, not showing any dipolar splitting with down to 2^0K a Curie-
type temperature dependence, its intensity relative to the
dipolar split line appears to depend on the sample and usually
varies between 0.1 and 2% of the total number of spins the
TCNQ$^-$ ions would show, if there were no interaction pairing them.
Clark et al (8) have recently shown that the g-tensor of this
signal corresponds to the g-tensor of the TCNQ$^-$ion, which
already shows that the "extra" spins are not due to an impurity
but must derive from TCNQ. In contrast to many inorganic systems
one does not expect organic impurities to have a spin; TCNQ$^-$-ion
is a sufficiently rare case.

In a very well executed and careful study of the impurity
signal at temperatures down into the millikelvin region Clark
et al (3) also showed that the behavior of the spin-
susceptibility can be perfectly understood on the basis of a
random anti-ferromagnetic exchange interaction. The fact that
it was random is essential in his interpretation. What then is
the nature of this impurity signal ? Only at very low temperatures,
where all temperature activated spins have been frozen out one
calls the remaining spins contributing to the Curie "tail",
impurities. At those temperatures all these systems will have

gone through some kind of (spin) Peierls in stability leading
to a diamagnetic ground state. The temperature of this transition T_c, however, is finite and the free energy $F = U - TS$ will
therefore be lowest if some entropy S can remain in the spin
system, even if this costs some energy U. One would therefore
expect the system not to become completely diamagnetic at T_c,
but rather to arrange itself in such a manner that some spins
remain free. The Peirls transition, however, involves nuclear
displacements and once they have occurred it is naieve to
suppose that a further nuclear rearrangement could still take
place at yet lower temperatures. Therefore, the spins remain
"frozen in". Apart from the spin entropy there is some entropy
in the random choice of where to leave free spins. In general
one therefore expects a random distribution of them, which
will lead to the small random exchange interaction, which is
exactly what Clark et al (8)   found. For a half-filled band,
which dimerizes in the Peierls transition the situation can be
easily visualized and understood.   In systems
with other band fillings it will be more complicated, but the
general argument will hold.

6.   Speculations on the value of U

As we have pointed out in an earlier contribution to
this series (9), a reasonable estimate of U can be obtained
from a consideration of the ionization energy I, the electron
affinity A of the TCNQ⁻ion and of the polarization energies
P_+ and P_-, around a hole and an electron, respectively:
$$U = I - A - P_+ - P_-$$
For purposes of this discussion it is advantageous to separate
the energies P into two parts: $P = P^{core} + P^{val}$, where P^{core} is
the polarization energy arising from the "core" (the paired)
electrons of the solid and P^{val} the polarization energy arising
from the valence (i.e. unpaired) electrons of the TCNQ=ions.
We then define $U_o = I - A - P^{core} - P_-^{core}$
$$U = U_o - P_+^{val} - P_-^{val}$$
In a half-filled band, with an electron at every site, the
contribution of P^{val} will not be excessively large. If U_o is
finite to start with, it takes a lot of energy to shift an
electron from one site to another, polarization is limited,
P^{val} is small.

A different situation obtains, however, when we consider a
partially filled band. Now, there are sites not containing
electrons even in the ground state. The hole and electron
created by the U-type excitation can really be surrounded by
electrons or holes, respectively, since a considerable rearrangement is possible without having to apply the energy U.
The polarization (like solvation in electrochemistry) is now
considerable, P^{val} will be large and the resulting U will be
much smaller than in the case of the half-filled band. This

appears to be the mechanism by which a large number of
partially charge transferred solids have such a high conductivity, as is by now well-known (10)

In this context it is worthwhile noting the work of
Sawatzky et al ($_{11}$), who have   succeeded in measuring U of the
transition metals by an Auger - ESCA technique, and for instance
shown that polarization changes U from 23 eV in Zn-atom, to 11
eV   in ZnO to 9.5 eV   in Zn-metal.

7. Conclusions

It was our purpose to show what contributions Electron
Spin Resonance as performed by our group in Groningen can make
to the understanding of the simple alkali TCNQ salts. It was
shown that in RbTCNQ the Hubbard U $\approx$ 1.1 eV   , the transfer
integrals t' and t'' are 0.35 eV and 0.10 eV, respectively and
that the anisotropy of triplet exciton motion is $\nu_1/\nu_t \approx 10^5$.

Acknowledgements

The work reviewed here is a summary of a considerable
effort over by Drs.Vegter, Hibma, and Kuindersma for which I am
very grateful.
I am indebted to Mr.Huizinga for commenting on this manuscript.
The cooperation of Dr.G.A.Sawatzky in all of this work has been
essential and I'm happy to be able to acknowledge it here.

References

1.a. J.G.Vegter, T.Hibma and J.Kommandeur
     Chem.Phys.Lett.3, 427 (1969)
  b. J.G.Vegter and J.Kommandeur
     Mol.Cryst.Liq.Cryst. 30, 11 (1975)

2.   S.Huizinga, J.Kommandeur, G.A.Sawatzky and B.T.Thole
     Phys.Rev.B 19, 4723 (1979)

3.   See f.i. Z.G.Soos and R.C.Hughes,
     J.Chem.Phys. 46, 253 (1967)

4.   T.Hibma, G.A.Sawatzky and J.Kommandeur
     Phys.Rev.B 15, 3959 (1977)

5.   A.Hoekstra, T.Spoelder and A.Vos, Acta Crystallogr.
     B 28, 14 (1972)

6.   T.Hibma, G.A.Sawatzky and J.Kommandeur
     Chem.Phys.Lett. 23, 21 (1973)

7.   T.Hibma and J.Kommandeur, Phys.Rev.B 12, 2608 (1975)

8.   W.G.Clark, J.Hammann, J.Samry and L.C.Tippie
     Lecture Notes in Physics 96, Quasi-one dimensional
     conductors II, Proceedings Dubrovnik 1978 ed.Barisic
     et al, p.255, Springer-Verlag 1979

9.   J.Kommandeur, Low-Dimensional Cooperative Phenomena
     Ed. H.J.Keller New York, p.65, (1975)

10. See f.i. J.B.Torrance in Molecular Metals, ed. by
     W.E.Hatfield, Plenum Press 1979, p.7.

11.a. E.Antonides and G.A.Sawatzky
      J.Phys.C. 9, L547 (1976)
   b. E.Antonides, E.C.Janse and G.A.Sawatzky
      Phys.Rev.B 15, 1669 (1977)

PHONON SCATTERING IN QUASI 1-d CONDUCTORS: TTF-TCNQ

Esther M. Conwell

Xerox Webster Research Center, Webster, N.Y., 14580, USA

In the first part of this paper I derive expressions for conductivity σ and thermopower Q arising from 1- and 2-phonon
scattering in a model quasi 1-d conductor. It is demonstrated
that the usually neglected term in Q, arising from the energy-
dependence of the scattering time, is as large as the "band" term.
The theory can account for the experimental data for TTF-TCNQ
in the range 150-300K with reasonable values of the parameters.
It is shown that the 2-libron theory of transport is in disagreement with the data on several important points.

1.   INTRODUCTION: THE MODEL

The model to be used for the calculation of the transport coefficients is the tightbinding band model with single-particle scattering by the phonons. It neglects electron-electron interactions. (U $<<$ t). This model should be applicable from 300K
down to 150 or perhaps 100K. Below $\sim$ 100K the data of Andrieux
et al (1), showing a dip in σ as commensurability is approached
(2), may indicate that collective transport makes a significant
contribution. It is also possible that phonon drag is important
for some range of temperature below 150K. (3)

Since the lattice vibration spectrum is quite complicated in
the organic solids we consider, I shall initially make a number
of simplifying assumptions in order to gain some physical insight. The first assumption is that there is one molecule per
unit cell, the molecules being planar and stacked as shown in
Fig. 1. The lattice vibrations are conveniently divided into
external and internal modes. The latter are based on the

L. Alcácer (ed.), The Physics and Chemistry of Low-Dimensional Solids, 213–222.

molecular vibrations and there are 3m - 6 of them, m being the
number of atoms in a molecule. The remaining 6 modes arise from
the 3 translational and 3 rotational degrees of freedom. In a
lattice of high enough symmetry (4) the translational and rotational modes do not mix at the center of the Brillouin zone (q=0)
and perhaps at other high symmetry points. In the theoretical
development I shall assume that there are 6 distinct branches, 3
based on translations and 3 on rotations, that do not mix anywhere
in the zone. The 3 rotations, or librations, are assumed to take
place about the 3 inertial axes, indicated on the top molecule in
Fig. 1, although this is not in general true. (4) Their quanta
will be called librons, while those of the acoustic modes will be
called translons. In the quasi 1-d case the important acoustic
modes for scattering the electrons are those that propagate along
the stacking axis. These consist of longitudinal (LA) modes
with displacements along the b axis and two transverse (TA) modes
with displacements that we take to be in the directions indicated
c* and a* in Fig. 1. Neutron scattering (5),(6) indicates that
the TA vibrations are primarily in these directions in TTF-TCNQ.

We proceed now to calculate the scattering of electrons by the
phonons just described, and later to apply the results to TTF-TCNQ.

2. THE MATRIX ELEMENTS FOR SCATTERING

The Hamiltonian that we use to describe the 1-d system of electrons and phonons is (7)

$$H=\sum_{j}\mathcal{E}_j a_j^+ a_j + \sum_{n,q} (b_{q,n}^+ b_{q,n}+1/2)\hbar\omega_{q,n} + \sum_{j,i=j\pm 1}(t_{ij}a_i^+ a_j+h.c.)\quad(1)$$

The first term, to be summed over all N molecules in the chain,
gives the energy of electrons on the j'th molecule. The second
term gives the energy of the phonons, the summations being over
the n branches of the phonon spectrum and all values of the wave
vector. The coupling of the electrons to the external modes
arises from the third term. The transfer integral t_{ij} is a
function of the spacing and relative orientation of the neighboring molecules i and j, which are affected by the external
vibrations. To obtain the coupling constants, we: (1) expand
t_{ij} as a Taylor series in the displacement u of a molecule from
its equilibrium position; (2) insert this into eq. (1); and (3)
make the usual normal mode expansions of u in terms of the b's
and a_j^+ in terms of the a_k^+, the creation operator for an electron
with wave vector k. We then obtain from the linear terms in the
expansion of t_{ij} the perturbing Hamiltonian for a one-phonon,
(1-p), scattering process,

$$H'_{1-p,n} = N^{-1/2} \sum_{q,k} g_{1-p,n} a^+_{k+q} a_k (b_{q,n} + b^+_{-q,n}) . \qquad (2)$$

For elastic scattering by one translon, we obtain (7),(8)

$$g_{1-t} = \pm 4i \left(\partial t / \partial u_n \right)_o (\hbar/2M\omega_{2k_F,n})^{1/2} \sin k_F b \qquad (3)$$

where M is the mass of a molecule, k_F the Fermi wave vector and b the lattice constant in the stacking axis direction. Similarly, for 1-libron scattering we obtain

$$g_{1-\ell} = 4i(\partial t / \partial \theta_n)_o (\hbar/2I\omega_{2k_F,n})^{1/2} \sin k_F b, \qquad (4)$$

where θ_n represents the angular displacement about the axis appropriate to the n'th librational mode and I the moment of inertia about this axis. For the internal vibrations g in eq. (2) may be written (7)

$$g_{1-i} = g_n \hbar \omega_n, \qquad (5)$$

where g_n is the dimensionless. coupling constant defined and calculated in reference 9.

We see from eqs.(3) and (4) that a vibration for which $(\partial t/\partial u_n)$ or $(\partial t/\partial \theta_n)$ vanishes does not scatter electrons to first order. Examples of such vibrations are the TA(a*) mode and the and η, ξ librations.(8)

By the same procedure as described above for the linear term in the t_{ij} expansion, we obtain from the quadratic term the perturbing Hamiltonian for 2-p processes (10):

$$H'_{2-p,n} = N^{-1} \sum_{k,q,q'} g_{2-p,n} a^+_{k+q+q'} a_k (b_{q,n} + b^+_{-q,n})(b_{q',n'} + b^+_{-q',n'}) \qquad (6)$$

For the translons

$$g_{2-t} = (\partial^2 t/\partial u_n \partial u_{n'})(\hbar/2M)(\omega_{q,n}\omega_{q',n'})^{-1/2} f(k,q,q') \qquad (7)$$

where, for elastic scattering $(q'=-q-2k_F)$

$$f(k,q,q')=2(\cos k_F b - \cos(k_F+q)b). \qquad (8)$$

A similar expression holds for the 2-ℓ case with $u_n u_{n'}$ replaced by $\theta_n \theta_{n'}$ and M by I. Two-phonon scattering is expected to occur for any mode when its amplitude is large enough.

For the calculation of the scattering time due to any of the processes above the quantity required is the absolute square of the matrix element of H' for that process between the initial and final states. For the 1-p processes this is

$$\left|\langle k'|H'_{1-p,\eta}|k\rangle\right|^2 = N^{-1}|g_{1-p,\eta}|^2 \delta(k'-k_\mp q) \begin{Bmatrix} n_q & \text{absorption} \\ n_{q+1} & \text{emission} \end{Bmatrix} \quad (9)$$

while for the 2-p processes

$$\left|\langle k'|H_{2-p,n}|k\rangle\right|^2 = N^{-2}|g_{2-p,n}|^2 \delta(k'-k_\mp q \mp q') \begin{Bmatrix} n_q \\ n_q+1 \end{Bmatrix} \begin{Bmatrix} n_{q'}' \\ n_{q'}'+1 \end{Bmatrix} \quad (10)$$

In (10) one factor is chosen from each pair of braces according to whether the corresponding phonon is absorbed or emitted. The signs in the argument of the δ function must be chosen accordingly to satisfy conservation of crystal momentum.

3. THE TRANSPORT COEFFICIENTS

On the evidence of the measured mobilities, scattering rates are low enough (over at least most of the temperature range for the materials of interest) so that first order perturbation theory can be used to find the scattering probability/unit time. It is then found that for the 1-p processes, as described above, a relaxation time τ exists (11) and the Boltzmann equation can be solved very simply. The resulting conductivity is given by

$$\sigma = -\int_0^{\mathcal{E}_0} \sigma(\mathcal{E})(\partial f_0/\partial \mathcal{E}) d\mathcal{E}, \quad (11)$$

where $\mathcal{E}_0$ is the bandwidth, equal to 4t, f_0 the Fermi-Dirac distribution, and

$$\sigma(\mathcal{E}) = \eta e^2 v^2 \tau \rho(\mathcal{E}), \quad (12)$$

η being the number of chains/cm^2, e the charge on the electron, v the electron speed ($=(b/\hbar)[\mathcal{E}(\mathcal{E}_0-\mathcal{E})]^{1/2}$) and $\rho(\mathcal{E})$ the number of states/unit energy range ($=(2N/\pi)[\mathcal{E}(\mathcal{E}_0-\mathcal{E})]^{-1/2}$). It is also found that the thermopower for a single chain is given by

$$Q = -(1/eT\sigma) \int_0^{\mathcal{E}_0} (\mathcal{E}-\mathcal{E}_F)\sigma(\mathcal{E})(\partial f_0/\partial \mathcal{E}) d\mathcal{E}, \quad (13)$$

where $\mathcal{E}_F$ is the Fermi energy. For two chains $Q=(\sigma_1 Q_1 + \sigma_2 Q_2/(\sigma_1+\sigma_2)$

It is informative to evaluate (11) and (13) for single scattering processes. This is particularly easy to do when the scattering is elastic, as is the case for acoustic and libron modes when $k_B T \gg \hbar\omega(2k_F)$. For the 1-t and 1-$\ell$ processes, when equipartition is valid for the phonons, (12)

$$\sigma \propto \omega_{2k_F}^2 \mathcal{E}_0^2/T(\partial t/\partial u)^2.$$

For the 2-p processes, under equipartition, (12)

$$\sigma \propto \omega^4 \mathcal{E}_0^2 / T^2 .$$

From eq. (13) we derive, just as in the 3-d case,

$$Q = \frac{\pi^2}{3} \frac{k_B^2 T}{e} \left[\frac{1}{v^2 \rho} \frac{d}{d\mathcal{E}} (v^2 \rho) + \frac{1}{\ell} \frac{d\ell}{d\mathcal{E}} \right]_{\mathcal{E}_F} \quad (14)$$

For the 1-t case, if we allow for the dispersion of the acoustic modes, $\tau \propto [\mathcal{E}(\mathcal{E}_0 - \mathcal{E})]^{1/2}$. This is precisely the same $\mathcal{E}$-dependence as shown by the first term (the "band" term) in eq. (14) and the two terms therefore make equal contributions to Q. The result is

$$Q_{1-t} = \frac{\pi^2}{3} \frac{k_B}{e} \frac{k_B T}{\mathcal{E}_0} \frac{1 - 2 \mathcal{E}_F / \mathcal{E}_0}{(\mathcal{E}_F / \mathcal{E}_0)(1 - \mathcal{E}_F / \mathcal{E}_0)} \quad (15)$$

It is easily seen that this is still correct when more than one acoustic phonon is scattering. In the 1-ℓ case, if we neglect the dispersion of the libron branches since they are optical, $\tau \propto [\mathcal{E}(\mathcal{E}_0 - \mathcal{E})]^{1/2}$, which results in $Q_{1-\ell} = 0$.

4. APPLICATION TO TTF-TCNQ

With 4 molecules per unit cell in TTF-TCNQ there are 24 external mode branches rather than the 6 considered in the simplified model. At q=0 these break down into 3 acoustic branches and 21 optical branches. From the lattice structure of TTF-TCNQ it is not likely that any of the latter have the character of η or ξ rotations, although the ζ rotation might survive because the ζ axis corresponds to a symmetry direction of the crystal. Away from q=0 mixing of many of the branches, including translations and rotations, can take place. In particular neutron scattering data show a mixing of TA(a*) and TA(c*). (6) One consequence of the lattice vibrations being more complicated than the model we have been considering is the blurring of the distinctions between branches so far as the possibility of 1-p or 2-p processes is concerned. Thus 1-p scattering should actually be possible from most, if not all, branches. It is reasonable to expect, however, that the proportion of 1-p to 2-p scattering is not much different from the predictions of our earlier simplified model.

For numerical calculations, for the LA branch, since it does not obviously mix with any other and its dispersion appears normal, (8) I chose $(\partial t / \partial u_b) = - 0.2$ eV/Å, in reasonable agreement with theoretical calculations (13),(14) and the measured

shift of the plasma edge under pressure.(15) The 1-p scattering
arising from all other branches was lumped together with that
from TA(c*) and the quantity $(\partial t /\partial u_{_{"c*"}})$ treated as a
parameter in the calculations. The phonon energy $\hbar\omega_{2k_F}$ was taken
as 85K for the LA branch, 57K for the "TA(c*)"
branch. (5,6) Phonon softening was not taken into account in
the calculations since they are probably not valid anyway below
$\sim$100K. For the internal modes the frequencies and coupling
constants were those from reference 9, also used in reference 11.

The bandwidth of TCNQ was taken as 0.5 eV, or 6000K, while the
TTF bandwidth was treated as a parameter. Since the volume-
dependences of the various quantities entering into the transport coefficients are not well known, the calculations were done
for constant volume.

The results of two different sets of numerical calculations for
the resistivity at constant volume, ρ_v , vs T are shown in
Figs. 2A and 2B. In the calculation of Fig. 2A only 1-p processes are included, while in that of Fig. 2B 2-p external modes
were included as well. As is seen, in both cases the major
contribution is made by the external mode 1-p processes. The
300K value for "intrinsic" conductivity was taken to be 800-900
ohm^{-1} cm^{-1}. To fit this for case A, under the assumption that
$\partial t/\partial u$ values are the same for TTF as for TCNQ, requires $(\partial t/\partial u_{_{"c*"}})=$-
0.24 eV/A, a reasonable value. If 2-p scattering is allowed,
it is apparent that the value of $(\partial t/\partial u_{_{"c*"}})$ required to match
σ at 300K will be smaller. For the amount (chosen arbitrarily)
of 2-p scattering included in case B the value of $(\partial t/\partial u_{_{"c*"}})$
was-0.20 eV/A. In both the cases of Fig. 2A and Fig. 2B ρ_v varies
approximately linearly with temperature, in agreement with the
experimental data corrected to constant volume.(16),(17) In the
case of Fig.2A, on the average $\rho \propto T^{1.23}$, while in the case of
Fig. 2B $\rho \propto T^{1.43}$. The average exponent found by Friend et al(16)is
1.29. They have under estimated the steepness, however, since
they assumed that ρ was affected only by the b axis length changes,
whereas it is almost as much affected by changes in the transverse
dimensions. (18),(17) Thus the case including 2-p scattering
is in better agreement with ρ_v vs T. It is also in better agreement with the changes of ρ with pressure.(19).

In fitting ρ, the value of ε_o(TTF) is not important so long
as σ_{TTF} is considerably smaller than σ_{TCNQ}, a condition required by the fact that Hall mobility (20) and thermopower (21)
have values close to what is expected for the TCNQ chain alone.
When $(\partial t/\partial u)$'s are assumed the same for the two chains, this
condition must be satisfied by choosing ε_o (TTF) smaller than
that of TCNQ. In that case, to obtain the correct 300K thermo-power,
-28 μ V/K(21), ε_o (TTF) must be chosen as 3000K. The ratio
(σ_Q/σ_F) is found to be 4.3. It is clear that σ and Q can

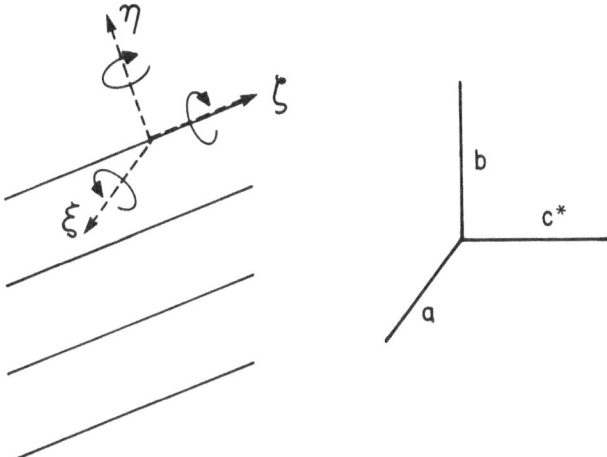

Fig. 1: The chain of molecules, showing inertial axes, and the coordinate system giving polarization directions of the acoustic vibrations.

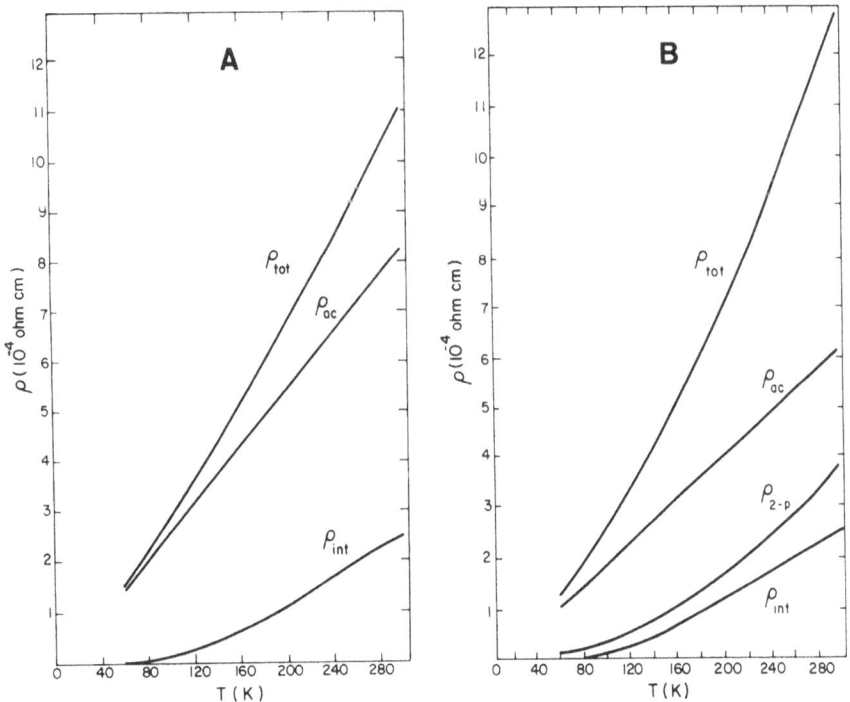

Fig. 2: Calculated ρ_v vs T with (A) including only 1-p processes and (B) including 2-p processes as well.

be fit equally well with larger $\mathcal{E}_o$ (TTF) provided the values
of $(\partial t/\partial u)$ are increased correspondingly. For example, to
fit Q with $\mathcal{E}_o$ =4500 the $(\partial t/\partial u)$ values for TTF must be increased about 50%. According to theoretical estimates (13),(15)
$(\partial t/\partial u)$ for TTF is smaller, rather than larger, than
$(\partial t/\partial u)$ for TCNQ, lending support to the smaller values of
$\mathcal{E}_o$ (TTF).

Whatever the $\mathcal{E}_o$ value the calculated Q is found to vary linearly
with T, in fairly good agreement with experiment down to $\sim$ 100K.
This reflects the fact that Q should be much less sensitive to
volume changes than $\mathcal{G}$. Most of the volume-dependent factors
in $\mathcal{G}$ are common to $\mathcal{G}_Q$ and $\mathcal{G}_F$, and thus their effect will
cancel out in Q. Evidence for this is the finding that at 300K
under 20 kb of pressure Q is reduced by about 70%, whereas $\mathcal{G}$
increases by a factor 7 or 8.(22).

I conclude that the behavior of $\mathcal{G}$ and Q can be accounted for
by phonon scattering, at least in the range $150 < T < 300K$, with
1-p processes more important than 2-p, for reasonable values
of the parameters.

5. COMPARISON WITH 2-LIBRON THEORY

In contrast to the above, proponents of the 2-λ theory claim
that the dominant scattering from 300K down to 60K is by 2-p
processes . (3) They postulate that 1-p scattering is ineffective
because the interaction between electrons and $2k_F$ phonons is
so strong that momentum lost by electrons to these phonons is
returned to the electrons before it can be dissipated in p-p
scattering. Since p-p scattering increases rapidly with T, it
is questionable that it can be so ineffective, particularly above
$\sim$ 150-200K. To counter this objection, WGK contend that the
2-p processes are intrinsically more effective at the higher
temperatures. They estimate that at 300K the scattering frequency
for a 1-p process is 3×10^{14}/sec while that for a 2-p process
is 3×10^{15}/sec. The scattering times, which are the reciprocals
of these frequencies, are then 3×10^{-15} sec and 3×10^{-16} sec, respectively. Insertion of the latter time into the Uncertainty
Principle gives an uncertainty in energy for the 2-p scattering
of 2 eV, several times the bandwidth. The 1-p scattering time,
significantly, is in good agreement with what we have calculated
for 1-p processes and thus with the scattering time deduced from
the measured

Since $\mathcal{G}_{2-p} \propto T^2/\omega^4$, to account for ρ_v being quasilinear
in T, WGK postulate that ω increases strongly with T at constant
volume, 20% in the range $60 < T < 300K$. (3) The change in this
same temperature range due to the change in volume is estimated
as -25%. The net change in ω as T goes from 60 to 300K is

then -5%. (3) This is in disagreement with the observation from
neutron scattering that ω decreases by -17% in this temperature
range for the TA(a*) mode(6), which shows no softening. Further,
it is in disagreement with the behavior of anthracene, naphthalene
and pyrene, for which the change in ω at constant volume is
much less than that at constant pressure. (23) To counter the
latter argument WGK, agreeing that TTF-TCNQ should be compared
with these organics, claim that the comparison should be made
at 10-12 kb of pressure, where the compressibility of TTF-TCNQ
matches that of anthracene and naphthalene. This argument is
totally unconvincing. The measured ratio of expansivity to
compressibility for TTF-TCNQ equals tht for pyrene <u>at ambient</u>
<u>pressure</u>. (23) Thus, it would be even more appropriate to
compare TTF-TCNQ with pyrene at ambient, which, as noted before,
shows a much smaller change in ω with T at constant volume than
is required to account for the quasilinear f_v.

Another serious difficulty for the 2-ℓ theory comes from the
measured 300K P-dependence of the Hall constant, R, (20) and
Q. (23) According to WGK, the observed 300K P-dependence of
σ, $\sim 25\%$/kb, is accounted for by the P-dependence of ω^4,
since ω increases by 5 to 6%/kb for low frequency phonons in
organic crystals. (23) This gives no way of explaining R and
Q growing less negative with P, indicating a decrease in the
ratio σ_Q/σ_F. The objection that was originally posed to
this explanation(20) was that the plasma frequency is only weakly
dependent on P. If σ_Q is several times σ_F, however, the
motion of the plasma edge at low pressures reflects mainly the
change in $\mathcal{E}_0$ of TCNQ. Since, as indicated earlier, there is
no reason to believe $\partial t/\partial$ u larger for TTF than TCNQ, the
relatively large changes in σ_Q/σ_F required to explain the
P-variation of R and Q point to a small value of $\mathcal{E}_0$ (TTF), $\sim$
$(1/2)\,\mathcal{E}_0$(TCNQ). The same value of ($\partial t/\partial$ u) on both chains
would then make the percent increase in σ_F at a given pressure
twice that in σ_Q.

REFERENCES

(1)   Andrieux, A., Schulz, H.J., and Jerome, D.: 1979, Phys.
         Rev. Lett. 43, pp. 227-229.
(2)   Conwell, E.M. : 1979, submitted for publication.
(3)   Gutfreund, H., Kaveh, M. and Weger, M. : 1979, "Quasi
         One-Dimensional Conductors I," Proc. Dubrovnik, 1978
         (Springer-Verlag, Berlin) pp. 105-129. This also gives
         earlier references. These authors will be referred
         to as WGK.
(4)   Kitaigorodsky, A. I. : 1973, "Molecular Crystals and
         Molecules" (Academic Press, New York)
(5)   Shirane, G., Shapiro, S.M., Comes, R., Garito, A.F.
         and Heeger, A.J. : 1976. Phys. Rev. B14, pp.2325-2334.

(6)   Shapiro, S.M., Shirane. G., Garito, A.F. and Heeger,
         A.J. : 1977, Phys. Rev. B15, pp. 2413-2415.
(7)   See, for example, Rice, M.J., Duke, C.B. and Lipari,
         N.O. : 1975, Solid State Comm. 17, pp. 1089-1092.
(8)   Conwell,E.M. : 1979,"Quasi One-Dimensional Conductors
         I", Proc. Dubrovnik, 1978 (Springer-Verlag, Berlin)
         pp. 270-278.
(9)   Lipari, N.O., Rice, M.J., Duke, C.B., Bozio, R., Girlando,
         A. and Pecile, C. : 1977, Inter. Jour. of Quantum Chem:
         Quantum Chemistry Symposium 11, pp. 583-594.
(10)   Gutfreund, H. and Weger, M. : 1977, Phys. Rev. B 16,
         pp. 1753-1755.
(11)   Conwell, E.M. : 1977, Phys. Rev. Lett. 39, pp. 777-
         780.
(12)   Detailed results are given in a manuscript in prepara-
         tion.
(13)   Berlinsky, A.J., Carolan, J.F. and Weiler, L. : 1974,
         Solid State Comm. 15, pp. 795-801.
(14)   Herman, F., Salahub, D.R. and Messmer, R.P. : 1977,
         Phys. Rev. B16, pp. 2453-2465.
(15)   Welber, B., Seiden, P.E. and Grant, P.M. : 1978, Phys.
         Rev. B18, pp. 2692-2700.
(16)   Friend, R.H., Miljak, M., Jerome, D., Decker, D.L.
         and Debray, D. : 1978, J. Phys. Lett. 39, pp L-134-
         L-138.
(17)   Cooper, J.R. : 1979, Phys. Rev. B19, pp. 2404-2408.
(18)   Bouffard, S. and Zuppiroli, L. : 1978, Solid State
         Comm. 28. pp. 113-117.
(19)   Conwell, E.M. : 1979, Phys. Rev. B19, pp. 2409-2410.
(20)   Cooper, J.R., Miljak, M., Delplanque, G., Jerome, D.,
         Weger, M., Fabre, J.M. and Giral L. : 1977, Jour. de
         Phys. 38, pp. 1097-1103.
(21)   Chaikin, P.M., Kwak, J.F., Jones, T.E., Garito, A.F.
         and Heeger, A.J. : 1973, Phys. Rev. Lett. 31, pp. 601-
         604.
(22)   Weyl, C., Jerome D. and Chaikin, D.M. : 1979, ASI,
         Tomar, Portugal.
(23)   Zallen, R. and Conwell, E.M. : 1979, Solid State Comm.
         31, pp. 557-561.

ELECTRONIC PROPERTIES AND NEW FORMS OF INSTABILITIES
IN TCNQ-SALTS WITH INTERMEDIATE CONDUCTIVITY

Jean-Pierre FARGES

Laboratoire de Biophysique, U.E.R.D.M.,
Université de Nice-Valrose, 06034 NICE Cedex, FRANCE

I. ELECTRONIC PROPERTIES OF TEA(TCNQ)$_2$

$TEA^+(TCNQ)_2^-$ is one of the first $TCNQ$ salts synthesized in the 1960's (1,2,3,4) (TEA^+ = triethylammonium$^+$). It is a good semi-conductor obtainable in the form of superb parallelepipedic black crystals, as large as 10x4x0.5 mm^3, offering a special opportunity for investigating the electrical anisotropy. Extensive work has been devoted to this salt, part of it by our group, readily promoting the salt to the rank of a model compound.

I.1 X-ray data

The crystal structure of TEA(TCNQ)$_2$ is shown on Figure 1 (5). It consists of parallel and well separated TCNQ columns forming TCNQ layers which alternate with layers of TEA cations. Such a molecular segregation into columns and layers accounts for the pronounced tri-dimensional anisotropy of the electron system in this material.

The TCNQ columns are all equivalent, consisting of TCNQ dimers with an intradimer spacing of 3.22 Å. There are two alternating interdimer spacings of 3.32 and 3.34 Å. There are also three slightly different modes of overlap between the TCNQ molecules, one intradimer and two interdimer. The unit cell, which has a centre of symmetry (space group P1̄), contains two dimers of the same TCNQ column and two TEA cations. As a result of the 1:2 stoichiometry and of a complete donor-acceptor charge transfer, there is formally one electron per dimer, thus producing the dimer-ion (TCNQ)$_2^-$ which is basically associated with conduction and magnetism in the solid state. X-rays do not show evidence for a localization of the

L. Alcácer (ed.), The Physics and Chemistry of Low Dimensional Solids, 223–232.

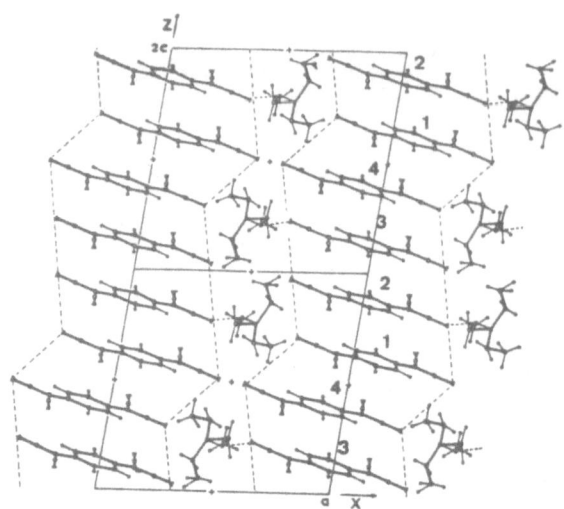

Figure 1. X-ray structure of TEA(TCNQ)$_2$, after (5).
Note the tetradic superstructure.

transferred electron preferentially on one molecule of the dimer
(5,6). This contradicts a previous assertion (7). The organization
of the TCNQ molecules into tetrad units along the columns is made
clear on Figure 1. It can be noted that tetrads are regularly
staggered with respect to each other in the column.

All the data considered below will be referred to the following orthogonal frame. Axis 1 is parallel to the TCNQ columns and
coincides with the crystallographic Z-axis and with the long axis
of crystals. Axis 2 is in the TCNQ layers, perpendicular to the
columns, and axis 3 is perpendicular to the alternate layers of
TCNQ and TEA ions.

I.2 DC conductivity

The temperature dependence of the three components, σ_1, σ_2
and σ_3, of the DC conductivity is shown on Figure 2 (8,9,10,11,12).
Similar results have been obtained by other groups (2,3,4,13,14).
The material exhibits a good-intermediate-conductivity, together
with a remarkable anisotropy. The largest component is σ_1 measured,
as expected, along the TCNQ columns. At room temperature:
$$\sigma_1 = 7.4 \ (\Omega \text{xcm})^{-1} \text{ and } \sigma_1/\sigma_2 = 164 \ , \ \sigma_1/\sigma_3 = 2850$$

Thus, TEA(TCNQ)$_2$ is almost a 1D semiconductor, with an essentially isotropic activation energy obtainable from the low temperature data: $E_a = 0.12$ eV (12). However, deviations towards higher
conduction at higher temperatures are observed. This is especially

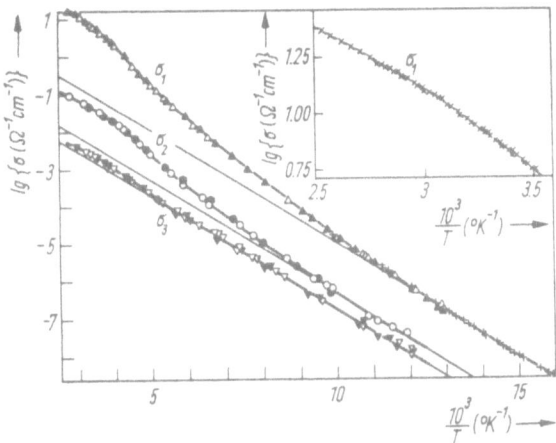

Figure 2. DC conductivity of TEA(TCNQ)$_2$. The slope of
the straight lines is E_a/k with E_a = 0.12 eV.

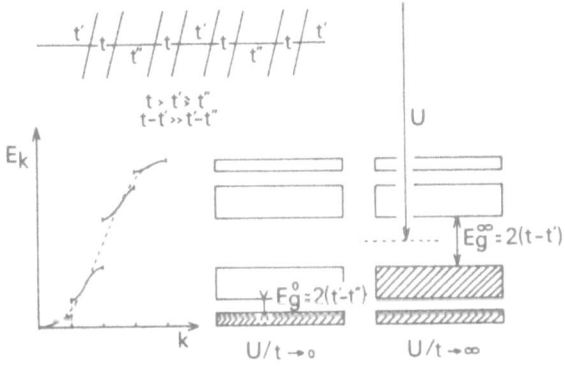

Figure 3. Schematic band structure for TEA(TCNQ)$_2$.

true for the largest component σ_1 (see Figure 2). These deviations
have been assigned to a temperature dependence in the mobility (12).
However, another possibility will also be discussed in the second
part of this lecture.

The measured activation energy for conduction, E_a = 0.12 eV,
is considerably larger than the characteristic energy associated
with magnetic excitations, $E_m \simeq 0.04$ eV (15). This is one argument
in favor of the existence of strong electron-electron interactions
in this material (12,16). However, as conduction does not require
highly energetic TCNQ$^=$ states, the conduction gap is not a HUBBARD
gap ($\simeq U$) but a much smaller-band-gap, $E_g = 2E_a$, resulting from the
underlying tetradic modulation of the TCNQ columns (Figure 3). As
for other TCNQ salts, U is evaluated to $\simeq$ 1 eV, in particular from
CT bands in crystalline optical spectra (16). The X-ray data imply

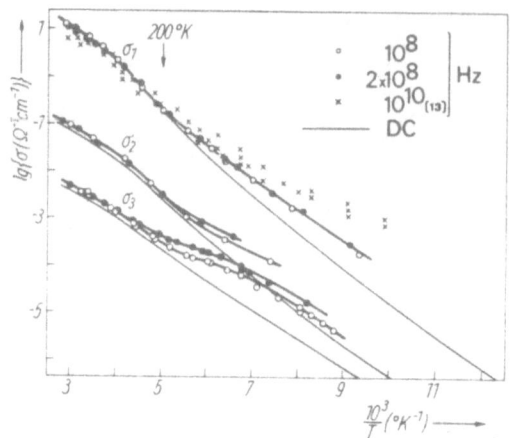

Figure 4. AC conductivity of $TEA(TCNQ)_2$.

three transfer integrals, t, t' and t", in the columns, such that
$t > t' \gtrsim t"$ and $t - t' >> t' - t"$ (Figure 3 (12)). In the strong coupling
limit, $U/t \to \infty$, thus, $E_g \simeq 2(t-t')$ and (17) $E_m \simeq 2t'^2/U$. In addition,
the small difference between t' and t" is able to account for the
EPR fine structure observed in $TEA(TCNQ)_2$(18).

I.3 AC conductivity and dielectric constant

The temperature dependence of the three components of the AC
conductivity at 10^8 and 2×10^8 Hz is shown on Figure 4 (19). Radio-
frequencies are shown to produce significant deviations towards
higher conduction below 200 °K. These effects may be attributed to
some kind of disorder in the material (such as cation disorientation).
They could also be an early manifestation of the coupling of the
conduction electrons to an intermolecular phonon mode of particularly low frequency (see below, I.5). Evaluations of the dielectric
constant at 77 °K gave:

$\varepsilon_1 \simeq 70 \pm 20$ and $\varepsilon_3 \simeq 6.2 \pm 0.7$

whereas SCHEGOLEV reported $\varepsilon_1 = 35$ at 10^{10} Hz and 4.2 °K (13).

I.4 HALL effect and thermopower

The HALL effect has been observed at room temperature in
$TEA(TCNQ)_2$, Figure 5 (20,21). In this experiment (repeated on
several crystals) the current I ($\simeq$ 1 mA) was flowing parallel to
the TCNQ columns (axis 1) whereas the HALL voltage was detected
perpendicular to the columns in the TCNQ layers (axis 2). The
magnetic field B ($\simeq$ 8 KG) was applied parallel to axis 3.

As a result of the strong anisotropy, the shorting effect of
the current electrodes proved to be considerable. After appropriate
correction, the HALL mobility has been evaluated to be
$\mu_H \simeq 8.7$ cm^2/V.s (22).

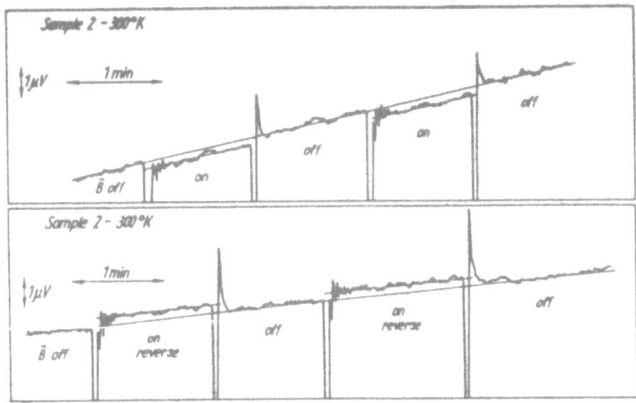

Figure 5. HALL effect in TEA(TCNQ)$_2$.

Unexpectedly, the sign of the HALL constant R_H was positive, opposite to the sign known at this time for the thermopower S from measurements parallel to the TCNQ columns (2,4). In fact, a simple formulation for anisotropic intrinsic semiconductors shows that the two results $R_H > 0$ and $S_1 < 0$ are reconcilable only if the transverse thermopower is positive, thereby implying a remarkable sign anomaly of the thermopower (22). The existence of such an anomaly has been firmly established by subsequent experimental work.

The three components S_1, S_2 and S_3 of the thermopower in TEA(TCNQ)$_2$ are plotted versus temperature on Figure 6 (23,21). These data also indicate a considerable anisotropy. In particular,

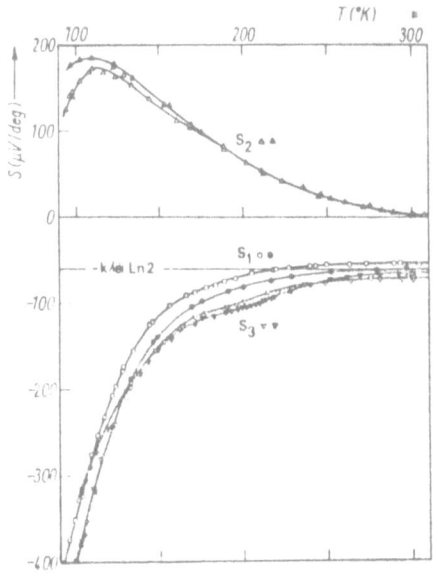

Figure 6.
Thermopower of TEA(TCNQ)$_2$.

we remark that S_1 and S_2 have opposite signs, and no inversion, over the entire temperature range investigated, from 100 to 300 °K. Curiously, the behaviour of S_3 is very close to S_1. S_3 exhibits a small anomaly near 200 °K. Room temperature values are:

$$S_1 = -59 \ \mu V/°K, \quad S_2 = +1.5 \ \mu V/°K \quad \text{and} \quad S_3 = -69 \ \mu V/°K.$$

The saturation of the longitudinal thermopower S_1 near to −60 $\mu V/°K$ (Figure 6) is just as would be expected from strong electron-electron interactions. It has been shown that, in the strong coupling limit of the HUBBARD model ($U/t \rightarrow \infty$), the high temperature thermopower is entirely due to the $kLn2$ spin entropy of the carriers (24). For a system such as TEA(TCNQ)$_2$ consisting of N/2 carriers over N (TCNQ) sites, the result is $S = -(k/|e|)Ln2 = -59.8 \ \mu V/°K$ (24,25). However, this saturation value ought to be independent of the direction in the crystal. The very different behaviour observed for S_2 might be therefore a major objection to the reliability of the considered model, at least in this specific case.

I.5 Polarized optical reflection spectra

Reflection spectra for TEA(TCNQ)$_2$ are shown on Figure 7 (26, 12). They make it obvious that a considerable anisotropy also exists at optical frequencies. For the two polarizations perpendicular to the TCNQ columns (R_2- and R_3-spectra), the reflection is very low and nearly structureless. In contrast, when the polarization is parallel to the columns (R_1-spectrum), a broad band corresponding to absorption by the conduction electrons is observed, commencing at an energy of ≃ 0.2 eV which is comparable to twice the activation energy E_a for DC conduction. In addition, co-existing with this electronic band in the R_1-spectrum, is a remarkable series of strong and sharp infrared bands quite identifyable as molecular vibrations of the TCNQ molecules (26,27,12).

However, the polarization of this vibrational spectrum, exclusively in the direction of the columns, is highly surprising since most of the molecular vibrations are polarized in the plane of the TCNQ molecules and hence almost perpendicular to the TCNQ columns. Even more confusing is the fact that not only is the polarization of these bands abnormal, but also their optical activity. In fact, a detailed analysis indicates that the frequencies of these bands are closely conform to the ten totally symmetric A_g modes of the TCNQ molecule (27) which are RAMAN active modes but formally non-active in the infrared. No theory was appropriate to help clarify these puzzling results, however the polarization of the vibrational bands exclusively parallel to the highly conducting axis was soon recognized as indicating strong interactions between the 1D electron system and the intramolecular TCNQ vibrations (26,27,12).

A decisive step towards the understanding of these phenomena has been made recently by RICE et al. (29,31) and by RICE (30,32).

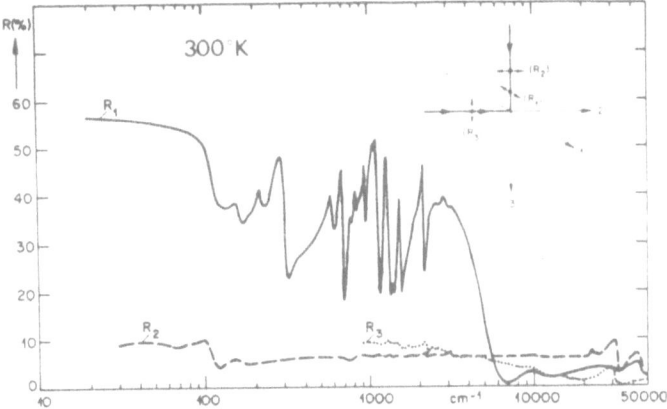

Figure 7. Polarized reflection spectra of TEA(TCNQ)$_2$.

The molecular orbital of the conduction electron of TCNQ is b_{2g} and, for symmetry reasons, linear coupling, with constants g_n, is only possible to the ten, n= 1 to 10, A_g vibrational modes of the molecule (33). As a result of this electron-phonon (e-p) coupling, in TEA(TCNQ)$_2$, the A_g modes are renormalized as collective or "phase phonon" modes, thereby acquiring the abnormal oscillator strength and polarization observed in experiments.

In absence of coupling, the optical conductivity is simply the result of one electron (1e) excitations accross a gap 2V:

$$\sigma^\circ(\omega) = \sigma_{1e}^{2V}(\omega) \qquad g_n = 0$$

In RICE's theory, the e-p coupling introduces a new, collective term σ_c, together with a renormalization of the 1e gap:

$$\sigma(\omega) = \sigma_{1e}^{2\Delta}(\omega) + \sigma_c(\omega) \qquad g_n \neq 0$$

σ_c describes a series of ten absorption bands and provides an enhancement $\Delta\varepsilon_c$ over the PENN dielectric constant:

$$\varepsilon_s = \varepsilon_{1e}^{2\Delta} + \Delta\varepsilon_c \text{ , with } \varepsilon_{1e}^{2\Delta} = 1 + 2/3 \ (\omega_p/2\Delta)^2$$

RICE deduced "experimental" values of the 10 g_n from a fit of his theory to the curve $\sigma(\omega)$ obtained from a KRAMERS-KRONIG analysis of the R_1-spectrum of Figure 7. He also employed the experimental values $\omega_p \simeq 5800$ cm^{-1}, $\varepsilon_s \simeq 70$ and $2\Delta(=2E_a) \simeq 0.2$ eV. The deduced g_n values are in good semiquantitative agreement with pure quantum-chemistry-calculations. The e-p coupling provokes considerable stabilization of the semiconducting gap: $V/\Delta \simeq 0.13$. Finally, any intermolecular phonon mode in the very far infrared should also make contribution to σ_c and $\Delta\varepsilon_c$.

II. NEW FORMS OF INSTABILITIES IN TCNQ SALTS

MeI-NEtBz$^+$(TCNQ)$_2^-$ is a highly anisotropic salt of intermediate conductivity formed from the cation methyl-1-N-ethyl-benzimidazolium (Figure 8)(4). As for TEA(TCNQ)$_2$, delicate needles and large crystals are obtained in two successive stages of crystallization. We recently found that the axis-conductivity σ_1 of needles undergoes on cooling strong discontinuities in which it drops by a factor which can be as large as 300, either in a large single step or in

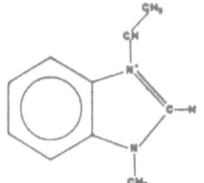

Figure 8.
Methyl-1-N-ethyl-benzimidazolium.

two or more closely spaced steps (Figure 9)(34,35). The room temperature value is $\sigma_1 = 23.7$ $(\Omega xcm)^{-1}$. The temperature at which the (first) discontinuity occurs, of the order of 250 °K, varies from needle to needle and, for a given needle, from one thermal cycle to the next. By re-heating the needle, just after the transition is achieved, it is possible to gradually recover the initial room temperature value, to better than 5% in general, although with pronounced hysteresis. Discontinuities are never observed with the larger specimens produced by re-crystallization. In this later case, strong but continuous slope-anomalies in the σ_1 versus T data of the large crystals are observed as a residual effect (21, 35). The origin of the instability responsible for these phenomena

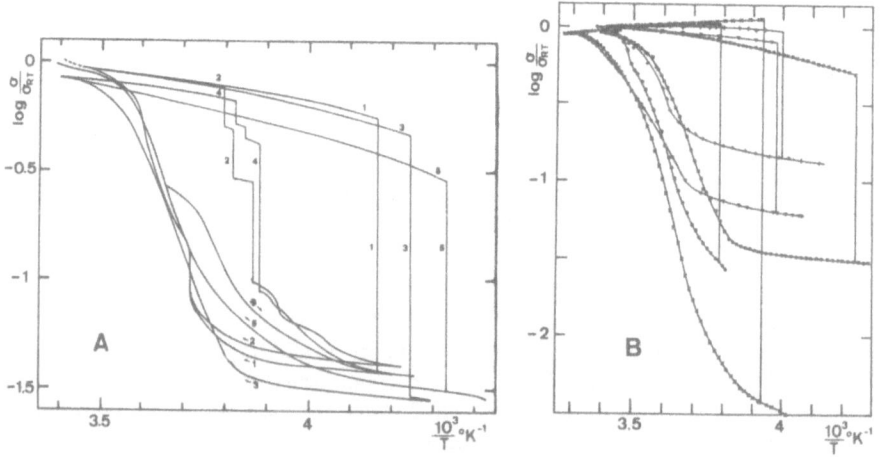

Figure 9. Examples of σ_1-steps in MeI-NEtBz(TCNQ)$_2$.
A. Several thermal cycles for one needle.
B. One thermal cycle for several needles.

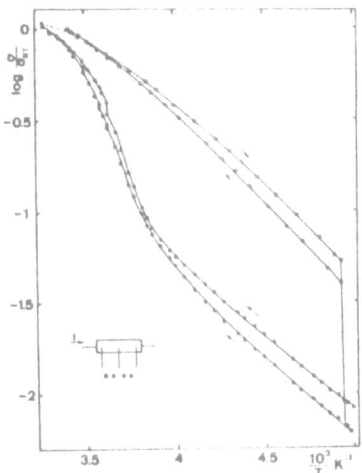

Figure 10. σ_1-discontinuities in a needle of $TEA(TCNQ)_2$.

is presently unclear and the investigation is continuing. Detailed X-ray data for needles and large crystals are needed, part of them are yet available (36). One additional point, however, is worthy to mention. σ-discontinuities have also been found recently in needles of $TEA(TCNQ)_2$, Figure 10 (37), thus establishing that such instabilities do also exist in other TCNQ salts not having direct chemical relation to $Mel-NEtBz(TCNQ)_2$. Lastly, we believe that the σ-deviations discussed in I.2 for the case of large re-crystallized specimens of $TEA(TCNQ)_2$ (see also Figure 2), are also related, at least partly, to these instabilities.

ACKNOWLEDGMENTS

    I am indebted to my colleagues A.BRAU and P.DUPUIS for their continuous and valuable collaboration, and also to H.GRASSI for his more recent contribution to our work. Part of this work was supported by the D.R.M.E. (PARIS).

REFERENCES

1) D.S.ACKER, R.J.HARDER, W.R.HERTLER, W.MAHLER, L.R.MELBY, R.E. BENSON and W.E.MOCHEL, J.Amer.Chem.Soc.82, 6408(1960)
2) R.G.KEPLER, P.E.BIERSTEDT and R.E.MERRIFIELD, Phys.Rev.Letters 5, 503(1960)
3) W.J.SIEMONS, P.E.BIERSTEDT and R.G.KEPLER, J.Chem.Phys.39, 3523(1963)
4) P.DUPUIS and J.NEEL, C.R.Acad.Sci.(PARIS)C265, 688(1967); C265, 1297(1967)

5) J.JAUD, D.CHASSEAU, J.GAULTIER and C.HAUW, C.R.Acad.Sci.(PARIS)
   C278, 769(1974)
6) A.T.McPHAIL, G.M.SEMENIUK and D.B.CHESNUT, J.Chem.Soc.A, 2174
   (1971)
7) H.KOBAYASHI, Y.OHASHI, F.MARUMO and Y.SAITO, Acta Crys.B26,
   459(1970)
8) J.P.FARGES, A.BRAU and F.GUTMANN, J.Phys.Chem. Solids 33,
   1723(1972)
9) A.BRAU and J.P.FARGES, Phys.Letters A41, 179(1972)
10) J.P.FARGES, Phys.Letters A43, 161(1973)
11) A.BRAU and J.P.FARGES, phys.stat.sol.(b)61, 257(1974)
12) J.P.FARGES, thesis, Université de NICE, 1974(unpublished)
13) I.F.SCHEGOLEV, phys.stat.sol.(a)12, 9(1972)
14) R.M.VLASOVA, Y.G.NURULLAEV, L.D.ROZENSHTEIN, V.N.SEMKIN, S.K.
    KOSIMI, K.S.KARIMOV and V.D.ERMAKOVA, Fiz.Tverd.tela 17, 1169
    (1975); Sov.Phys. Solid State 17, 749(1975)
15) See for instance S.FLANDROIS, J.AMIELL, F.CARMONA and P.
    DELHAES, 4th Int.Symp. Organic Solid State, BORDEAUX (1975)
16) Z.G.SOOS and D.J.KLEIN in Treatise on Solid State Chemistry 3,
    pp.679-767, N.B.HANNAY ed., PLENUM, N.Y.(1976)
17) P.PINCUS, Sol.Stat.Comm.11, 305(1972)
18) S.FLANDROIS, J.AMIELL,F.CARMONA and P.DELHAES, Sol.Stat.Comm.
    17, 287(1975)
19) J.P.FARGES and A.BRAU, phys.stat.sol.(b)61, 669(1974)
20) J.P.FARGES, A.BRAU, D.VASILESCU, P.DUPUIS and J.NEEL, phys.
    stat.sol.37, 745(1970)
21) A.BRAU, thesis, Université de NICE, 1976(unpublished)
22) J.P.FARGES and A.BRAU, phys.stat.sol.(b)92, K131(1979)
23) J.P.FARGES and A.BRAU, phys.stat.sol.(b)64, 269(1974)
24) G.BENI, J.F.KWAK and P.M.CHAIKIN, Sol.Stat.Comm.17, 1549(1975)
25) E.M.CONWELL, Phys.Rev.B18, 1818(1978)
26) A.BRAU, P.BRUESCH, J.P.FARGES, W.HINZ and D.KUSE, phys.stat.sol.
    (b)62, 615(1974)
27) M.D.KAPLUNOV, T.P.PANOVA and Y.G.BORODKO, phys.stat.sol.(a)13,
    K67(1972)
28) A.GIRLANDO and C.PECILE, Spectrochim.Acta 29A, 1859(1973)
29) M.J.RICE, C.B.DUKE and N.O.LIPARI, Sol.Stat.Comm.17, 1089(1975)
30) M.J. RICE, Phys.Rev.Letters 37, 36(1976)
31) M.J.RICE, L.PIETRONERO and P.BRUESCH, Sol.Stat.Comm.21, 757
    (1977)
32) M.J.RICE, Sol.Stat.Comm.25, 1083(1978)
33) N.O.LIPARI, C.B.DUKE, R.BOZIO, A.GIRLANDO, C.PECILE and A.
    PADVA, Chem.Phys.Letters 44, 236(1976)
34) A.BRAU, J.P.FARGES and H.GRASSI, phys.stat.sol.(a)51, K45(1979)
35) J.P.FARGES, H.GRASSI, A.BRAU and P.DUPUIS, phys.stat.sol.
    (a)55, n°1(1979)
36) D.CASTAGNE, D.CHASSEAU, J.GAULTIER, C.HAUW, P.DUPUIS, J.NEEL
    and A.FILHOL, 4th Int.Symp. Organic Solid State, BORDEAUX
    (1975)
37) H.GRASSI, A.BRAU and J.P.FARGES, submitted to phys.stat.sol.

COHESIVE ENERGY AND IONICITY

Robert Melville Metzger

Department of Chemistry, The University of Mississippi,
University MS 38677, U.S.A.

The calculation and experimental determination of the cohesive energy of organic ionic and partially ionic crystals will
be reviewed, with particular emphasis on the role of partial
ionicity.

I.   INTRODUCTION

Despite the rapid growth of experimental data and theoretical models for organic quasi-one-dimensional metals, as exempli
fied by the contributions to these Proceedings, there has not
been, to date, a detailed and complete understanding of the various contributions to the crystal lattice energy, or cohesive
energy, of these materials.   Sofar, the cohesive energy
calculations, such as will be presented below, have been <u>after</u>
<u>the</u> <u>fact</u> calculations, i.e. the organic crystals have first been
synthesized with certain experimentally characterized static and
dynamic (transport) properties, and theoretical calculations
have tried to justify why these crystals did form with their
particular mode of intermolecular overlap, stacking, and formal
charge.

Of course, the ultimate goal of such calculations is far
more ambitious: to <u>predict</u>, from the properties of the isolated
molecular or ionic constituents, what sort of a crystalline
lattice will be formed, with what amount of formal charge
transfer; then a reasonable guess could be made as to what solid-
state properties (conductivity, etc.) may be expected. Unfortunately, we are still rather far from this goal. Since, however,
the synthetic chemist interested in preparing new and exciting

233

L. Alcácer (ed.), The Physics and Chemistry of Low Dimensional Solids, 233–246.
Copyright © 1980 by D. Reidel Publishing Company.

compounds needs some approximate guiding principles, therefore
many such rules-of-thumb have been proposed (see the contributions
by Bechgaard, Torrance, and Wudl, elsewhere in these Proceedings).

We propose to search for the energetic and crystallographic
criteria for the formation of neutral, ionic, and partially ionic
organic electron donor-electron acceptor crystals (1). For
neutral hydrocarbon molecular crystals C_xH_y (and less well for
neutral crystals containing also N, O, and S) there exists a vast
thermochemical data base of experimental lattice energies, and
fairly sophisticated crystal packing energy computer programs,
which use ad hoc van der Waals and repulsion parameters to fit
the experimental lattice energy and packing geometry for certain
model compounds, and then can predict them for new compounds (2).
The goal is to do as well, if not better, with organic ionic and
partially ionic crystals, and to find predictive criteria for the
formation of organic metals. The theoretical program is to
continue the previous efforts to calculate Madelung energies (1)
and to compute polarization, hybridization, van der Waals, and
repulsion energies; the experimental program is to create an
adequate thermochemical data base by measuring the relevant
energies in the Born-Haber cycle, so as to guide the overall effort
with benchmark experimental data.

II. BORN-HABER CYCLES

The Born-Haber cycle for a neutral donor-acceptor (DA)
crystal is given in Fig. 1 (not drawn to scale). The enthalpies
of formation of $D(c)$, $A(c)$, and $DA(c)$ are available from combustion calorimetry, while the enthalpies of sublimation can be
obtained from vapor-pressure measurements. Neglecting terms of
the order of RT, where R is the gas constant and T is the absolute
temperature, U^n_{exp} is the experimental cohesive energy, which is
usually matched by computing an attractive van der Waals energy
and a repulsive energy.

Fig. 2 shows the Born-Haber cycle for a fully ionic D^+A^-
crystal. Now, in addition to what is needed for neutral crystals,
we must obtain the adiabatic gas-phase ionization potential I_D
from mass spectrometry or X-ray photoemission spectroscopy, and
the adiabatic electron affinity of the acceptor, A_A, from ion
cyclotron resonance spectroscopy for small molecules (3) or cesium
beam electron attachment experiments (4) for the larger molecules
of interest in these Proceedings. One may discuss the experimental
cohesive energy U_{exp} (which is negative, by convention, if the
$D^+A^-(c)$ energy level is lower than the $D^+(g),A^-(g)$ energy level),
and compare it with the cost $I_D - A_A$ of ionizing all $D(g)$ and $A(g)$
molecules, or else one may consider the ionic stabilization energy
E_I. For the usual "model" ionic solid, sodium chloride,

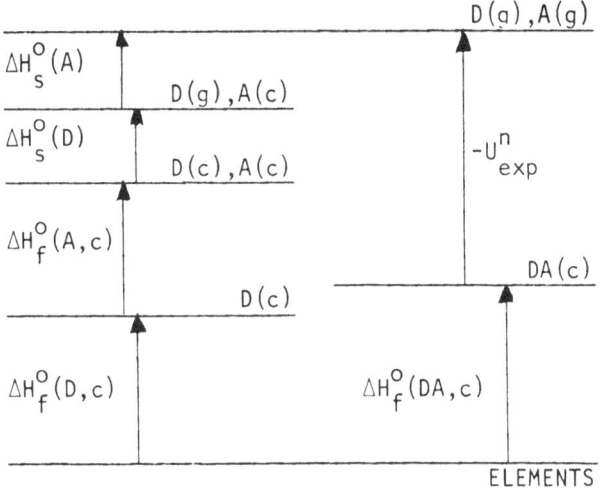

Fig. 1.   Born-Haber cycle for neutral DA crystal (not to scale).

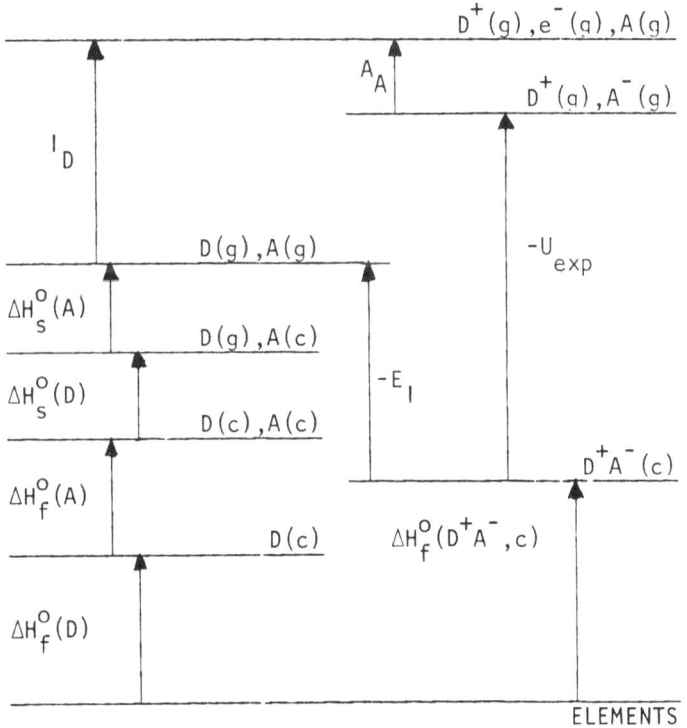

Fig. 2.   Born-Haber cycle for ionic D^+A^- crystal (not to scale).

U_{exp} =-764 kJ/mol (5), I_D-A_A= 147 kJ/mol, whereas the Madelung, or classical Coulomb lattice energy E_M=-860.8 kJ/mol, so that for such a salt $E_M \simeq U_{exp}$, and one may say that NaCl is ionic because $E_M + I_D - A_A$ is negative, i.e. one obtains far more lattice energy than it costs to ionize the lattice.   Thus a simple criterion for ionicity is (6):

$$E_M + I_D - A_A < 0; \qquad\qquad (1)$$

the symmetric statement is that if $E_M + I_D - A_A > 0$, then the crystal remains neutral. The Born-Haber cycle for a partially ionic crystal $D^\rho A^{-\rho}$ (charge transfer ρ, $0 < \rho < 1$) is given in Fig. 3.   One way of describing the binding energy is to compare

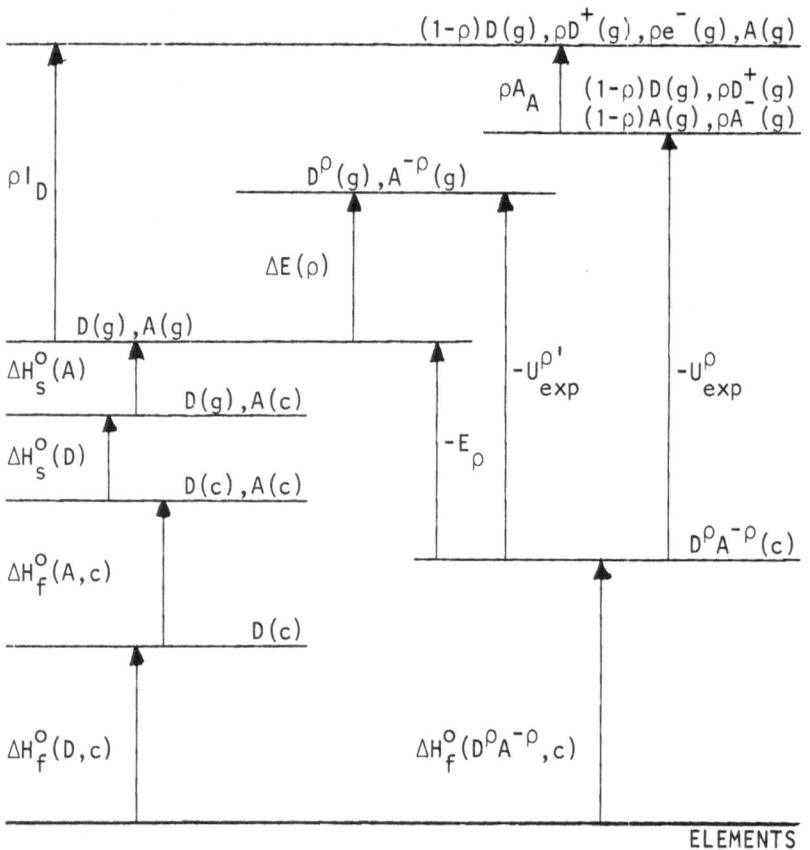

Fig. 3. Born-Haber cycle for partially ionic $D^\rho A^{-\rho}$ crystal.

the experimental crystal $D^\rho A^{-\rho}$(c) state with a reference state where a fraction ρ of the donor and acceptor molecules in the gas phase has been ionized; then the lattice energy is U^ρ_{exp}, and the crystal is stable if $U^\rho_{exp} + \rho(I_D-A_A) < 0$. This approach requires that the theoretical lattice energy calculation must include all solid-state corrections (polarization, etc.) to the cost of ionization. This however places a heavy burden on the theory: if a linear interpolation is used for the lattice of formal charge ρ between the neutral lattice and the fully ionic lattice for the formal charges, polarizabilities, etc., then the various lattice energies tend to scale as ρ^2, with maybe some higher-order contributions, while the cost of ionization scales as ρ. Thus a quadratic form in ρ is obtained, which has no minimum at intermediate ρ. This difficulty can in part be traced to the molecular-field arguments used when Eq. (1) was first introduced for organic charge-transfer crystals (6): the energy $E_M + \rho(I_D-A_A)$ has a minimum at either $\rho=0$ or $\rho=1$, and a collective-ionization scheme was used to justify full charge transfer.

An alternate idea is to define a fictitious state of partial-ly ionized $D^\rho(g)$ and $A^{-\rho}(g)$ species (7): this state then includes solid-state ionization energy lowering effects (polarization) and should justify a polynomial fit (8):

$$\Delta E(\rho) = a_o + a_1\rho + a_2\rho^2 + \ldots \qquad (2)$$

to the cost of ionization: $\Delta E(\rho=0)=0$, $\Delta E(\rho=1)=I_D-A_A$; to obtain a smooth $\Delta E(\rho)$ as a function of ρ, one may choose either $\rho=-1$ (8) or $\rho=2$ (9), or any set for which adequate data can be obtained. This smooth $\Delta E(\rho)$ replaces the connected-segment approach of $\rho(I_D-A_A)$, and will be shown to provide, for both TTF TCNQ and TMPD TCNQ an energy minimum at intermediate charge transfer.

III. CONTRIBUTIONS TO THE THEORETICAL LATTICE ENERGY.

The Madelung energy E_M is defined as the classical lattice sum of all the formal charge-charge interactions:

$$E_M = \sum_{i}^{MN/Z} \sum_{j}^{MN/Z} z_i z_j / r_{ij} = \frac{N}{Z} \sum_{m=1}^{M} z_m \phi_m \qquad (3a,b)$$

of atoms i, j with charges z_i, z_j at crystal lattice sites r_i, r_j; N is Avogadro's number, $r_{ij} = |\vec{r}_i - \vec{r}_j|$; there are M atoms and Z molecules per unit cell. The Madelung site potential ϕ_m can be differentiated to yield the Madelung electric field:

$$\vec{F}_m = -\nabla \phi_m. \qquad (4)$$

In the work reported here, for the large organic molecules the fractional point-charge approximation is used. For example, for the TCNQ⁻ ion in the room-temperature TTF TCNQ lattice (10) the point charges given by a Mulliken population analysis of an INDO (11) calculation is shown in Fig. 4. It should be emphasized that the fractional atomic point charge is a fictitious concept, but in previous E_M calculations it has worked rather well (1).

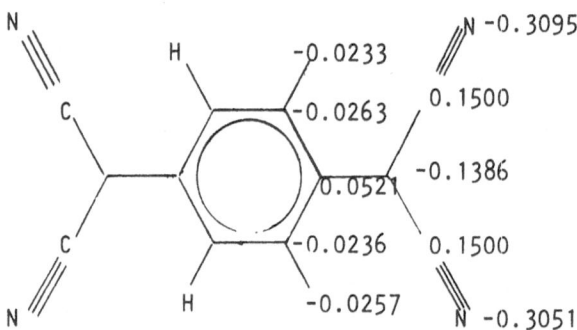

Fig. 4. INDO charges for TCNQ⁻ in TTF TCNQ.

Recently the finite-perturbation method of calculating INDO and MINDO/3 molecular polarizabilities α (12,13) by adding a weak electric field F and computing the molecular dipole moment μ in the presence and in the absence of that field:

$$\underset{\sim}{\alpha} = (\underset{\sim}{\mu}(\underset{\sim}{F}) - \underset{\sim}{\mu}(0))/\underset{\sim}{F} \tag{5}$$

has been adapted (14) to obtain atom-in-molecule polarizability tensors $\underset{\sim}{\alpha}_m$ (such that $\Sigma_m \underset{\sim}{\alpha}_m = \underset{\sim}{\alpha}$). For example, the scalar averages of the $\underset{\sim}{\alpha}_m$ atom polarizability tensors for TTF⁺ in TTF TCNQ are given in Fig. 5.

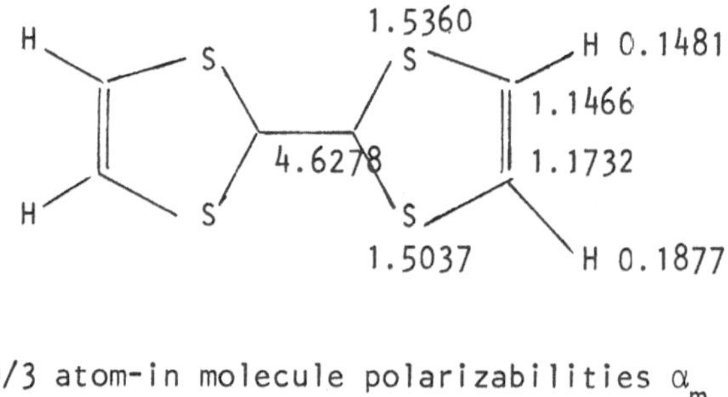

Fig. 5. MINDO/3 atom-in molecule polarizabilities α_m (A³) for TTF⁺ in TTF TCNQ.

The polarizability perpendicular to the molecular plane is badly underestimated for all atoms, because of a lack of polarization functions (15), and this, or an arbitrary decomposition of the α_m for fused-ring atoms, may explain why the α_m for sulfur is lower than expected. The effect of formal charge on α and α_m is large but not dramatic (16). Using the $\underset{\sim}{F}_m$ (Eq.(4)) $\overset{\sim}{}$ and $\overset{\sim}{}^m$ the $\underset{\sim}{\alpha}_m$ (Eq.(5)) one obtains the polarization energy:

$$E_{pol} = - \frac{1}{2\,Z} \sum_{m=1}^{M} \underset{\sim}{F}_m \cdot \underset{\sim}{\alpha}_m \cdot \underset{\sim}{F}_m \; . \qquad (6)$$

Another contribution to the theoretical cohesive energy is the hybridization energy (16):

$$E_{hyb} = - \frac{1}{Z} \sum_{m=1}^{M} \underset{\sim}{\mu}_{hyb_m} \cdot \underset{\sim}{F}_m \qquad (7)$$

where $\underset{\sim}{\mu}_{hyb_m}$ is the one-center hybridization correction to the INDO or MINDO/3 molecular dipole moment. For centro-symmetric molecules like TTF or TCNQ the dipole moment for the whole molecule vanishes, but the local atomic moments do have a non-zero interaction energy with the Madelung electric field (16). In fact, for TTF TCNQ and for TMPD TCNQ E_{hyb} is somewhat smaller than, and opposite in sign to, E_{pol} (see below).

The exchange Coulomb lattice energy E_{ex} (1) is, alas, not accessible by classical lattical sum methods, but must await a more serious tight-binding Hartree-Fock-Fwald treatment of the quantum mechanical crystal binding energy.

The kinetic energy binding correction for organic metals is rather small for these narrow-band systems (1).

The van der Waals or dispersion energy can be computed ad hoc:

$$E_{vdW} = - \frac{N}{Z} \sum_{m=1}^{M} \sum_{j=1}^{NM/Z} \frac{B_m B_j}{r_{mj}^6} \qquad (8)$$

using Kitaigorodskii or Williams non-bonded parameters (2), or using the London expression:

$$E_{vdW} = - \frac{3}{4} \frac{N}{Z} \sum_{m=1}^{M} \sum_{j=1}^{NM/Z} \frac{\alpha_m \alpha_j}{r_{mj}^6} \frac{E_m E_j}{(E_m + E_j)} \qquad (9)$$

where α_m, α_j are scalar atom-in-molecule polarizabilities and E_m, E_j are some characteristic averages of the molecular excitation spectrum.

Finally, the repulsion energy is needed to keep the crystal from collapsing to a point; its physical origin lies with the finite overlap between the wavefunctions of adjacent molecules; it can be represented in a Born-Lande form:

$$E_{rep} = \sum_{i>j} \sum \frac{A_i A_j}{r_{ij}^n}, \quad n = 9 \text{ to } 12, \qquad (10)$$

or in the form:

$$E_{rep} = \sum_{i>j} \sum C_i C_j \exp(-r_{ij}(D_i + D_j)), \qquad (11)$$

where the A_i, C_i, and D_i are ad hoc parameters. Recently the cohesive energy of aromatic crystals, and even that of TTF TCNQ (2,17) has been treated by matching with experiment only $E_M + E_{vdW} + E_{rep}$, but one wonders to what extent the latter two terms are really "lumped" parameters which mask the combination of several lattice forces (16).

IV. CALCULATED LATTICE ENERGIES.

Our computer programs CELMAP and EWALD have been described previously (1): crystal structure data are input into the former, and the quantum-chemical programs CNDO/2, INDO, and MINDO/3 (11) produce charges z_i (H to Cl for CNDO/2, Ge,As,Se,Br for CNDO/2 (18), H to Cl for MINDO/3, H to F for INDO) and polarizability tensors α_m (H to F for INDO or CNDO/2, H to Cl for MINDO/3); finally, EWALD can compute E_M, E_{pol}, E_{hyb}, E_{vdW}, and E_{rep}. Typically, the α_m for TCNQ$^-$ require about 10 hours of DECsystem 1077 computer time, and the lattice energies for TTF TCNQ require about 2 hours.

Some representative Madelung energies for ionic and neutral organic insulating and semiconducting crystals are given in Table 1. The E_M values are fairly independent of charge model, i.e. of the quality of the quantum-chemical single-molecule calculation (19), and, wherever Eq.(1) is applicable, it is satisfied by the data of Table 1. However, the full ionicity of TMPD TCNQ and TMPD p-chloranil has recently been called into question by Soos (8,20); this point is discussed further below.

For organic metals, however, E_M depends rather strongly on the charge model used (21), and Eq.(1) is not satisfied, for $\rho = 0$,

Table 1.
Some Madelung Energies for Neutral and Ionic Insulators
and Semiconductors (Ref.(1)).

Compound	ρ_{exp}	Stacking	E_M (kJ/mol)	$I_D - A_A$ (kJ/mol)	Is Eq.(1) Rule OK?
Naphthalene TCNE	0	DA, Reg.	-415.8	530	Yes
Hexamethylbenzene p-chloranil	0	DA, Reg.	-425.5	540	Yes
TMPD p-chloranil	1 ?	D^+A^-Reg.	-427.4	371	Yes
TMPD TCNQ	1 ?	D^+A^-Reg.	-380.1	333	Yes
TMPD Iodide	1	D^+D^+Reg.	-411.0	286	Yes
TMPD Perchlorate	1	D^+D^+Reg.	-396.6	-	-
TMPD Perchlorate	1	D^-D^-Alt.	-398.5	-	-
Na TCNQ	1	A^-A^-Alt.	-510.4	227	Yes
Rb TCNQ(I)	1	A^-A^-Alt.	-456.4	133	Yes
Rb TCNQ(II)	1	A^-A^-Reg.	-447.7	133	Yes
Rb TCNQ(III)	1	A^-A^-Alt.	-455.0	133	Yes
Cs_2TCNQ_3	1,0,1	A^-AA^-	-886.7	213	Yes
TEA $TCNQ_2$	$\frac{1}{2},\frac{1}{2}$	A_2^-, Alt.	-248.0	-	-

$\rho=1$, or any intermediate value of ρ (1). The relevant E_M values
for organic metals are given in Table 2 (1), which also give (22,23)
recently obtained Madelung energies for two insulators (TMTSF
TCNQ red, and HMTSF $TCNQF_4$) for $\rho=0$, 1, and 2. As for NMP TCNQ,
E_M essentially vanishes at ρ_{exp} for the two triclinic salts
HMTSF TNAP and TMTSF TCNQ black. The two compounds HMTSF TCNQ
and HMTSF $TCNQF_4$ are isostructural (24), and the E_M values reflect
this, and yet the former is a metal and the latter is an insulator.
The red and black forms of TMTSF TCNQ crystallize under the same
conditions, and yet one is an insulator with large E_M, and the
other is a metal with vanishing E_M. All these data, however, are

Table 2.
Madelung Energies for Some Organic Metals and Two Insulators(*)

Compound	ρ_{exp}	$\rho=0$ E_M	$\rho=1$ E_M	I_D-A_A	$\rho=2$ E_M	other ρ ρ	E_M
NMP TCNQ	0.94?	-	0 ± 30	-	-	0.50	-173 W^a
TTF TCNQ	0.59	-2.242	-225.470	400[b]	-812.144	0.50	-154 W_b
TTF $Br_{0.79}$	0.79	-	-376[b]	311	-	0.74	-312 W^b
HMTSF TNAP	-	16.175[c]	-20.003[c]	-	163.274[c]	-	-
HMTSF TCNQ	0.74	3.487[c]	-248.806[c]	-	-959.672[c]	-	-
TMTSF TCNQ black	0.38	11.985[c]	-31.550[c]	-	82.655[c]	-	-
TMTSF TCNQ red*	0.17	-8.110[c]	-297.068[c]	-	-1236.230[c]	-	-
HMTSF $TCNQF_4$*	1.0	41.221[c]	-246.353[c]	-	-912.452[c]	-	-

W=Wigner crystal; (a) Ref.(25); (b) Ref.(26); (c) Ref.(22).

crucial to a cohesive energy theory recently elaborated by Bloch
(27) and discussed in this Advanced Summer Institute, where two
molecular chemical potentials, $\Delta\mu_0$ and $\Delta\mu_1$ are introduced that
help explain and classify insulators and conductors in this
family of compounds and also stabilizes the partially charge-
transferred state.

In Table 3 are given values of E_M, E_{pol}, E_{hyb} obtained
recently for TTF TCNQ (16), TMPD TCNQ (28), and Rb TCNQ(III)
(29). For TTF TCNQ the experimental value of E_ρ (30) is also
listed, and should be compared with $E_M + E_{pol} + E_{hyb} + I_D - A_A$.
It is clear that without van der Waals and repulsion energies
the experimental cohesive energy cannot be matched. As mentioned
above, Govers (17) has in fact matched U_{exp}^ρ with $E_M + E_{vdW} + E_{rep}$
but has not explained the stability of the partially charge-
transferred state. The point of performing laborious $E_{hyb} + E_{pol}$
calculations is, however, to obtain SCF quantum-mechanical
corrections to the lattice energy picture afforded just by the
fractional point charges used in the Madelung energy calculation.

Table 3.
Madelung, Polarization, and Hybridization Energies, compared
with $I_D - A_A$ and with experiment, for the fully ionic lattice
of one metal and two semiconductors.
(all energies in kJ/mol)

	TTF TCNQ	TMPD TCNQ	Rb TCNQ(III)
E_M	-205.551	-313.149	-454.993
E_{hyb}	184.269	104.703	-87.365
E_{pol}	-259.991	-226.251	-65.886
$I_D - A_A$	400.4	332.9	133.2
$E_M + E_{hyb} + E_{pol} + I_D - A_A$	119.2	-101.8	-475.0
E_ρ	-235 ± 6	-	-

Calculations of $E_M + E_{hyb} + E_{pol}$ as a function of ρ for $\rho=0$,
0.1, 0.2, ..., 1.0 for both TTF TCNQ and TMPD TCNQ (16,28) have
shown no energy minimum for intermediate ρ, and, a fortiori, no
minimum in $E_M + E_{hyb} + E_{pol} + \rho(I_D - A_A)$ as a function of ρ. Thus
either the polarizabilities were not sufficiently large or sensitive to formal charge to yield the hoped-for minimum, or the scheme of interpolation did not permit the location of a minimum at
partial charge transfer, even though for TTF TCNQ $\rho_{exp} = 0.59$ is
very well established, and for TMPD TCNQ Soos and van Duyne have
shown (8) theoretical and experimental evidence that $\rho = 0.7\pm.1$.
However, if $\Delta E(\rho)$, computed as suggested in Ref.(8), is added to
$E_M + E_{hyb} + E_{pol}$ in place of $\rho(I_D - A_A)$, then a minimum is observed

at $\rho=0.4$ for TTF TCNQ, and at $\rho=0.8$ for TMPD TCNQ (16,28); the location of the minimum is very sensitive to the choices of the relevant I_D and A_A values, and an uncertainty of 0.2 or 0.3 is not unreasonable. The crucial point is that the use of $\Delta E(\rho)$ does indeed stabilize the partial charge transfer state for $E_M + E_{hyb} + E_{pol}$ (16,28) as well as just for E_M (8).

V.   EXPERIMENTAL COHESIVE ENERGIES.

Although the experimental cohesive energy U_{exp}^ρ for TTF TCNQ was measured using an existing combustion calorimeter (31) at the Bartlesville, Oklahoma Energy Technology Center (Department of Energy), about 7 grams of tetrathiafulvalene had to be burned. Clearly, such large samples, perfectly reasonable for normal petrochemicals, are too large for determining enthalpies of formation for organic metals, where large samples are not usually practical.

Therefore, with the generous support of the U.S. National Science Foundation (32) a new semi-micro platinum-lined rotating bomb combustion calorimeter was built on the Bartlesville model. The design features are summarized in Table 4. With this new calorimeter, sample sizes of the order of 50 mg per run may be burned, albeit with some reduction in precision.

Table 4.
A Semi-Micro Platinum-Lined Rotating-Bomb Combustion
Calorimeter.

	Bartlesville Design (1954)	Mississippi Adaptation (1979)
Calorimeter operation	Isothermal Jacket	Adiabatic (Parr Model 1243)
Calorimeter constant (J/deg)	16783.2 ± 1.3	4153.0 ± 4.0[a]
Bomb internal volume (cm³)	340	29
Sample size (benzoic acid, g)	1.2	0.050
Temperature rise	2°	0.35°
Thermometry	Pt resistance (now Quartz crystal)	Quartz crystal
Data acquisition	Müller bridge (now Minicomputer)	Minicomputer

(a) Precision limited by unusual final-period heating curve

At present, the Mississippi calorimeter and the data reduction programs can process compounds of C, H, O, N, S, I, and Na. Air-sensitive and moisture-sensitive compounds can be handled by encapsulation in Mylar bags sealed in an inert atmosphere.

Recently obtained data for Na TCNQ and for a TTF pseudohalide, TTF $(SCN)_{0.545}$, are presented in Table 5 (33).

Table 5.

Born-Haber Cycle Data for Three Salts

(All energies in kJ/mol)

Compound	$\Delta H_f^0(c)$	E_ρ	ρ_{exp}	U_{exp}^ρ	$E_M(\rho=1.0)$
Na TCNQ[a]	305 ± 12	-573 ± 20	1.00	-590 ± 20	-510
TTF TCNQ[b]	918 ± 2	-235 ± 6	0.59	-471 ± 16	-225
TTF SCN[a,c].545	265 ± 40	-	0.545[c]	-448 ± 40[c,d]	-

a: Mississippi calorimeter

b: Bartlesville calorimeter, Ref.(30).

c: Sample was labelled as $TTF_{11}(SCN)_6$, and was so analysed. In fact, it should be reanalysed as $TTF_{12}(SCN)_7$, and will be in the near future.

d: Using $\Delta H_f^0(SCN^-,g)=-63.4$ kJ/mol, MNDO theoretical value.

The experimental cohesive energy for Na TCNQ is very close to the Madelung energy, and if the $E_{hyb} + E_{pol}$ values for Rb TCNQ given in Table 3 are a good indication of the possible corresponding values for Na TCNQ, then the correlation between theory and experiment is even closer. For the TTF pseudohalide, there is strong resemblance between its experimental lattice energy and the value of the TTF $Br_{0.74}$ Madelung energy quoted in Table 2. Although at the moment the thermochemical data given in Table 5 show large errors, which will hopefully decrease with better technique and better data reduction algorithms, still the beginnings of an adequate experimental data base for cohesive energies of organic ionic crystals can be seen. (34)

REFERENCES

1. R.M.Metzger, Ann. N. Y. Acad. Sci. 313, 145 (1978).
2. A.I.Kitaigorodskii, "Molecular Crystals and Molecules", Academic Press, New York 1973; K. Mirsky, Acta Cryst. A32, 199 (1976); D. E. Williams and T. L. Starr, Computers and Chem. 1, 173 (1977).
3. Cf. e.g. K.C.Smith, R.T.McIver,Jr., J.I.Brauman, and R.W. Waller, J. Chem. Phys. 54, 2758 (1971).
4. C.E.Klotz, R.N.Compton, and V.F.Raaen, J. Chem. Phys. 60, 1177 (1974).

5. 1 eV/molecule = 96.487 kJ/mol. For a binary compound, 1 mole is here one mole of dimers (DA pairs).

6. H. M. McConnell, B. M. Hoffman, and R. M. Metzger, Proc. Natl. Acad. Sci. U. S. $\underline{53}$, 46 (1965).

7. Cf. discussion of ΔE_{cov} in R. M. Metzger and A. N. Bloch, J. Chem. Phys. $\underline{63}$, 5098 (1975).

8. J. Hinze and H. H. Jaffe, J. Am. Chem. Soc. $\underline{84}$, 540 (1962); Z. G. Soos and S. Mazumdar, Chem. Phys. Letters, in press.

9. A. N. Bloch, paper presented at ACS-CSJ Chemical Congress, Honolulu, Hawaii, March 1979.

10. T. J. Kistenmacher, T. E. Phillips, and D. O. Cowan, Acta Cryst. B $\underline{30}$, 763 (1974).

11. INDO, CNDO/2, MINDO/3, MNDO computer programs: Quantum Chemistry Program Exchange, Department of Chemistry, Indiana University, Bloomington, Indiana USA.

12. J. A. Pople, J. W. McIver, Jr., and N. S. Ostlund, J. Chem. Phys. $\underline{49}$, 2960, 2965 (1968).

13. M. J. S. Dewar, R. C. Haddon, and S. H. Suck, J. Chem. Soc. Chem. Commun. 611 (1974).

14. R. M. Metzger, "INDO-level Atom-in-Molecule Polarizabilities", J. Chem. Phys., submitted.

15. J. J. C. Teixeira-Dias and P. J. Sarre, J. Chem. Soc. Faraday Trans. II, 906 (1975).

16. R. M. Metzger, "Crystal Binding Energies. I. The INDO-Level Madelung, Polarization, Hybridization and Dispersion Binding Energies of TTF TCNQ", J. Chem. Phys., submitted.

17. H. A. J. Govers, Acta Cryst. A $\underline{34}$, 960 (1978).

18. H. L. Hase and A. Schweig, Theor. Chim. Acta $\underline{31}$, 215 (1973).

19. R. M. Metzger, J. Chem. Phys. $\underline{57}$, 1870, 1876 (1972).

20. Z. G. Soos and S. Mazumdar, Phys. Rev.B18, 1991 (1978).

21. R. M. Metzger and A. N. Bloch, J. Chem. Phys. $\underline{63}$, 5098 (1975).

22. R. M. Metzger, F. M. Wiygul, and A. N. Bloch, unpublished results.

23. In each case, except for TMTSF TCNQ black, INDO charges were computed for the geometry of the molecules (ions) in the crystal lattice.

24. T. J. Kistenmacher, "Partial Charge Transfer and Charge Density Wave Modulation in the TTF-TCNQ Family of Quasi-One-Dimensional Organic Materials", in "Modulated Structures-1979 (Kailua Kona, Hawaii), J.M.Cowley, J.B.Cohen, M.B.Salamon, and B.J. Wuensch, Editors, AIP Conference Proceedings No. 53, American Institute of Physics, New York 1979, page 193.

25. I.I.Ukrainski, V.E.Klymenko, and A.A.Ovchinnikov, "Electrostatic energy and electronic structure of donor-acceptor molecular crystals based on tetracyanoquinodimethan", Preprint ITP-75-89E, Institute of Theoretical Physics, Ukrainian Academy of Sciences, Kiev, USSR (1975).

26. J. B. Torrance and B. D. Silverman, Phys. Rev. B15, 788 (1977).

27. A. N. Bloch, to be published.

28. R. M. Metzger, F. M. Wiygul, and Z. G. Soos, unpublished results.
29. R. M. Metzger and F. M. Wiygul, unpublished results.
30. R. M. Metzger, J. Chem. Phys. $\underline{66}$, 2525 (1977).
31. W. N. Hubbard, D. W. Scott, and G. Waddington, J. Phys. Chem. $\underline{58}$, 142 (1954).
32. Grant No. NSF-DMR-77-09314
33. We are indebted to F. Wudl and G. A. Thomas for the sample of TTF (SCN)$_{0.588}$.
34. Thanks are also due to C. S. Kuo and E. S. Arafat for the construction and patient operation of the calorimeter, to F. M. Wiygul for extensive programming, and to Z. G. Soos and A. N. Bloch for valuable suggestions.

MOLECULAR PROPERTIES OF THE MOLECULES USED IN CONDUCTING ORGANIC SOLIDS

Klaus Bechgaard
H. C. Ørsted Institutet, København, Danmark

Jan R. Andersen
Forsøgsanlæg Risø, Roskilde, Danmark

In this paper we shall try to communicate some of the ideas the organic chemist may utilize when trying to develop new conducting organic solids. First, we want to point out some important properties of typical organic molecules and try to show that if we want to form conducting solids, we are forced into working with certain classes of molecules. This in turn determines the special properties of the resulting solids.

In a typical organic molecule such as methane the energy of the highest occupied molecular orbital (HOMO) is -12.70 eV, and the lowest unoccupied molecular orbital (LUMO) is several eV's above. Since methane is a closed shell molecule, at best we can hope for a semimetallic or a small gap semiconducting solid. However, since the LUMO is too far from the HOMO, an insulating solid results.

The same general features are found in most other organic solids based on closed shell molecules. In order to obtain metallic solids, one might then employ open shell organic molecules, which could form half filled band conductors in the solid state. There are actually a few chemically stable free radicals, for example nitroxides, DPPH, and several sulphur containing heterocycles. Most of these, however, dimerize in the solid state or interact so weakly that insulating solids are formed.

The other possibility is to use open shell organic ions (ion radicals). This appears quite promising as many chemically stable systems are known, but of course there has to be enough energy to keep the open shell situation in the solid. Thus we must look for molecules giving rise to easily accessible oxidized or reduced ion radicals.

L. Alcácer (ed.), The Physics and Chemistry of Low Dimensional Solids, 247–263.
Copyright © 1980 by D. Reidel Publishing Company.

One possibility, and in fact the only one which has been utilized so far, is to use molecules having extended π-systems made from carbon p_z or heteroatom p_z orbitals. As a few examples we take:

σ-type	π-types		
CH_4	Benzene	Naphthalene	Furan
Ip 12.70 eV	Ip 9.24 eV	Ip 8.2 eV	Ip 8.9 eV

Although the delocalized π-systems give rise to lower Ip's or higher E_a's than σ-type molecules, there are many problems if we want to build a conducting solid from, for example, benzene cation radicals. In principle it should be possible, but in practice the benzene cation is very reactive and can only be handled under extreme conditions.

Stable benzene cation radical salts and related salts have recently been reported[1] and are in most cases simple paramagnets or semiconductors.

More success could be expected using higher homologues of benzene, as well as heteroaromatics or just extended π-systems preferentially containing heteroatoms (N, P, O, S, Se and Te) in the basic framework. In fact it seems that the latter type of molecules are the only ones which have given rise to metallic solids.

We do, however, pay a price when we are limited to rather large, flat molecules. Let us again use benzene as a model:

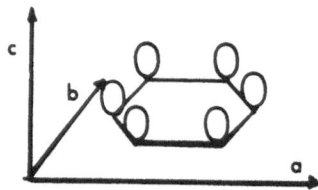

If we want to form an open energy band using benzene cation radicals, we will have to place the benzenes on top of each other to make the π-HOMO's interact strongly.[2] We may thus obtain bonding interaction along c, whereas the interaction along a and b must be substantially lower. In the "predicted" conducting solid we then get strongly anisotropic interactions, in fact a quasi one-dimensional situation.

Thus, one dimensional physics which has benefitted greatly from the occurrence of molecular conducting solids, has had this vitamine injection, not as a result of a well planned effort, but because the experimental systems are built from a particular kind of molecules.

It would probably be too definitive to claim that molecular organic conductors will always be more or less 1-D, since one might include graphite (the ultimate organic molecule) or stacked linear polymers (polybiphenylene, polyacetylene, etc.) among organic molecular conductors.

In resumé, the "possible" candidates for molecular conducting solids then are[3]: Planar, extended π-cation or anion radicals.

In binary compounds we have 3 combinations:

I	II	III
odd cation, closed anion	odd cation and anion	closed cation, odd anion
TTF-halides	TTF-TCNQ	$A_x(TCNQ)_y$
TTT and TSeT-halides	and derivatives	
1 chain	2 chains	1 chain

In the second part of the paper we focus on organic conductors derived from TTF-TCNQ, since these constitute the only larger group of similar materials. We will try to examine some of the molecular properties and try to find some general features which can be utilized in the search for new highly conducting systems.

TTF   TCNQ

TTF-TCNQ is a simple 1:1 charge-transfer compound crystallizing in segregated stacks.[4] The charge-transfer from donor (TTF) to acceptor stack (TCNQ) is incomplete as demonstrated by diffuse X-ray scattering experiments.[5]

Single crystal room temperature conductivities range from 500-1000 (ohm cm)$^{-1}$ rising to more than 10^4 (ohm cm)$^{-1}$ at 59K.[6] At 53, 47 and 38K three consecutive phase transitions occur, ultimately leading to 3-D order and an insulating state. The electro-

FORMULAS

TTF   TSeF   MTTF

TMTTF   TTTF   HMTTF

cis - and trans - DEDMTSF

TCNQ   MTCNQ   DMTCNQ

DETCNQ $TCNQCl_2$ $TCNQF_4$

TNAP

nic properties are highly anisotropic, and the conductivity is electronlike as indicated by the negative thermopower.

The mechanism of the metallic conductivity has so far remained a mystery, but it has been suggested that Frölich type charge density waves could contribute.[7] However, the most important property from an experimental point of view is that TTF-TCNQ is by now relatively easy to synthesize, and that single crystals of good quality can be obtained. Thus TTF-TCNQ can and has been subjected to a vast number of studies during the past six years, but it should be mentioned that in most systems derived from TTF-TCNQ, synthesis and crystal growth are not feasible.

An obvious way to evaluate the specific molecular influence on the properties of TTF-TCNQ is to prepare derivatives of the parent system.

We wish to point out some simple effects of substituting atoms or atom groups in TTF or TCNQ before elaborating on the molecular properties. Substitution of the hydrogens in TTF makes the molecules longer and create electronic perturbations.

In TCNQ, substitution of hydrogens basically leads to a wider molecule and in most successful experimental examples also breaks the symmetry of the parent molecule:

We expect that in most new systems the substitution will induce both different crystal structure (i.e. interstack interactions) and different intrastack interactions relative to the parent system.

It is thus possible to evaluate the effects of tuning the molecular properties, not a priori, since we cannot predict actual crystal structure, but a posteriori by careful examination of the resulting solids. We may then extract some key features, but so far no clear-cut picture has emerged.

TTF and derivatives

TTF is a planar, symmetrical (D_{2h}) molecule, having 14 π-electrons in a 10 nuclei framework. In solution as well as in the gas phase it is easily oxidized to the cation radical and the dication:

$$TTF \; \rightleftarrows \; TTF^{+\cdot} \; \rightleftarrows \; TTF^{2+}$$

The ionization potential is 6.83 eV in the gas phase and +.27 V vs SCE in acetonitrile solution. $TTF^{+\cdot}$ is rather stable towards nuclephilic attack (H_2O) as compared to most other radical cations. This is of relevance to crystal growth. Although part of the TTF is ionized, extensive decomposition is not observed in reasonably dry solvents.

The TTF skeleton was first synthesized in 1926 as part of dibenzo-TTF. This compound has first recently found some use as a building block in organic conductors.[8] A massive effort in the area of TTF derivatives was, however, first started when the preparation of TTF itself and TTF-chloride[9] was reported.

A few years ago a comprehensive review covering the synthesis of most of these derivatives was published,[10] hence we shall not consider synthetic details. Instead we classify some of the changes made of the TTF skeleton and briefly discuss the impacts on the resulting TCNQ complexes.

Symmetrical substitution: TTF has been changed to yield symmetrical congeners. The introduction of Selenium gave TSeF[11], from which a variety of substituted TSeF's can be derived. Among the successful new molecules, as judged by the ability to form highly conducting compounds with TCNQ and analogous acceptors, are the alkylated TTF's (TMTTF,[12] HMTTF[13] and others), all of which carry weakly electron donating substituents and are slightly better donors than TTF. Numerous derivatives having electron attracting substituents (CF_3, CN, aryl-, etc.) have also been synthesized, but tend to form DA structures rather than the intended segregated stacks. Still missing among TTF and TSeF derivatives are the structures substituted with strongly electron donating groups (CH_3O-, R_2N-).

Unsymmetrical substitution: Rather few unsymmetrically substituted TTF's have been made, mainly because of preparative difficulties. It is, however, possible to cross-couple TTF precursors
to yield mixtures of TTF's, which can be separated by fractional
crystallization or chromatography. Examples are MTTF[14] and TTTF[15].
In the TSeF series, to our knowledge, no pure unsymmetrical derivatives have been obtained so far.

A few years ago a very useful synthetic procedure for obtaining monosubstituted TTF's was developed (see below), but in term
of obtaining highly conducting solids the pay off has so far been
low.

Finally, the "poor man's" unsymmetrical TTF's and TSeF's can
be obtained in mostly unseparable mixtures of cis and trans forms
by coupling of unsymmetrical dithioles or diselenoles. These
mixtures in their TCNQ compounds give the expected effects: dis-
order-smearing of the phase transitions.

Three examples may serve in demonstrating the effects of
employing TTF derivatives (see fig. 2): 1. TSeF and derivatives
exhibit one anomaly (phase transition) rather than two (or three)
seen in most TTF derivatives. Furthermore, the thermopower is
hole-like indicating donor stack dominance. 2. Disorder, as for
example in DEDMTSeF,[16] gives rise to complete smearing of the
metal to insulator transition. 3. By substitution, size matching
of donor and acceptor stack can be obtained. HMTTF-TCNQ[13] and
HMTSF-TCNQ[17] have a chessboard-like structure, are less anisotropic than usual and probably undergo a metal to semimetal transition at lower temperature.

In conclusion the synthesis of TTF derivatives has reached a
rather sophisticated level and the main effects of substitution
on the physical properties are to some extent evaluated. More
exotic TTF derivatives such as polymeric TTF[18] and mixed metal-
bisdithiolene-TTF polymers[19,20] have been made but few details
are known since single crystals are not yet available.

Chemistry: The chemistry of TTF and derivatives has not yet
been investigated to any larger extent. The main objective for
most chemists has so far been the preparation of new materials.
However, a few reactions of TTF are known. In 1977 D. C. Green
published[21] the first examples of derivatives made from TTF itself. The procedure is low temperature lithiation followed by
reaction with various electrophiles. New molecules obtained were
mono-alkylated TTF's, TTF-carbaldehyde, TTF-carboxylic acid and
several other derivatives.

Another important reaction of TTF and alkylated TTF's is the
very facile reaction with oxygen to form TTF-S-oxides.[22] TSeF does

not seem to react similarly. TTF-S-oxide has been shown to inter-
fere in the crystallization process of TTF-TCNQ, a fact which to
some extent has been neglected so far. Thus exposure of TTF to
oxygen should be avoided.

Physical properties

The Ip's of a series of heterofulvalenes are given in Table
I. Only a few comparative studies have been reported, among these
the photoelectron spectra of a few TTF's (and TSeF's).[23] Mass
spectra of a large series[24] have been reported, clearly demonstrating the fragility of C-Se bonds relative to C-S bonds. In that
study field ionization spectra were shown to be a convenient and
sensitive tool in monitoring impurities.

Table I. Donor ionization potentials and electrochemical half-
wave potentials.

	I_p (eV)	$E_{\frac{1}{2}}^{1}$[*]	$E_{\frac{1}{2}}^{2}$
TTF	6.83	0.27	0.64
TMTTF	6.42	0.24	0.61
HMTTF	–	0.27	0.60
TSF	7.21[§]	0.42	0.70
TMTSF	6.58	0.35	0.59
HMTSF	–	0.35[‡]	0.59
DEDMTSF	6.45	0.35	0.59

[*]Volts vs SCE at Pt-button electrode in CH_3CN/n-Bu_4NBF_4 (0.1M)

[§]Appearance potential from mass spectroscopy (under the same
conditions TTF gives a value of 6.95 eV).

[‡]HMTSF is insufficiently soluble in CH_3CN for a cyclic volta-
mogram to be recorded. As TMTSF and HMTSF give exactly the
same values in CH_2Cl_2, however, the CH_3CN values obtained for
the former are taken to be representative of the latter as
well.

TCNQ and derivatives

TCNQ was   first prepared in 1960.[25] Cyclohexane-1,4-dione
was condensed with malonitrile and oxidized.[26] Unfortunately,
most derivatives of TCNQ cannot be prepared by similar simple reactions, but are obtained via long, difficult sequences. In fact
most synthetic work on TCNQ derivatives can be credited to Wheland and Martin, who in an excellent paper[27] reported 21 TCNQ
derivatives.

The parent molecule is a planar D_{2h} electron poor structure and is in solution easily reduced to the radical anion and dianion. The anion radical is remarkably stable in many solvents. TCNQ in solution exhibits a strong tendency to collect cations from glassware and must be sublimed onto teflon and/or handled in quartz to ensure high purity.

The electron affinity of TCNQ in the gas phase is 2.8 ± 0.1 eV.[28] Since electron affinity measurements in the gas phase are difficult, solution reversible redox potentials are often taken as indicative of relative electron affinities (see Table II).

Electrochemistry: Electrochemically TCNQ and most derivatives are reduced in two reversible one electron steps:

$$TCNQ \rightleftarrows TCNQ^- \rightleftarrows TCNQ^{2-}$$

It has been suggested, most recently by Addison et al.,[28] that the free energy from the equilibrium:

$$2\ TCNQ^- \rightleftarrows TCNQ + TCNQ^{2-}$$

is related to the on-site Coulomb repulsion, U, in a conducting TCNQ stack. From the relation $U \sim \Delta G^0 = -n \cdot F \cdot \Delta E$, where ΔE is the difference between the first and second reduction wave, U is estimated. We wish to point out that ΔE depends strongly on solvent and supporting electrolyte. In fact ΔE measured in CH_3CN using $LiClO_4$ or $n\text{-}Bu_4NBF_4$ (1 M) is found to be 0.30 V and 0.52 V, respectively. The reason is that Li^+ forms a 2:1 ion pair with the dianion, whereas $n\text{-}Bu_4N^+$ forms a 1:1 ion pair.[30] There is to our knowledge no argument supporting either a 2:1 or a 1:1 ion pair as leading to the "better" U, and we conclude that neither necessarily represent the real U of the solid. Similarly we find that E_1 (~electron affinity) depends somewhat on the nature and concentration of the supporting electrolyte. Thus for estimates of relative electron affinities identical experimental conditions are a necessity.

Table II

Compound	E_1, V^*
2,5-Dicyano TCNQ	+0.65
Tetrafluor TCNQ	+0.53
2,5-Dibromo TCNQ	+0.41
TCNQ	+0.17
2-Methyl TCNQ	+0.15
2,5 Diethyl TCNQ	+0.12
2,5-Dimethoxy TCNQ	-0.02

*Volts versus SCE in CH_3CN.

TCNQ reactions: TCNQ is a rather reactive molecule. In long lasting crystal growth experiments this adds extra problems. A variety of addition reactions can occur. A recent review lists several.[31] Thus careful selection of solvents and exclusion of light and moisture are necessary to ensure uncomplicated processes.

Spectroscopy: The spectroscopic properties of TCNQ, TCNQ⁻, $(TCNQ)_2^-$ etc. have been investigated in great detail. The above mentioned review[31] gives a comprehensive account.

Classification: Relatively few TCNQ derivatives have so far been used to any great extent. High electrical conductivity has been found in TCNQ salts of heteroaromatics, of TTF and its analogues. TCNQ derivatives have been used only with TTF- and TTT derivatives with success.

TCNQ's used with TTF-type donors are normally simple 1:1 solids, although examples of 1:2, 3:2 and 4:3 are known.[32] So far, the vast majority of work has been done with TCNQ itself, but a few conducting solids from methyl-TCNQ,[33] 2,5-dimethyl-TCNQ,[16,34] 2,5-diethyl-TCNQ,[35,36] 2,5-dichloro-TCNQ[8] and TNAP[37,38] have been reported. Several other TCNQ's have been reported to form conducting powders, but in most cases it has not been possible to obtain single crystals.

We strongly encourage that more detailed investigations of TCNQ derivatives are undertaken. Much attention was given to the isomorphous series TTF, TSeF and DTDSeF-TCNQ, and also the presumably isomorphous series $TTF-TCNQhal_2$ (hal = F, Cl, Br and I) may provide very important information, as these compounds exhibit dramatic effects,[39] as judged from preliminary measurements. The conductivities seem to change by orders of magnitude rather than the "small" changes seen when going from S to Se in TTF-TCNQ.[11,16]

Bandfilling: The variety of TCNQ's available has recently provided important experimental evidence as to which effects the degree of bandfilling (i.e. charge transfer) can have on the transport properties of TTF-TCNQ salts. HMTSeF-TCNQ and HMTSeF-$TCNQF_4$ are isomorphous, and exhibit 0.74 e/mole[40] and presumably 1.0 e/mole charge-transfer, respectively. HMTSeF-TCNQ has a room temperature conductivity of 1500 or more $(ohm \cdot cm)^{-1}$ whereas HMTSeF-$TCNQF_4$ is a semiconductor; $\sigma_{RT} \sim 10^{-3}$ $(ohm \cdot cm)$. This result has been interpreted in terms of HMTSeF-$TCNQF_4$ as a Mott-insulator.[41] Similar results are obtained for HMTTF-$TCNQF_4$[42] and it is hoped that alloying HMTSeF-TCNQ and HMTTF-TCNQ with $TCNQF_4$ may provide some important answers.

TCNQ conclusions: Many TCNQ derivatives have not yet been investigated. Available systems range (relative to TCNQ) from stronger to poorer electron acceptors. Most derivatives are 2,5-disubstituted, thereby breaking the completely uniform overlap in the TCNQ stacks. The investigation of derivatives of TCNQ in more detail will presumably provide much important information.

Classification of TTF-TCNQ compounds

We now proceed to classify some of the TTF-TCNQ's, which have been reported. The insulating mixed stack compounds are not included. In fact, we only list some of the well investigated systems, characterized by uniform segregated stacks and relatively high electrical conductivity. The compounds have been divided into 5 (or 6) classes of materials according to their transport (conductivity) properties.

Class 1. Semiconductors. Examples are HMTSF-TCNQF$_4$,[40] HMTTF-TCNQF$_4$,[41] TTF-TCNQI$_2$.[39] $\sigma_{RT} \ll 1$ (ohm·cm)$^{-1}$. In this class the use of very strong acceptors probably causes full charge transfer, leading to Mott insulators.

Class 2. Intermediate semiconductors (high T$_c$). Examples: TSeF-DETCNQ,[35] TTF-DETCNQ,[36] TTF-MTCNQ,[33] TTF-TNAP[37] and Dibenzo-TTF-TCNQCl$_2$.[8] The materials are characterized by high transition temperatures for the metal-insulator transition. $\sigma_{RT} > 30$ (ohm·cm)$^{-1}$. Careful examination of X-ray structures reveals asymmetric overlap in the stacks, caused by the asymmetric acceptors. This can give rise to narrow bands, which in turn could lead to high T$_c$'s.

Class 3. Metals-insulators (low T$_c$). TTF-TCNQ,[6] TSeF-TCNQ,[11] TMTSF-TCNQ,[43] TMTSF-DMTCNQ (amb. pressure),[34] DEDMTSF-TCNQ,[16] HMTTF-TCNQ (amb. pressure)[13] and TMTTF-TCNQ.[12] In this class room temperature conductivities range from 200 - 1500 (ohm·cm)$^{-1}$. T$_c$ is found below 100K, and the solids behave metallic from room temperature to well below 100K.

Class 4. Metals-semimetals (still highly conducting below T$_c$). HMTSF-TCNQ,[44] HMTSF-TNAP[38] and HMTTF-TCNQ (high pressure).[45] In the materials special features (disorder, low anisotropy and/or a characteristic structure) have been used to rationalize the experimental data.

Class 5. Metals. Recently it was shown that TMTSF-DMTCNQ at pressures above 10 kbar is metallic throughout the accessible temperature range. Conductivities above 10^5 (ohm·cm)$^{-1}$ were reproducibly observed below 15K, where the conductivity seems to saturate.[46] It is at present not clear whether a distinct structural change, or just increased intra- and interstack interactions, stabilize the metallic state.[47]

The high pressure phase of TMTSF-DMTCNQ is, however, the first example of a solid, which behaves as a metal down to very low temperatures.

(Class 6. Superconductors). It has been suggested that the high pressure state of TMTSF-DMTCNQ could be a special 1-D superconducting state.[46] More experimental evidence is needed to clarify the situation, but this intriguing possibility should of course be evaluated in detail.

Conclusion: After classifying the materials we should be able to draw some general conclusions. The present authors, however, feel that we, at this point, are not confident in trying to "explain" why a certain material goes into a certain class, except for materials, which go into class 1 due to full charge transfer. We can point out obvious features, as for example the effects of the detailed structure of a particular material, but how this influences the band structure and thereby the transport properties is in our opinion still in the shade, in spite of the numerous transport models which have been proposed.

Guidelines: After examining the new TTF-TCNQ derivatives which have appeared during the last six years, it is felt appropriate to extract a few general results of interest for future work. From a chemical point of view several important reactions in sulphur and selenium chemistry have been developed and are synthetically useful. In principle any desired TTF or TCNQ structure can be made in the search for new types of TTF-TCNQ organic conductors, and "special effects" can be built into the molecules.

A few results of interest for the solid state properties which are directly related to molecular properties are worth mentioning.

1. Substitution of sulphur with selenium raises the RT conductivity, lowers the metal to insulator transition and in most cases makes the donor stack dominate the transport properties. TTF and TSeF salts with the same acceptor are normally isomorphous and homogenous alloys can be made.

2. Bulky substituents on the constituent molecules push the stack apart (rise the anisotropy).

3. Disorder (permanent dipole moments, or cis-trans mixtures) smears the phase transition(s).

4. The I_p-E_a value of the DA pair is not very meaningful in determining the final charge transfer in the solid. Large values, however, seem to give rise to Mott insulators.

5. It is not possible to predict crystal structure (i.e. the effect of substituents) or when a TTF-TCNQ pair is more stable as a mixed-stack structure. Often it is, however, possible to favor the segregated stack structure by changing the crystal growth conditions.

6. It seems to be possible to prevent the metal to insulator transition.

Since it may be an ultimate goal for work in this field to avoid the metal to insulator transition some features of the few known systems which retain high conductivity to low temperature are of interest. HMTSeF-TCNQ and HMTTF-TCNQ (under pressure) retain conductivities in excess of 10 $(ohm \cdot cm)^{-1}$ at 1K. Their structure is unique and characterized by 4 short (Se-N or S-N) contacts for each DA pair. Also the structure is slightly disordered[17] and relatively low anisotropy is observed. The high conductivity at low temperature appears to result from a crossover from a 1D metallic state at high temperature to a 3D semimetallic state at lower tenperature. Recently, however, in crystals which appear less disordered from X-ray studies an anomaly has been detected at 24 K indicating that also this system undergoes a Peirls transition.[48]

The other example, TMTSF-DMTCNQ,[46,47] (mentioned above) has a quite "ordinary" structure at ambient pressure, and does not seem to enter a state which can unambiguously be described as semimetallic[47] when hydrostatic pressure is applied.

A few other systems which are highly conducting at low temperature are known. HMTSeF-TNAP[36] and $(TSeT)_2Cl[49]$ have both recently been investigated and more examples will hopefully appear as a result of future work.

One chain materials based on TTF's

Finally, we want to mention that one chain materials based on TTF and derivatives have until recently not been very exciting from the point of view of obtaining high conductivity and low T_c.[50] However, several new materials exhibiting very high conductivity have now been reported.[53] The rationale for this work was the observation that a selenium containing compound, TMTSF-DMTCNQ, although commensurate[48] (C.T. = $\frac{1}{2}$), is metallic at ambient pressure down to 42 K. Likewise, $(TSeT)_2Cl[49]$ is metallic down to 24 K. Delhaes et al.[51], however, reported a series of stoichiometric sulphur compounds, $(TMTTF)_2X$. This series exhibits class 2 properties as does a series of incommensurate $(TMTSF)_2^+$ halides and pseudohalides reported by Somoano et al.[52]

The new series,[53] (TMTSF)$_2$X, (X = PF$_6^-$, AsF$_6^-$, SbF$_6^-$ and NO$_3^-$),
consists of four commensurate salts which all exhibit T$_c$'s well
below 20 K. Conductivities greater than 10^5 are found for two
materials (X = PF$_6^-$ and NO$_3$) below 20K, the highest reported so
far for organic materials at ambient pressure. Moreover, the resistivity results (ρ decreases 200x from room temperature to 20K)
indicate that these simple one chain materials can supply some answers as to which resistive mechanisms are important in organic
conducting solids.

From this result and the above mentioned results on TMTSF-
DMTCNQ we extract that heavy atom effects are important in stabilizing molecular metals to low temperatures. Thus new donors (and
acceptors) containing heavy atoms should be prepared and investigated. Especially interesting would be the preparation of di- or
tetratellurofulvalenes, both for use in one- and two-chain materials.

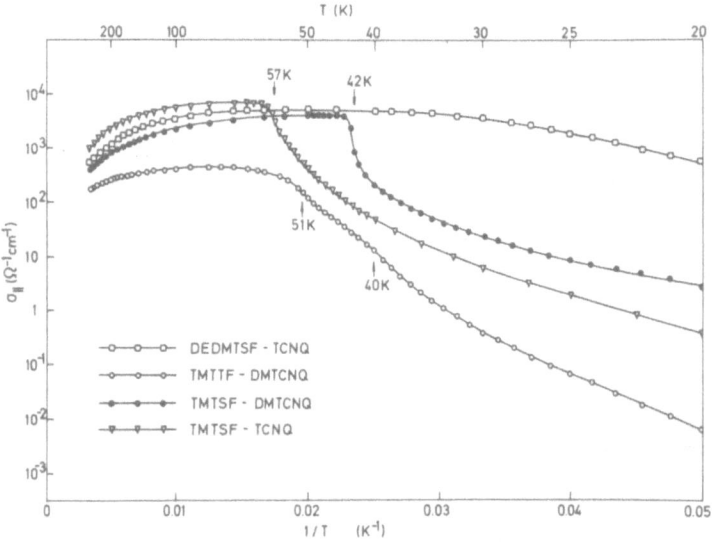

Figure 1. Examples of multiple phase transitions in
a sulphur compound (TMTTF), vs. a single transition
with a selenium donor (TMTSF), and of disorder-smea-
ring (DEDMTSF).

References

1. N. Bartlett, R. N. Biagioni, G. McCarron, B. McQuillan, and F. Tanzella in Molecular Metals, W. E. Hatfield ed. Plenum Press 1979.
2. J. B. Torrance, this report.
3. F. Wudl, this report, pp. 265.
4. T. J. Kistenmacher, T. E. Phillips, and D. O. Cowan, Acta Cryst. B 30, 763 (1973).
5. F. Denoyer, R. Comes, A. F. Garito, and A. J. Heeger, Phys. Rev. Lett. 35, 445 (1975).
6. L. B. Coleman, M. J. Cohen, D. J. Sandman, F. G. Yamagishi, A. F. Garito, and A. J. Heeger, Sol. State Commun. 12, 1125 (1973).
    J. P. Ferraris, D. O. Cowan, V. Walatka, and J. H. Perlstein, J. Amer. Chem. Soc. 95, 498 (1973).
7. D. Allender, J. W. Bray, and J. Bardeen, Phys. Rev B 9, 119 (1974), and D. Jérome, this report, pp.123, and M. Weger, this report, pp. 77.
8. C. S. Jacobsen, K. Mortensen, H. Pedersen, and K. Bechgaard, to be published, and R. I Shibaeva, Lecture Notes in Physics 95, Springer 1979.
9. F. Wudl, G. M. Smith and E. J. Hufnagel, J. Chem. Soc., Chem. Commun. 1979, 1453 (1979).
    D. L. Coffen, J. Q. Chambers, D. R. Williams, P. E. Garret, and N. D. Canfield, J. Amer. Chem. Soc. 93, 2258 (1971).
10. M. Narita and C. U. Pittman, Jr. Synthesis 1976, 489 (1979).
11. E. M. Engler and V. V. Patel, J. Amer. Chem. Soc. 96, 7376 (1974).
12. J. P. Ferraris, T. O. Pochler, A. N. Bloch, and D. O. Cowan, Tetrahedron Lett. 1973, 2553 (1973).
13. R. L. Greene, J. J. Mayerle, R. R. Schumaker, G. Castro, P. M. Chaikin, S. Etemad, and S. J. LaPlaca, Sol. State Commun. 20, 943 (1976).
14. F. Wudl, unpublished.
15. D. Chasseau, J. Gaultier, C. Hauw, J. M. Fabre, L. Giral, and E. Torreilles, Acta Cryst. B 34, 2811 (1978).
16. C. S. Jacobsen, K. Mortensen, J. R. Andersen, and K. Bechgaard, Phys. Rev. B 18, 905 (1978).
17. T. E. Phillips, T. J. Kistenmacher, A. N. Bloch, and D. O. Cowan, J. Chem. Soc., Chem. Commun. 1976, 334 (1976).
18. Y. Ueno, Y. Masugama, and M. Okawara, Chem. Lett. 1975, 603 (1975).
19. N. Martinez Rivera, E. M. Engler, and R. R. Schumaker, J. Chem. Soc., Chem. Commun. 1979, 184 (1979).
20. J. R. Andersen, V. V. Patel, and E. M. Engler, Tetrahedron Lett. 1978, 239 (1978).
21. D. C. Green, J. Chem. Soc., Chem. Commun. 1977, 161 (1977) and J. Org. Chem. 44, 1476 (1979).
22. L. Carlsen, K. Bechgaard, C. S. Jacobsen, and I. Johansen, J. Chem. Soc., Perkin II 1978, 862 (1978).

23. R. Gleiter, M. Kobayashi, J. Spanget-Larsen, J. P. Ferraris, A. N. Bloch, K. Bechgaard, and D. O. Cowan, Ber. Bunsenges. Phys. Chem. 79, 1218 (1975).

24. J. R. Andersen, H. Egsgaard, E. Larsen, K. Bechgaard, and E. M. Engler, Org. Mass Spectroscopy 13, 121 (1978).

25. D. S. Acker, R. J. Harder, W. R. Hertler, W. Mahler, L. R. Melby, R. E. Benzon, and W. E. Mockel, J. Amer. Chem. Soc. 82, 6408 (1960).

26. D. S. Acker and W. R. Hertler, J. Amer. Chem. Soc. 84, 3370 (1962).

27. R. C. Wheland and E. L. Martin, J. Org. Chem. 28, 3101 (1975).

28. C. E. Klots, R. N. Compton, and V. F. Raaen, J. Chem. Phys. 60, 1177 (1974).

29. A. W. Addison, J. P. Barnier, V. Gujval, Y. Hoyano, S. Huizinga, and L. Weiler, Molecular Metals, W. E. Hatfield ed. Plenum Press 1979.

30. B. S. Jensen, K. Bechgaard, and J. R. Andersen, to be published.

31. B. P. Bespalov and V. V. Titov, Russ. Chem. Rev. 44, 1091 (1975).

32. K. Bechgaard, unpublished.

33. C. S. Jacobsen, J. R. Andersen, K. Bechgaard, and C. Berg, Solid State Commun. 19, 1209 (1976).

34. J. R. Andersen, C. S. Jacobsen, G. Rindorf, H. Soling, and K. Bechgaard, Acta Cryst. B 34, 1901 (1978).

35. J. R. Andersen, R. A. Craven, J. E. Weidenborner, E. M. Engler, J. Chem. Soc., Chem. Commun. 1977, 526 (1977).

36. A. J. Schultz, G. D. Stucky, R. Craven, M. J. Schaffman, and M. B. Salamon, J. Amer. Chem. Soc. 98, 5191 (1976).

37. P. A. Berger, D. J. Dahm, G. R. Johnson, M. G. Miles, and J. D. Wilson, Phys. Rev. B 12, 4085 (1975).

38. K. Bechgaard, C. S. Jacobsen, and N. H. Andersen, Sol. State Commun. 25, 825 (1978).

39. R. C. Wheland and J. L. Gillson, J. Amer. Chem. Soc. 98, 3916 (1976).

40. C. Weyl, E. M. Engler, S. Etemad, K. Bechgaard, and G. Jehanno, Sol. State Commun. 19, 925 (1976).

41. A. N. Bloch and D. O. Cowan, private communication.

42. J. B. Torrance, J. J. Mayerle , K. Bechgaard, B. D. Silverman, and Y. Tomkiewicz, to be published.

43. K. Bechgaard, D. O. Cowan, and A. N. Bloch, J. Chem. Soc., Chem. Commun. 1974, 937 (1974).

44. A. N. Bloch, D. O. Cowan, K. Bechgaard, R. E. Pyle, R. H. Banks, and T. O. Poehler, Phys. Rev. Lett. 34, 1561 (1975).

45. R. L. Friend, D. Jérome, J. M. Fabre, L. Giral, and K. Bechgaard, J. Phys. C. 11, 263 (1978).

46. A. Andrieux, C. Duroure, D. Jérome, and K. Bechgaard, Le Jour. de Phys. Lett. 40, L-381 (1979).

47. A. Andrieux, P. M. Chaikin, C. Duroure, D. Jérome, C. Weyl, K. Bechgaard, and J. R. Andersen, Le Jour. de Phys. Lett., submitted.

48. J. P. Pouget, private communication.

49. I. F. Schegolev and R. B. Lubovskii, Lecture Notes in
     Physics 95, 39, Springer 1979.

50. B. A. Scott, S. J. LaPlaca, J. B. Torrance, B. D. Silverman,
     and B. Welber, J. Amer. Chem. Soc. 99, 6631 (1977).

51. P. Delhaes, C. Coulon, J. Amiell, S. Flandrois, E. Toreilles,
     J. M. Fabre, and L. Giral, Mol. Cryst. Liq. Cryst. 50, 43
     (1979).

52. R. B. Somoano, private communication.

53. K. Bechgaard, C. S. Jacobsen, K. Mortensen, H. J. Pedersen,
     and N. Thorup, Solid State Commun., submitted.

CHEMISTRY OF ORGANIC CONDUCTORS: A REVIEW OF STRATEGIES

F. Wudl

Bell Laboratories, Murray Hill, NJ 07974 USA

INTRODUCTION

The purpose of this lecture is to present a review of the strategies which have evolved over approximately the past six years in the design of organic materials which will be metallic over a long temperature range. Known organic conductors exhibit a plethora of solid state transitions as a function of temperature. These transitions are usually accompanied by dramatic (or subtle) changes in structural, electric and magnetic properties. It is these variations as a function of temperature which have captivated a large segment of the population of solid state physicists (but a considerably smaller fraction of chemists) and kept their attention over a period of approximately five years (1973-1978). Even with the small number of chemists involved, a symbiosis between physicists and chemists has almost always led to the discovery of novel materials and/or properties. Design must therefore have input from both groups. With that in mind, this lecture would not have been possible without input from and collaboration with A. N. Bloch, F. J. DiSalvo, W. E. Geiger, R. C. Haddon, M. L. Kaplan, D. Nalewajek, D. Moncton, S. G. Soos, G. A. Thomas, W. M. Walsh and E. T. Zellers.

The goal of most chemists involved in the materials search end of the subject of this conference has been, and is, to stabilize the organic metallic state. From what the theorists and experimental physicists tell us, in order to achieve this goal, one needs disorder and an increase in dimensionality.

L. Alcácer (ed.), The Physics and Chemistry of Low Dimensional Solids, 265–279.
Copyright © 1980 by D. Reidel Publishing Company.

Other chemists have attempted to improve on the one
dimensional organic metals and have evolved a set of guidelines for design which will be the subject of K. Bechgaard's
and J. Torrance's lectures. In this lecture those guidelines
will be implied. Possible new donors containing sulfur,
selenium or tellunium will be proposed, while the design of
new acceptors will not be covered.

Strategies

The strategies to achieve a truly metallic organic solid,
an event which has not yet occurred at atmospheric pressure,
are (I) design of donors with sites of maximum spin at the
periphery of the molecule to increase interchain interactions,
(II) substitution of sulfur by selenium and tellurium,
(III) reduction of on-site coulomb repulsions in acceptors,
(IV) incorporation of a certain amount of disorder to thwart
the Peierls distortion, (V) design of polyconjugated species
with two donor, two acceptor, and donor and acceptor sites on
the same molecule, (VI) design of polyconjugated donors or
acceptors, (VII) emulation of alkali metals, and (VIII) emulation and extrapolation based on $(SN)_x$ and $(CH)_x$.

(I) Design of Donors with Peripheral Spin

The donor TTT has the spin bearing chalcogens on part of
its periphery but intermolecular interactions

are somewhat hindered by the peri hydrogens. Also, it is a
relatively large molecule whose size was assumed to be
incompatible with that of the popular acceptor TCNQ. We
designed TTN(2) with the thought that the sulphur will be more
exposed and with a molecular size closer to TTF, the possible

packing problems of TTT were expected to be ameliorated. In fact, TTT TCNQ is a semiconductor(1) and TTN TCNQ has a short temperature range metallic behavior(2) in its conductivity. But this result is misleading since TTT is a considerably stronger reducing agent(3) than TTN(2) and is probably fully charged in its TCNQ salt. In support of this hypothesis is the fact that the more conducting TCNQ salt of TTT is $TTT(TCNQ)_2$(1). The latter is, in fact, identical in its conductivity behavior to TTN TCNQ(1).

On the whole, TTN turned out to be an uninteresting donor with respect to other salts such as the iodide [$TTNI_2$(or TTN_3I_6)] which, contrary to TTT_2I_3 is a semiconductor. However, in collaboration with Professor W. Geiger and F. J. DiSalvo, we found that the insulating ($\rho > 10^6 \Omega$cm) TTN · Ni(tfd)$_2$ exhibits extremely unusual magnetic properties (cf Figure 1). Diffuse X-ray scattering and particularly neutron scattering measurements are so far precluded because of crystal size. In the absence of these and other measurements we cannot speculate on the cause for this unusual first order transition.

A more interesting approach was attempted by Cava, who prepared(4) compound 1, an isomer of TTF.

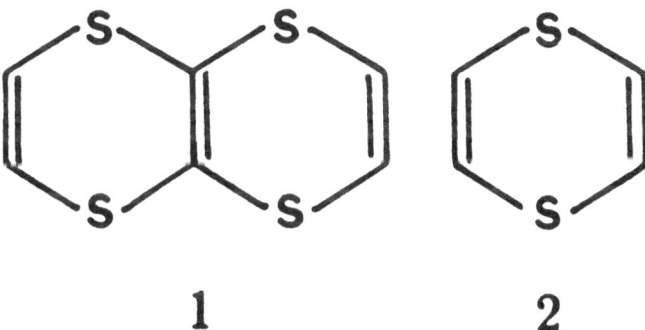

1   **2**

Quite surprisingly, while this molecule fits most of the other guidelines(cf tutorials by Bechgaard and Torrance), it turned out to be an absolute failure in regard to its donor characteristics and ability to form D-A complexes. While a priori it was inconceivable that the observed results would occur, it was known that dithiines (2)(5) are actually very stable, relatively poor donors.

These are reported cases of molecules belonging to this category. There remain some molecules (3-6) to be prepared which should have spin (or hole) density only in the periphery:

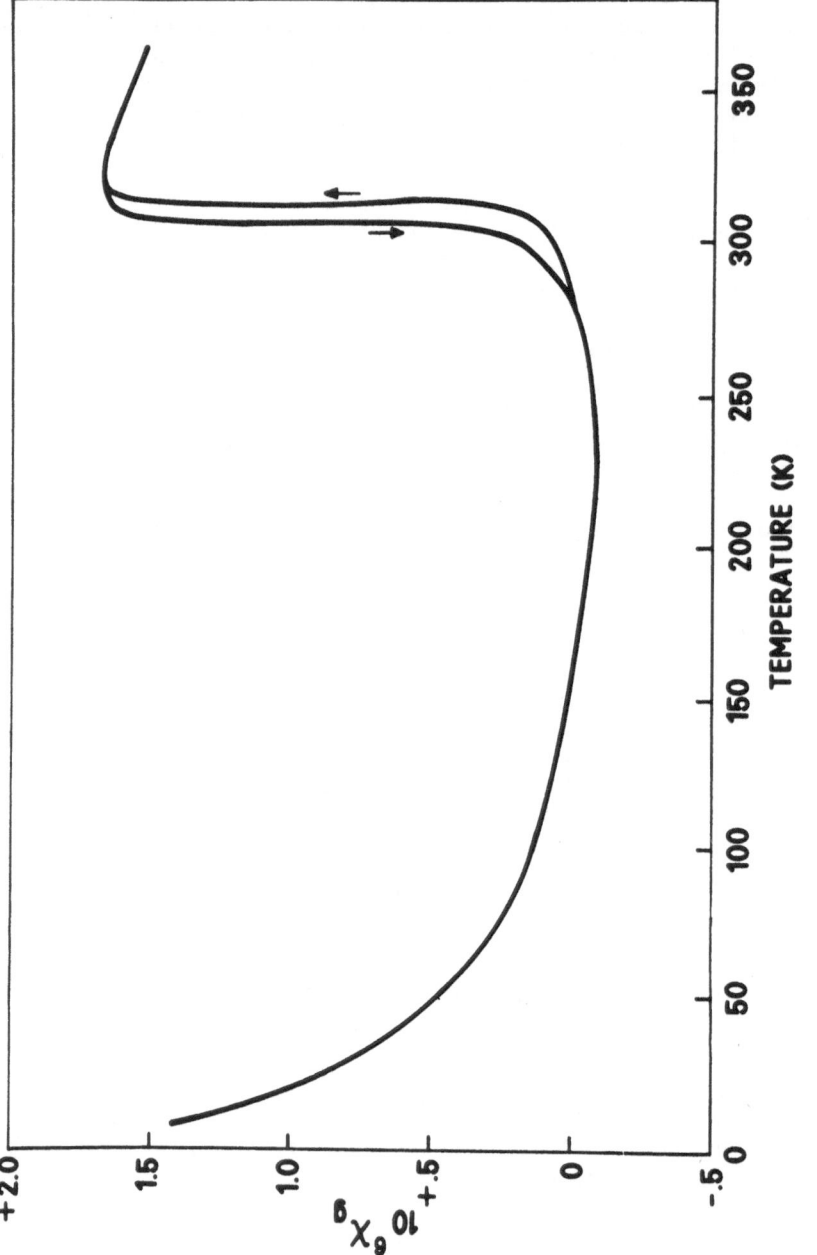

Figure 1.   Magnetic susceptibility of TTN Ni(tfd)$_2$ not corrected for diamagnetic contributions.

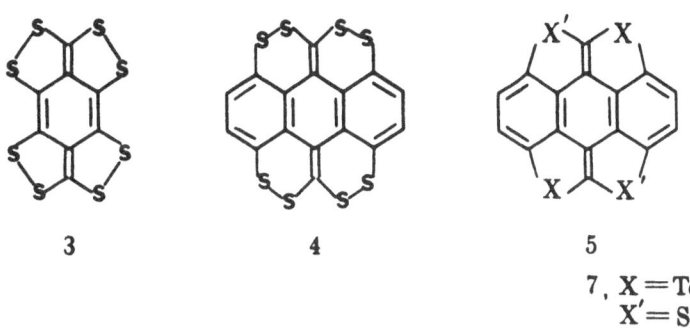

3   4   5

7, X = Te
   X' = S

A relative of 3, 6, has been reported in the literature(6) but its solid state properties have not yet been

6   8

investigated. Molecular models indicate that only large chalcogens such as Se and Te will produce 5 devoid of bond strains. Indeed, 5 promises to be the only reasonably stable symmetric, tellurium containing donor; however its potential insolubility may seriously limit its usefulness as a donor. In attempts to prepare 7, which is expected to be more soluble than 5, we prepared the precursor 8. The latter turned out to be very similar in its properties to perylene but had excess reactivity associated with the methylene carbons(7).

(II) Substitution of S and Se by Te.

Since Te is more electropositive than the other chalcogens and is considerably larger, incorporation into a donor would clearly tend to increase both intra- and inter-stack delocalization of electrons. Unfortunately the C-Te bond is extremely weak and synthetic procedures for its formation are not as well developed as those for the sulfur and selenium analogs. To date, the measurement of physical properties of single crystals of a tellurium containing D-A salt have not been reported. Compaction measurements are very encouraging, however, for the salt 9(8).

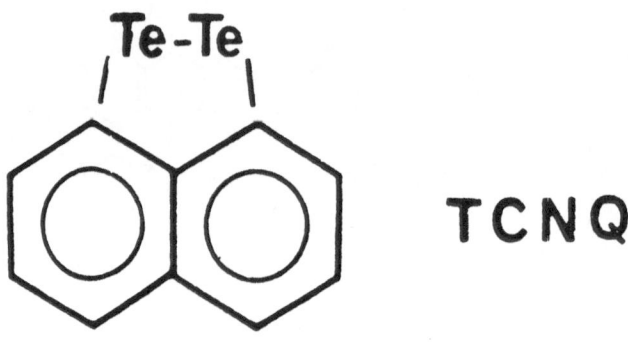

TCNQ

9

(III) Reduction of On-Site Coulomb Repulsion.

Before it was known that there is only partial charge transfer in TTF TCNQ, it was suggested that to minimize on-site coulomb repulsion, larger acceptors should be synthesized. This would spread the negative charge in the radical anion, and particularly the dianion over a larger area. The strategy here was to stretch out TCNQ as in the molecules A-E.

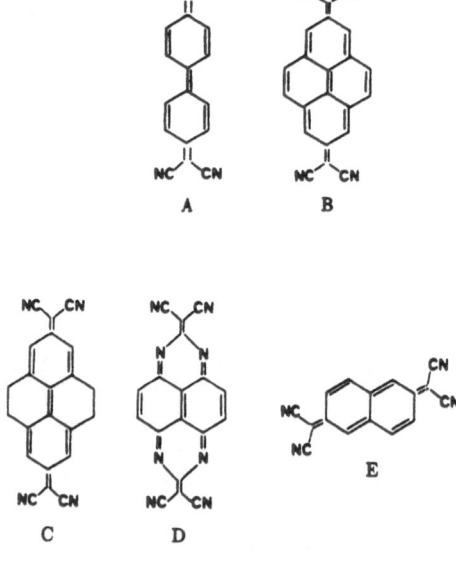

Of all these, only C & E have been isolated. Anions derived
from D and A were isolated but to date exhibited uninteresting
solid state properties(9)(10). It was reported that C did not
form a charge transfer salt with TTF(11). The TTF salt of E
is a poorer conductor than the TCNQ salt whereas HMTSF·E is a
better room temperature conductor than HMTSF·TCNQ.

(IV) Incorporation of Disorder.

Lectures in this ASI dealing with the physics of one
dimensional metals show that the Peierls transition
(responsible for converting organic metals to insulators) is
due to a periodic distortion of the lattice. If the periodicity
of the distortion could be interrupted, then the distortion
could be destroyed. One way to accomplish this would be to
introduce random disorder without disturbing the remainder
of the lattice. Some of the molecules that were designed
toward this end are shown below(12)

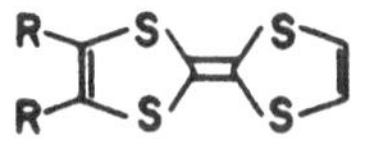

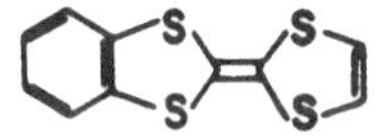

10 a R = CH₃, R'=H
    b R = CH₃, R'=CH₃
    c R = Ph, R'=H

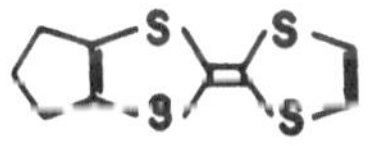

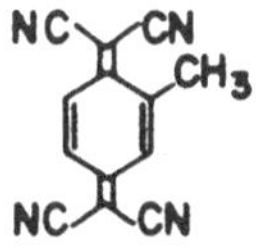

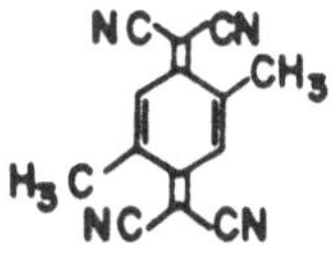

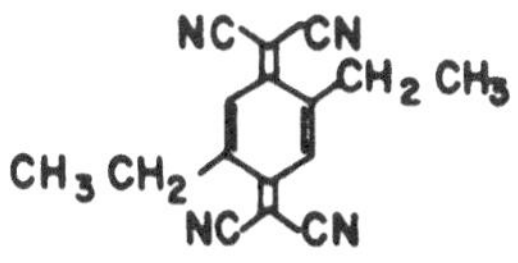

The first report on a successful smearing of the Peierls
distortion was by Engler and Bechgaard, et al(13) upon doping
of TSeF TCNQ with MTCNQ; however, the metal to insulator
transition was not destroyed and the room temperature
conductivity was observed to be lower compared to the undoped

material.   Successful attempts to prepare single crystals
with the unsymmetrical dimethyl TTF's have so far not been
reported.

(V) Design of conjugated D-D, D-A, and A-A Molecules.

Another strategy to increase dimensionality is to link
two donors or acceptors or a donor and acceptor together
with unsaturated bridges.   So far, only 11 and 12 have been
reported(14)(15).   Other linked TTF's

11   12

have been prepared but do not fit the criteria mentioned above.
Unfortunately neither of these molecules, to this day, produced
single crystals (with suitable acceptors) which exhibited
properties of a multidimensional solid.

Other research groups have tried to prepare compounds
where an acceptor such as TCNQ and another acceptor or a donor
such as TTF have been linked together but no results of these
synthetic efforts have appeared.

(VI) Polyconjugated Donors.

Polymers such as 13, 14, 15 and 16 belong to this family.

13   14   15

16

Of these, only 14 and 15 have been reported in the literature(16)(17) and 14 with M=Ni is the only one exhibiting relatively high compaction conductivity ($\sim 30 \ \Omega^{-1} cm^{-1}$)(16).

(VII) Emulation of Alkali Metals.

Because of the D_{3h} symmetry of the phenalenyl system (F), its HOMO is a non bonding orbital. As a result, the radical, anion, and cation oxidation states are all practically isoenergetic. If we were to suitably substitute the phenalenyl with strong electron donating substituents such as the chalcogens, particularly in the positions of high reactivity, then the ionization potential would be lowered at least to the level of most reactive metals (e.g. Zn).

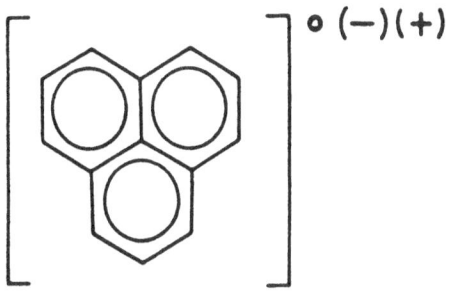

It was shown recently that one disulfide bridge in 2 out of the 6 reactive positions lowered the ionization potential of the radical in solution by ca 0.52 v.(18) and stabilized it toward dimerization relative to phenalene. Attempts to prepare more highly substituted phenalenes are currently in progress and assuming strong intermolecular interactions, these phenalenes may give rise to exciting organic solids where a conductivity mechanism will be other than that for TTF TCNQ or any other TTF type salt.

(VIII) Emulation of (SN)$_x$ and (CH)$_x$.

Because (SN)$_x$ is the first example of a superconducting material composed of elements which are themselves not even metallic, it has prompted a great deal of search for other solids whose components are linked in a manner identical to (SN)$_x$. This search has taken several directions: (a) attempted preparations of (SeN)$_x$, (b) oligomers of (SN)$_x$ with organic or silyl terminal groups(19), (c) replacement of N by P or carbon.

While no reports on the successful preparation of (SeN)$_x$ exist, it appears as though the precursor Se_4N_4 does not exhibit the chemical properties of S_4N_4 such that the reaction:

$$S_4N_4 + Br_2 -> (SN)_x \ Br_n$$

does not take place(20). Also, the critical reaction:

$$S_4N_4 \xrightarrow[\Delta]{Ag} S_2N_2 \rightarrow (SN)_x$$

remains <u>sui generis</u> to S_4N_4.

Some oligomers of $(SN)_x$ have been known for some time. A recent attempt at an alternate synthesis of $(SN)_x$ afforded higher molecular weight $R_2(SN)_n$. Their solid state properties, particularly after doping have not been reported(19).

It would appear on first sight that $(SP)_n$ should be similar to $(SN)_x$ or that $(PN)_x$ may exhibit similar properties. However, all the known (SP) compounds contain P^V and are cyclic or bicyclic. Furthermore, just as is the case with the $(X_2P=N)_n$ high molecular weight polymers with P^V, these (P^VS) polymers would be expected to be insulators. Also, the P-S bond is hydrolytically very unstable.

On the other hand, $P^{III}N$ has been generated in the gas phase. But unfortunately its condensation product is <u>white</u>, indicating no long-range extended interactions. Nonetheless if someone could prepare the unsubstituted cycle $P_2^{III}N_2$ it may undergo a <u>solid state</u> polymerization to an $(SN)_x$ equivalent.

On the other hand, replacement of N by C, or the addition of C to SN, could lead to potentially interesting polymers.

In order to prepare the exact analog of $(SN)_x$ one would have to prepare the heterocycle shown below either as an intermediate or actually matrix isolated.

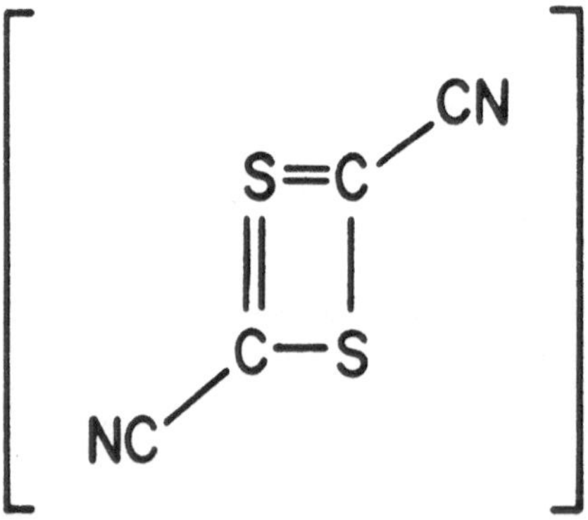

17

Unfortunately this family of heterocycles is unknown and efforts to prepare 17 may be fruitless since all valence electron MO calculations seem to indicate that the order of stability is as follows:

Other approaches to the preparation of carbon analogs of $(SN)_x$ involved the stabilization of the trivalent carbon as follows:

18   19

We have found highly conducting iodine doped precursors (or perhaps oligomers) of 18 but were unable to prepare 18 by itself(21). Similarly we have not been able to prepare undoped 19. Polymers derived from 19 and doped with pyridine, Bu_4N+ and TTF+ were prepared. Of these, the latter had a compressed pellet resistivity of 19 Ω cm(22).

The literature on $(CS)_n$ is confusing(23). It is generally prepared via the decomposition of CS_2. Variously colored films are produced. These may contain a variety of impurities including nickel.

We had found a novel approach to the preparation of 20. The mass spectrum of this compound revealed that it cleaved cleanly to C_2S_2:

$$\text{(structure)} \xrightarrow{70\,ev} C_2S_2 + I_2 + CO$$

20

Since usually photochemical reactions emulate mass spectrometer fragmentations, we attempted to prepare $(C_2S_2)_n$ from the above precursors. Unfortunately, and as is well known(24), the only reaction that took place was the loss of CO with concomitant formation of an insulating $(C_2S_2I_x)_n$ brown solid. It remains to be seen what happens when 20 is subjected to plasma discharge conditions.

The main problems with polymers such as 13-16, 18 and 19 is that they are prepared by solution polymerization. This process invariably leads to extremely insoluble, amorphous powdered materials as contrasted to crystalline $(SN)_x$ which is produced via a solid state polymerization reaction of a crystalline monomeric precursor.

While the results of oxidation and reduction doping of polyacetylene are exciting; again, only polyconjugated solids which already exist in the literature(28) exhibit redox properties similar to $(CH)_x$ (but not nearly as dramatic). Attempts to design materials based on $(CH)_x$ with possibly improved fabricability have so far not been fruitful(28).

Not all polyconjugated solids are subject to doping, for example, polydiacetylenes do not exhibit anywhere near the dramatic changes in transport properties upon doping as $(CH)_x$(29).

Conclusions

While we did not cover every new material that has been prepared and deemphasized physical properties as well as actual syntheses, the following are generalizations which can be made based on events of the past five years:

(1) All new donors and acceptors which have been designed ab initio (i.e. not simple variations on a theme e.g. TSeF) with the sole purpose to improve on TTF, TTT, TCNQ, and TNAP, have been failures(25)(26).

(2) All new polymers designed around $(SN)_x$ and TTF have been made with the wrong morphology i.e., not as thin, uniform films or single crystals.

In other words, the rate of return on pushing for systems beyond those whose unusual properties were discovered long after the materials were already in existence has been poor. One explanation for this result is that (1) we are essentially ignorant of most factors which are necessary for crystal growth (theory and practice), (2) the synthetic methodology for the generation of slightly more complicated molecules containing Se or Te is virtually non-existent.

The most important message is to keep an open mind. For example, TTF Cu(tfd)$_2$ was a totally uninteresting compound until its magnetic susceptibility as a function of temperature was determined. TTN Ni(tfd)$_2$ was equally uninteresting until its magnetic susceptibility above room temperature was determined. $(CH)_x$ was just another insulator even after it had been made in film form until it was doped. It had been doped previously(30) in amorphous powder form without attainment of the dramatic effects observed in thin sheets.

Finally, of the three approaches to research on the subject of this NATO ASI; namely, (a) "discover" an existing compound where crude measurements by others gave results which were encouraging enough to warrant careful work (e.g. TTF-TCNQ, $(SN_x Hg(SbF_6)_x, (CH)_x, TTTI_x, TSeTCl_x, PerI_x)$, (b) Perform variations on a theme and careful measurements on these (e.g. TSeF, HMTSeF), and (c) design novel systems, prepare them, and measure their properties [e.g. 1, TTN, 8, 21(27)], clearly the first and second are the ones that have been the most successful and the last the least.

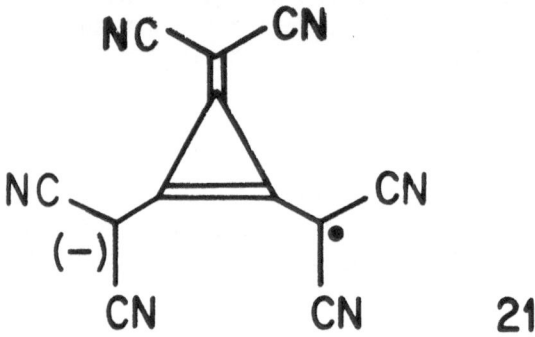

A more positive conclusion is that the syntheses of conducting polymers 18 and 19 required the development of molecules with unique sulfur and selenium hybridization containing new functional groups whose chemistry is still virtually unknown particularly in regard to transition metal reactions.

REFERENCES

1. L. I. Buranov, O. N. Eremenko, R. B. Lyubovskii,
   L. P. Rozenberg, M. L. Khidekel, R. P. Shibaeva,
   I. F. Shchegolev, and E. B. Yagubskii, J E T P Lett, 20,
   208 (1974).
2. F. Wudl, D. E. Schafer, B. Miller, J. Amer. Chem. Soc.,
   98, 252 (1976).
3. W. E. Geiger, Jr., J. Phys. Chem. 77, 1862 (1973).
4. M. Mizuno, M. P. Cava, A. F. Garito, J. Org. Chem., 41,
   1484 (1976).
5. W. E. Parham, H. Wynberg, W. R. Hasek, P. A. Howell,
   R. M. Curtis, W. L. Lipscomb, J. Amer. Chem. Soc., 76,
   4957 (1954).
6. L. K. Hausen, A. Hordvik, J. Chem. Soc. Chem. Commun.,
   800 (1974).
7. F. Wudl, R. C. Haddon, E. T. Zellers, and F. B. Bramwell,
   J. Org. Chem., 44, 2491 (1979).
8. J. Meinwald, D. Dauplaise, F. Wudl, J. J. Hauser, J. Amer.
   Chem. Soc., 99, 255 (1977).
9. A. W. Addison, N. S. Dalal, Y. Hoyano, S. Huizinga, and
   L. Weiler, Can. J. Chem., 55, 4191 (1977)
10. F. Wudl, M. L. Kaplan, B. K. Teo, and J. H. Marshall,
    J. Org. Chem., 42, 1665 (1977).

11. L. Weiler, ABSTRACTS, ACS/CSJ CHEMICAL CONGRES, PHYSICAL CHEMISTRY, April 2-6 (1979), D. O. Cowan, private communication.
12. F. Wudl, A. A. Kruger, M. L. Kaplan, R. S. Hutton, J. Org. Chem., 42, 768 (1977). M. P. Cava and M. V. Lakshmikantham, PROC. N. Y. ACAD. SCI. USA, 313, 355 (1978). D. C. Green, Ibid., 313, 361 (1978) and references therein.
13. E. M. Engler, R. A. Craven, Y. Tomkiewicz, B. A. Scott, K. Bechgaard, and J. R. Andersen, J. Chem Soc. Chem. Commun., 337 (1976).
14. V. Y. Lee, R. R. Schumaker, ABSTRACTS, ACS/CSJ CHEMICAL CONGRESS, HONOLULU, HAWAII, April 1979, ORGN 424.
15. M. L. Kaplan, R. C. Haddon, and F. Wudl, J. Chem. Soc. Chem. Commun. 388 (1977).
16. E. M. Engler, R. R. Schumaker, and F. B. Kaufman, in MOLECULAR METALS, W. E. Hatfield, Ed., NATO Conference Series VI, Plenum Press, N. Y. 1979, pp. 31-34.
17. B. K. Teo, F. Wudl, J. J. Hauser, and A. A. Kruger, J. Amer. Chem. Soc., 99, 4862 (1977).
18. R. C. Haddon, F. Wudl, M. L. Kaplan, J. H. Marshall, R. E. Cais, and F. B. Bramwell, J. Amer. Chem. Soc., 100, 7629 (1978).
19. J. Kuyper and G. B. Street, J. Amer. Chem. Soc., 99, 7848 (1977).
20. G. Wolmerhäuser, G. R. Brulet, and G. B. Street, Inorg. Chem., 17, 3586 (1978).
21. F. Wudl, A. A. Kruger, and G. A. Thomas, PROC. N. Y. ACAD. SCI. USA, 313, 79 (1978).
22. F. Wudl, ABSTRACTS, ACS/CSJ CHEMICAL CONGRESS, HONOLULU, HAWAII, April 1979.
23. Not fully characterized $(CS)_n$ and $(CSy)_n$ were reported: R. Steudel, Angew Chem. Int. Ed. Engl., 6, 635 (1967); B. Krebs, G. Gattow, Z. Anorg, Allg. Chem., 338, 225 (1965).
24. N. Jacobsen, P. DeMayo, A. C. Weedon, Nouv. J. Chim., 2, 331 (1978).
25. N. F. Haley, J. Chem. Soc. Chem. Commun., 207 (1977).
26. D. J. Sandman, A. P. Fisher, III, T. J. Holmes, and A. J. Epstein, Chem. Commun., 687 (1977).
27. T. Fukunaga, N. D. Gordon, P. J. Krusic, J. Amer. Chem. Soc., 98, 611 (1976).
28. R. H. Baughman, SYMPOSIUM ON THE STRUCTURE AND PROPERTIES OF HIGHLY CONDUCTING POLYMERS AND GRAPHITE, SAN JOSE, CALIF. March 29-30, 1979.
29. D. Bloor, C. L. Hubble, and D. J. Ando, in MOLECULAR METALS, W. F. Hatfield, Ed., NATO Conference Series VI, Plenum Press, N. Y. 1979 pp. 243-247.
30. D. J. Berets, D. J. Smith, Trans. Faraday Soc., 64, 823 (1968).

THE RADICAL CATION SALTS : A VIEW OF ORGANIC METALS

P. DELHAES

Centre de Recherche Paul Pascal (CNRS)
33405 TALENCE Cédex (France)

————————

There is a two-fold goal in this presentation : in the first part
an overview about this class of materials is given. During the second one new results on these one-chain compounds with modulated
structures are presented and analyzed.

————————

The radical-ions salts (RIS) are different from the charge transfer complexes (CTC) because they normally offer one efficient stack where the radical-ions are associated with a counter-
ion.

In this class of compounds the radical-cation salts (RCS)
complexed with halogens and pseudo-halogens have been investigated
recently : they are one-chain compounds which present a metallic
behavior as good as the CTC or two-chain compounds. To observe a
metallic conductivity, two conditions must be fulfilled :

- segregated stacks of cations D
- a fractional stoîchiometry for the halogen X : $D^{+}X_{x}^{-}$,
  $x \neq 1$.

This second point is leading us to the following remark :
the fractional stoîchiometry is equivalent to intermediate degree
of charge transfer in classical CTC. The degree of charge transfer
is deduced from stoichiometry as far as the halogen species is well
defined (for exemple Br^{-} or I_{3}^{-}, supposed to be the unique counter-
ion).

Neglecting the preparative difficulties to obtain single
crystals with a good quality and with defined and reproducible
stoîchiometries I will analyse the results with halide or pseudohalide salts from different series (TTF tétrathiafulvalene, TMTTF

281

L. Alcácer (ed.), The Physics and Chemistry of Low-Dimensional Solids, 281–291.
Copyright © 1980 by D. Reidel Publishing Company.

tetramethyl TTF, TTT tetrathiotetracene, their selenium analogs and DIPS ϕ_4 tetraphenyldithiopiranylidene).

There are two purposes to this presentation : on one hand to synthetise the knowledge about this class of materials, on the other hand to offer new results which confirm the interest we give them.

1. STRUCTURAL CLASSIFICATION

From the X-rays cristallographic structures the radical-cation salts can be classified regarding these two points :

(i) The molecular arrangement of cations along the stacking axis (C-axis) with the number of orientations (r) of the cation stacks in the (ab) plane.

(ii) The counter-ion position and the existence of a sublattice associated with it looking for the molecular arrangement in a stack, the following classification is put up :

- *eclipsed configuration*

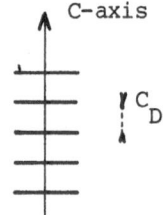

$$TTF-(SCN)_{0.588}$$
$$TTF-(SeCN)_{0.580}$$
$$TTF-(Br)_{0.74\ 0.79}$$
$$TTF-(I)_{0.71}$$
$$TSeT-(I)_{0.71}$$

tetragonal lattices (1)(2) (r=2)
(3)
monoclinic (4)
lattices (5) (r=2)

- *slipped configuration*

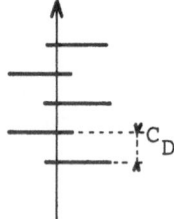

$$TTT_2(I_3)_{1+\delta}$$
$$DIPS\ \phi_4-(I_3)_{0.71}$$
$$TSeT-I_{0.50}$$
$$TSeT-Cl_{0.50}$$

orthorhombic lattices (6) (r=1)
monoclinic lattice (r=2) (7)

The molecules are tilted relative to the stacking axis as in TTF-TCNQ ("herring bone fashion").

- *alternate configuration*

$$TMTTF-X'_{0.5}\quad (X'=Br,BF_4\ ClO_4,PF_6,SCN)$$
triclinic lattices (r=1) (8)
(it is interesting to quote that ternary iodine-TCNQ salts present the same configuration too (9))

TABLE 1 - *INTERACTIONS BETWEEN CATION AND ANION SUBLATTICES*
(incommensurate and commensurate structures)

Compounds (RCS)	Commensurability of the two sublattices $\frac{c_D}{c_X}$	S(Se)-λ shortest distance in (ac) plane d_{exp}(Å)	V.d.W radii of S(Se) + ionic r. of X(X') $d_{calc.}$(Å)	Comparison $\frac{d_{exp}-d_{calc}}{d_{calc.}}$	X-rays diffuse scattering, surstructures
TTF-SCN$_{0.58}$ (1) (9)	0.58	3.40	2.00	0.70	2 KF satellite
TTF-SeCN$_{0.58}$ (1) (9)	0.58	3.38	2.30	0.47	" "
TTF-I$_{0.71}$ (4)	0.71	3.56	3.20	0.11	intermodulation of the two sub-
TTF-Br$_{0.79}$ (3)	0.79	3.36	3.00	0.12	lattices
TTT$_2$-(I$_3$)$_{1+\delta}$ (6)	1+δ	4.14	3.20	0.29	small intermodulation
TSeT-I$_{0.71}$ (5)	0.71	3.43	3.30	0.04	-
DIPS ϕ_4-(I$_3$)$_{0.76}$ (15)	0.76	3.82	3.20	0.13	-
TSeT-I$_{0.5}$ (6)	1	3.42	3.30	0.04	
TSeT-Cl$_{0.5}$ (7)	1	3.02	2.95	0.02	
TMTTF$_2$-X'(Br) (8)	0.5	4.90	3.00	0.63	(slightly dimerized)

The interstack distances C_D are always shorter than the
sum of Van der Waals radii, but the eclipsed configuration gives
the most effective overlap of the π-orbitals which are smaller in
the two other configurations. But no quantitative calculation has
been carried out to know in each situation the exact electronic
bandwidth. It is also useful to examine the counter-ion's positions.
In the channels between the cation stacks, these chains constitute
a second quasi-independent sublattice which is either commensurate
(x = 0.5 - 1 - 2...) or incommensurate.

On table 1 a few structural parameters have been reported to illustrate the interactions between the two sublattices.
The distance between chalcogen and halogen has been examined assuming that it plays a similar role as S(Se)---N contacts for donor-
acceptor interstack interactions in CTC. The experimental values
are compared to the calculated ones as indicated on table 2. Several comments are coming out from this table :

- for the commensurate structures this comparison shows
that there is a transverse electronic interaction because of the
possibility of an electron hopping specially in TSeT salts ; this
process must compete with the direct interstack interaction.

- for the incommensurate structures the analysis is not
so straightforward. I must indicate before that this distance is
just a projection measured in the (ac) plane and does not give the
exact distance for these compounds. The comparison between experimental and calculated values furnishes however a valuable information. Two sub-classes can be distinguished : these with a large
experimental value compared to the calculated one and those where
there are comparable. In the first case 2K$_F$ satellites in the dif-

fuse scattering have been observed (10) whereas in the second case intermodulations between the two sublattices are present (4).

2. GENERAL PHYSICAL PROPERTIES

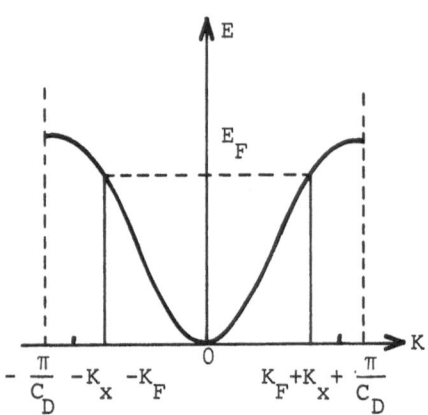

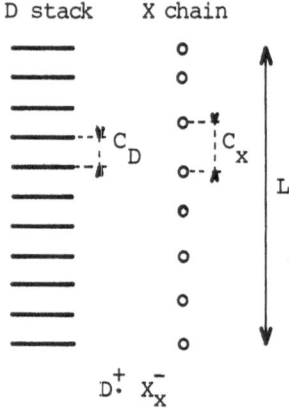

D electronic band
(1d tight binding model)
The Fermi wave rector value is:

$$|K_F| = \frac{(2-x)\pi}{2c_D}$$

The counter-ion periodic potential acting on the electronic chain :

$$|K_x| = \frac{\pi}{c_X} = \frac{\pi x}{c_D}$$

The relationship between them
(12)

$$g = 2K_F + K_x$$

(g : reciprocal lattice vector)
The phonon mode associated
with K_F :

$$q_F = 2K_F = \frac{(2-x)\pi}{c_D}$$

c_D, c_X are the lattice constants
M,N the number of donor and
acceptor stacks per unit cell
f_X, f_D fractional charge per
molecule

$$f_D \frac{L.M}{c_D} = f_X \frac{L.N}{c_X}$$

by assuming $f_X = 1$
  M = N (the gene-
ral case)

$$f_D = \frac{c_D}{c_X} = x$$

the stoichiometry defines the
band filling which is T and P
independent; for $x < 1$ the
free carriers are holes (the
thermoelectric power sign
must be positive).

These results are obtained from a standard 1d free electronic gas. However in a solid with two incommensurate sublattices
in coulombic interaction there is a tendency to restore a compromise periodicity to lower the system energy (12). A lattice modulation will appear with a wave vector value :

$$q_C = 2M(\frac{1}{c_D} - \frac{1}{c_X}) \qquad (q_C \neq q_F)$$

q_C can give rise to a displacive modulated structure. The rigid

TABLE 2 - *MAIN PHYSICAL PROPERTIES OF RADICAL CATION-SALTS*

Compounds RCS	Electrical conductivity			Paramagnetism		T_{M-I} (K)	EPR line-widths		$(\chi_p \cdot 10^4 \cdot \sigma")_{RT}$
	$\sigma"_{RT}$ $(\Omega^{-1}cm^{-1})$	$\dfrac{\sigma_{max}}{\sigma_{RT}}$	T_{max} (K)	$\chi_p \cdot 10^4$ (emu CGS mole^{-1})	T-dep.		$\overline{\Delta H}_{RT}$ (gauss)	T-dep.	
TMTTF$_2$-X' (13)	20-60	1.1-1.2	190 -245	5.8-6.5	slight low to T_{M-I}	41(ClO$_4$) 70(BF$_4$)	2.9-3.2	decrease	120-400
TTF-SCN$_{0.58}$	550	1.0-1.2	220	0.6	linear	175	~8	slight	350
TTF-SeCN$_{0.58}$ (1) (2)	750	.1-1.3	240	0.5	decrease	187	11	decrease	375
TTF-Br$_{0.79}$	~500	1.0	-	0.35	decrease	174	50	large	~175
TTF-I$_{0.71}$ (3) (11) (14)	~350	1.0	-	0.40	factor 2	230	200	decrease	~175
TTT$_2$-(I$_3$)$_{1+\delta}$ (15) (16)	~1000	up to 9	100 -38	2-3	decrease	29-42	200	decrease	2000 3000
TSeT-Cl(Br)$_{0.5}$ (17)	~2000	10	26	1.0	slight decrease	-	-	-	2000
TSeT-I$_{0.5}$ (5) (6)	10^3-3.10^3	~3-9	60-80	0.27	flat	-	30	-	270-510
TSeT-I$_{0.71}$ (6)	80	1.0	-	0.68	flat	-	~500	-	55
DIPS Φ_4-(I$_3$)$_{0.76}$ (18) (19)	250	1.2	200	0.30	flat	165 ?	5	decrease	85

band model described is perturbed and the above relationship has to be reconsidered. The position of the electronic gap can be different from $2K_F$ if a lattice distortion occurs.

A sketch of the physical properties is presented on table 2. In relation with the structural classification, an analysis of these results can be done. To start with these results, it is possible to analyze the longitudinal electrical conductivity ($\sigma"$) in terms of the mean $\sigma"$ free path of carriers (ℓ). Two situations occur :

- $\ell/C_D < 1$ The electric transport is diffusive. This is the case of TMTTF salts (13).

- $\ell/C_D > 1$ The phenomenon is coherent, we are in presence of a metallic behavior. This is the situation presented by the other salts.

Now by looking at the paramagnetic susceptibilities we observe that the TMTTF salts have the higher values, one order of magnetic is larger than the others except TTT$_2$-I$_3$ which is a special case (see next paragraph). As postulated recently (20) the enhancement of paramagnetism can be explained by a strong reduction of the charge carriers mean free path caused by a dynamic disorder. Whatever the microscopic model is, if we look for the product ($\chi_p \cdot \sigma"$) at room temperature, the same order of magnitude is roughly obtained except for TSeT$_2$- Cl$_{0.5}$ and still TTT$_2$-I$_3$.

The other evident observations about the general behavior of these compounds are :

- a very flat maximum of conductivity exists in most of the cases

- the EPR linewidth which can be narrow or large is always decreasing with temperature at the opposite of the observations on CTC that belong to TTF–TCNQ series (27).

        - the transition phase occurs at different temperatures often quite far from the maximum of conductivity (T_{max}). But only in few cases this is unambiguously a Peierls distorsion whereas in some other cases no metal-insulator transition is detected.

3. ELECTRONIC PROPERTIES OF RADICAL-CATION SALTS WITH A MODULATED STRUCTURE

        What are the characteristic properties of these compounds ?

        The only clear indication is the EPR linewidth. For this class of compounds namely TTF-bromine and iodine, TTT and TSeT-iodine a very large linewidth is observed (50-500 gauss) one order of magnitude larger than in the other materials (table 2). As suggested by Sugano and Kuroda (22) these linewidths must be due to a strong spin-orbit coupling caused by the interaction of the halogen atoms with the cations : this is in agreement with the structural analysis (table 1). One has to assume that a back charge transfer occurs : $X^- D^+ \rightarrow X° D°$. This is an activated process which causes the strong decrease of the linewidth with temperature (see figure 3). In correlation with this mecanism an analysis of the g-factor anisotropy has to be carried out. Now to illustrate this point I will present some new results on compounds prepared at CIBA-GEIGY laboratories by Drs. B. HILTI and C.W. MAYER.

        a)   TSeT - $I_{0.71}$

        This is a new compound which is isostoïchiometric but not isomorphous with TTF-$I_{0.71}$ (5) ; it presents also a strong intermodulation of these dual sublattices at room temperature.

        The thermal variation of electrical conductivity is peculiar, flat at room temperature ($\sigma_{//} \simeq 80 \ \Omega^{-1} \ cm^{-1}$) with a semiconductive behavior below 200 K and a strong hysteresis effect (see figure 1 where the results obtained for 4 successive cycles are presented).

        The same hysteresis   effect is observed on the EPR linewidth but not on thermal variation of paramagnetism (figure 2). A similar observation has been done by d.c. conductivity measurements on TTF-$I_{0.71}$ (14).

        We can infer from these observations that this is a dynamical effect due to the iodine chains. When the temperature is lowering there is an increase of the electrostatic interaction between the two sublattices with a correlation between tri-iodide chains. Elastic domains analogous to those observed in ferromagnetism should exist but a structural investigation at different temperatures is necessary.

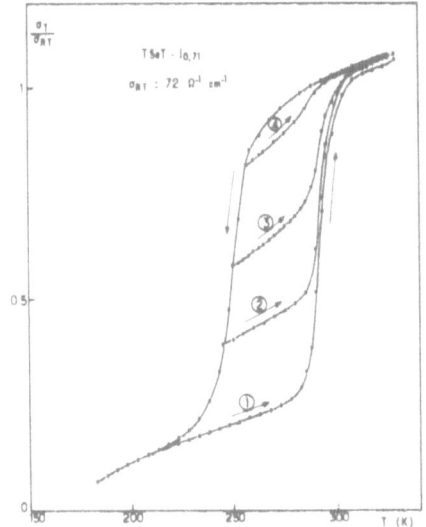
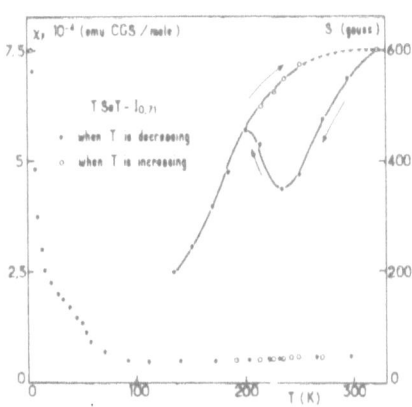

Fig.1 : *Thermal variations of d.c.*   Fig.2 : *Thermal variations of*
       electrical conductivity   *paramagnetism and EPR*
       measured along the needle   *line width on powder.*
       axis b.

b) $TTT_2 - (I_3)_{1+\delta}$

The tetrathiotetracene tri-iodide is of a special interest as we have seen by examining the table 2. It particular behavior comes from the two sublattices which are quasi-commensurate
along the stacking direction : $C_{I_3} = 2(1 - \delta)C_{TTT}$.

The first work by Kaminskii et al. (15) has allowed them
to distinguish four typical behavior in function of the iodine
concentration. The physical property changes are driven by δ the
excess of iodine. To get a better insight about the relation between iodine departure and the observed properties we have looked
upon two series of samples made by a cosublimation process and on
which the mean content of iodine has been defined ($\delta = 0.02$ and
0.04)(16). The electrical conductivity thermal variations are presented on figure 3 where the values along the needle axis C have
been reported but also the anisotropy ratios which have been determined using the Montgomery's method (23).

With an X-band EPR spectrometer we have been able to
examine the bandwidth thermal variations for the three principal
magnetic directions. For the first batch tne results are presented
on figure 4. A large T-dependence with a disappearance of the anisotropy is observed ; an inflexion point occurs at the same temperature observed in conductivity experiments. A similar observation
has been done on the second batch (23). Besides a dysonian EPR
line has been observed when the static magnetic field is parallel

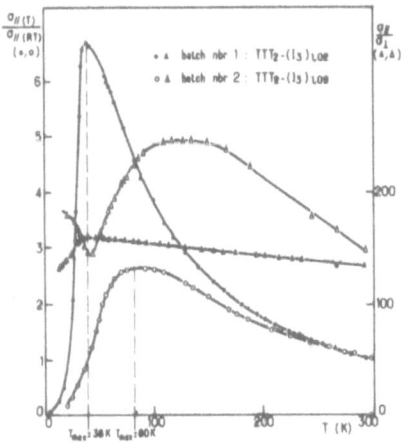

Fig.3 : *Thermal variations of d.c. electrical conductivity ani-sotropies and related values of longitudinal components.*

to c-axis. The asymmetric ratio $(\frac{A}{B})$ versus T is given on figure 5 where we ascertain a strong similarity with the conductivity curves (figure 3).

From these results the following analysis can be made :

- high-T regime : X-rays diffuse scattering study (24) has shown that the intermodulation of the two sublattices occurs progressively when the temperature is going down from 300 to 120 K. A tridimensional coupling between tri-iodide chains is arising simultanueously with the building up a new lattice. This structural evolution is accompanied with a conductivity increase (figure 3). As pointed out the ratio $(\frac{\sigma_{max}}{\sigma_{RT}})$ is very large in this material and increases with the iodine departure from ideal stoïchiometry (15). Superconductive or incommensurate CDW fluctuations (25) have been proposed to explain the conductivity increase. From the dysonian line analysis (figure 5) it is not possible to determine if either a single carrier process or a collective effect is predominant (24). For the first assumption the diffusion time accross the skin depth has to be considered but the usual calculation for metals does not work quantitatively in 1d systems ; in the second hypothesis a change of the microwave penetration depth with the temperature has to be invoked (23).

- Medium-T regime : A phase transition, function of δ, has been detected by several electronic properties, $\sigma_{d.c.}''$ (15) T.E.P. (26) and EPR linewidth (figure 4). No information is available from a structural point of view, the nature of this transition is unknown, it does not behave as a Peierls distortion however.

- low-T regime : it behaves as a disordered semi-conductor ; the electrical conductivity is governed by a phonon assisted hopping process and a huge isotropic and positive magnetoresistance is detected. In specific heat a linear time is found attributed to 1d phonons (16). In definite the nature of the ground

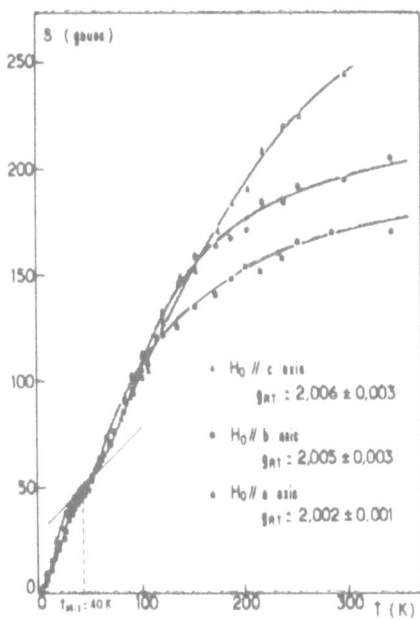

Fig.4 : *EPR line width thermal variations observed on a single crystal.*

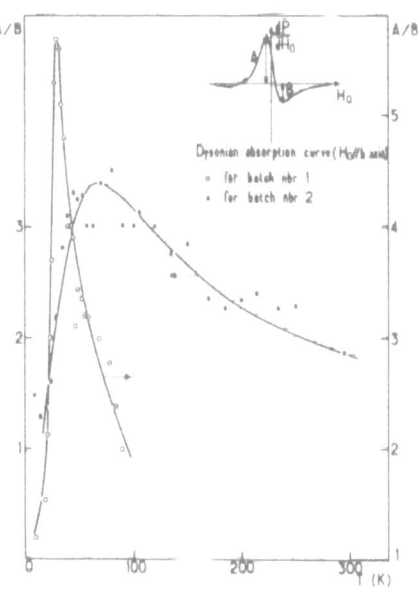

Fig.5 : *Temperature dependence of the observed dysonian line for a given orientation (c axis // H_0).*

state is not completely clear.

    To conclude a better insight could be obtained by comparing with the mercury chains compound (27). In the present case the incommensurate chain is not the conducting one. The iodine arrays perturb , because of their disorder, the TTT stacks ; nevertheless similarities about the structural properties are present.

4. CONCLUSION

    This class of organic conductors offers several attractive features. They are really monochain compounds, as KCP, on which the physics must be simpler than in two chains or CTC. Now several compounds which present a quite high conductivity have been presented, even more promising ones have been recently discovered (28).

    An attempt to classify these materials has been proposed :

    - Ordered structures (TSeT-halide salts) which present a slipped stacking. They present a quasi-metallic behavior down to low temperatures and they behave as almost anisotropic 3d electronic systems (5).

- Modulated structures with more or less effective coulomb interactions between the two substructures (table 1). In general they offer stacks with eclipsed configurations : in thiocynate and seleniocyanate compounds a weak interaction is present and a Peierls distorsion occurs with a lattice shear (10).

In iodide salts, the electrostatic interactions and the elastic energy of the chain compete. An energy gap can be opened at a wave vector different from $2K_F$ or even suppressed. Further investigations are necessary to understand this behavior.

BIBLIOGRAPHY

(1) Wudl, F., Schafer, D.E., Walsh, W.M., Rupp, L.W., Di Salvo, F.J., Waszczack, J.V., Kaplan, M.L., and Thomas, G.A., 1977, J. Chem. Phys. 66, p. 377.

(2) Somoano, R.B., Gupta, A., Hadek, V., Novotny, M., Jones, N., Datta, T., Deck, R., and Hermann, A.M., 1977, Phys. Rev. B 15, p. 595.

(3) Scott, B.A., La Placa, S.J., Torrance, J.B., Silverman, B.D., and Welber, B., 1977, J.A.C.S. 99, p. 6631.

(4) Johnson, C.K., and Watson, C.R., 1976, J. Chem. Phys. 64, p. 2271.

(5) Delhaès, P., Coulon, C., Manceau, J.P., Amiell, J., Flandrois, S., Rivory, J., Hilti, B., Mayer, C.W., and Rihs, G., (to be published).

(6) Hilti, B., MAYER, C.W., and Rihs, G., 1978, Helv. Ph. Acta 61, pp. 501 and 1462.

(7) Shibaeva, R.P., Lecture Notes in Physics Quasi 1d Conductors II (Proceedings Dubrovnik 1978),1979, 96, p. 167.

(8) Galigne, J.L., Liautard, B., Peytavin, S., Brun, G., Fabre, J.M., Torreilles, E., Giral, L., 1977, C.R. Acad. Sc. Paris, 285C, p. 475 ; 1978, Acta Crystallophica B 34, p. 620.

(9) Cougrand, A., Flandrois, S., Delhaès, P., Dupuis, P., Chasseau, D., Gaultier, J., and Miane, J.L., 1976, Mol. Cryst. and Liquid Cryst. 32, p. 165.

(10) Thomas, G., Proceedings of this ASI Nato School (Tomar, 1979).

(11) Chaikin, P.M., Craven, R.A., Etemad, S., La Placa, S.J., Scott, B.A., Tomkiewicz, Y., Torrance, J.B., and Welber, B., (to be published).

(12) Pouget, J.P., Ecole d'été sur les transformations de phase à l'état solide (Aussois, Sept. 1978).

(13) Delhaès, P., Coulon, C., Amiell, J., Flandrois, S., Torreilles, Fabre, J.M., Giral, L., 1979, Mol. Cryst. and Liquid Cryst. 50, p. 43.

(14) Warmack, R.J., Callott, T.A., Watson, C.R., 1975, Phys. Rev. B 12, p. 3336.

(15) Kaminskii, V.F., Khidekel, M.L., Lyubovski, R.B., Shchegolev, I.F., Shibaeva, R.P., Yagubski, E.S., Zvargkina, A.V., Zverana, G.L., 1977, Phys. St. Sol. A44, p. 77.

(16) Delhaès, P., Manceau, J.P., Coulon, C., Flandrois, S., Hilti, B., Mayer, C.W., 1979, Lecture Notes in Phys. Quasi 1d conductors II. (Proceedings Dubrovnik 1978), 96, p. 324.

(17) Schegolev, F.S., Lubrovskii, R.B., 1979, Lecture Notes in Phys. Quasi 1d Conductors II. Proceedings Dubrovnik 1978, 95, p. 39.

(18) Strzelecka, H., Weyl, C., Rivory, J., 1979, Lecture Notes in Phys. Quasi 1d Conductors II.(Proceedings Dubrovnik 1978), 96, p. 348, and unpublished results by Bordeaux's group.

(19) Lsett, L.C., Reynolds, G.A., Schneider, E.M., Perlstein,J.H., 1979, Sol. St. Comm. 30, p. 1.

(20) Marianer, S., Weger, M., and Gutfreund, H., sol. St. Comm. (to be published)

(21) Keryer, G., Delhaès, P., Amiell, J., Flandrois, S., Toreilles, E., Fabre, J.M., Giral, L., Lecture Notes in Phys. Quasi 1d Conductors I. (Proceedings of Dubrovnik 1978), 1979, 95, p. 65, and Sol. St. Comm. 26, p. 541.

(22) Sugano, T., and Kuroda, H., 1977, Chem. Phys. Letters 47, p. 92.

(23) Coulon, C., Thesis Bordeaux 1979, and Coulon, C., Amiell, J., Delhaès, P., (to be published).

(24) Megtert, S., Pouget, J.P., Comes, R., Fourme, R., 1979, Lecture Notes in Phys. Quasi 1d Conductors I.(Proceedings of Dubrovnik 1978) 96, p. 196.

(25) Abrahams, E., Gorkov, L.P., Kharadze, G.A., 1978, Sol. St. Comm. 25, p. 521.

(26) Khanna, S.D., Yen, S.P.S., Somoano, R.B., Chaikin, P.M., Lowe, M.A., Williams, R., Sanson, S., 1979, Phys. Rev. B 19, p. 655.

(27) Heeger, A., Proceedings of this ASI Nato School (Tomar 1979).

(28) Bechgard, K., Mortensen, K., Jacobsen, C.S., Pedersen, H., Communication at this ASI Nato School (Tomar 1979).

TRANSITION METAL TRICHALCOGENIDES

M. Renard

Centre de Recherches sur les Très Basses Températures, C.N.R.S., BP 166 X, 38042 Grenoble-Cedex, France

We give a review of the properties of transition metal trichalcogenides, and especially of $NbSe_3$. The structure and band properties, occurence of two charge density waves, and their correlation with anomaly in resistivity are reviewed. Special attention is given to non linear resistive effects associated with the depinning of the charge density wave.

We shall describe here properties of transition metals (say : Nb, Ta, Ti) trichalcogenides (say S, Se), and especially $NbSe_3$ which shows the more interesting properties.

STRUCTURE, BAND THEORIES

The samples look like whiskers, but of course, this is not a proof of the one dimensionality of the underlying physics.

On a structural point of view, every MX_3 is built with the same kind of elementary bricks : Triangular prism of X atoms, at the center of which, lies the M ion. These bricks are repeated along the b axis, giving a typical columnar structure.

The basic (Se) triangle has generally no high symmetry but shows two nearly equal sides and a shorter one. Every triangle appears two times in the unit cell, one of which being translated, by a displacement parallel to the b axis equal to the height of the elementary brick. So that the unit cell is twice the height of the elementary brick along b, and contains 2,4 or 6 columns, depending on the fact that there are : 1,2 or 3 different shapes of basic X triangles.

293

L. Alcácer (ed.), The Physics and Chemistry of Low Dimensional Solids, 293–303.
Copyright © 1980 by D. Reidel Publishing Company.

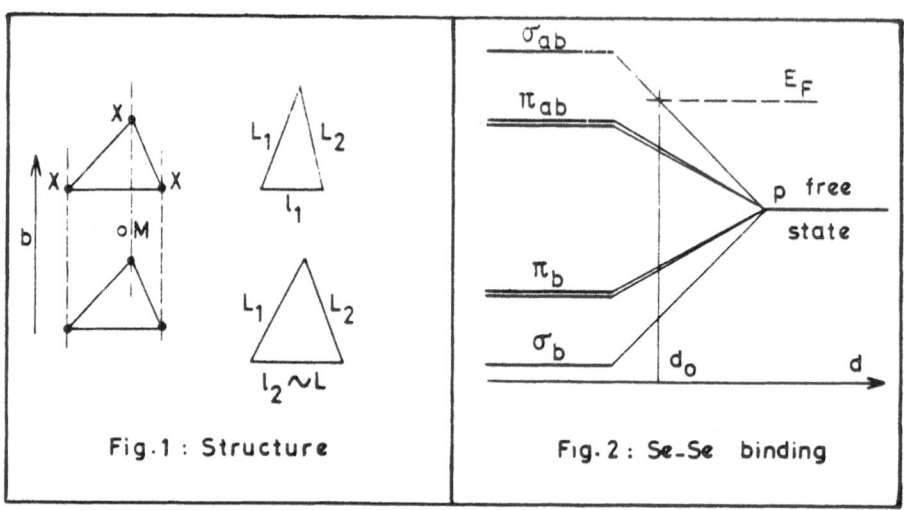

Fig.1 : Structure   Fig. 2 : Se-Se binding

Wilson pointed out the importance of the length. ℓ for a band theory of these materials : Since chalcogen atoms have 4 p electrons they can give X^{--} ions by filling the two empty p states. But for short distances of two ions they have a tendency to give the molecular ion (X_2^{--}). Neglecting electron-electron repulsion, and taking x axis along the direction Se-Se, we have the energy diagram of figure 2 : for bonding (b) or antibonding (ab) orbitals. At great distances every level degenerates in the common p level of the isolated atom. If the Fermi level falls at E_F, if the side ℓ of the triangle is : $\ell > d_o \rightleftarrows Se^{--}$ ions. If $\ell < d_o \sim (Se_2)^{--}$ ions. So, following the values of ℓ we can assume that Se prisms have the formulas :

$$\left\{ \begin{array}{lll} \ell \text{ short} & Se^{--}(Se_2)^{--} & \rightarrow 4 \text{ charges} \\ \ell \text{ long} & 3(Se^{--})^2 & \rightarrow 6 \text{ charges.} \end{array} \right.$$

For the metal atoms the z symmetry (z // b), will give a special role to the d_{z^2} orbital which, pointing to a positive ion, has to be lower in energy than the other quasi degenerate ($d_{x^2-y^2}$ d_{xy}) and (d_{xz} d_{yz}), transition metal bands.

A slight overlap of atomic M orbitals has to give rise to a finite width (d_{z^2}) band.

Taking the Nb atom : $(4d^4).(5s^1)$, we can assume that in $NbSe_3$, one type of chain with a spacing of Se atoms of 2.91 Å carry 6 charges, and the two types with 2.37 and 2.49 Å can be compared to pure solid Se : d = 2.32 Å and probably carry 4 charges. So that

3 Nb atoms have to give 6+4+4 = 14 electr. to Se ions, leaving one (d) electronic-state to share between two types of chains because the third one due to the electrostatic potential arising from the 6 charges, is empty of d electrons of Nb. The two low energy d_z2 sub-bands are nearly degenerate since their chains are supposed to carry the same charges. Since they can contain 4 electrons, we can think that in this picture Fermi level is at about one quarter of the zone boundary : π/b, and in this one dimensional model, wait for a Peierls transition with wave vector $q = 2k_F$, corresponding to $(1/4)(2\pi/b)$, and a wave length of about 4 unit cells along b axis.

Experimentally two CDW transitions occur, with wave vectors expressed in reciprocal lattice units :

$$T_c = 145 \text{ K} \quad q \; [0, 0.243, 0]$$
$$T_c = 59 \text{ K} \quad q \; [1/2, 0.263, 1/2]$$

A POSSIBLE MODEL FOR THE FERMI SURFACE OF $NbSe_3$

The complexity of these results is undoubtedly due to the complexity of the unit cell. For example, in a two dimensional model of chains linked by tunneling elements T_1 and T_2, the energy is given by :

$$= \frac{h^2k_b{}^2}{2m^*} \pm \sqrt{T_1^2+T_2^2+2T_1T_2 \cos k_a a_o}$$

giving for example the following Fermi surface which can fit with a q vector :

$$[0,q_o]$$

But if we have taken a single chain per unit cell and $T_2=T_1$ the Q vector will be :

$$[1/2,Q_o]$$

In the absence of a band calculation based on first principles, it is difficult to get more informations on the shape of Fermi-surface. Something is to be noted : By mixing two different sub-bands we loose the concept of pseudo-Brillouin zone, defined as the mediator plane from q_o vector. For a given k_a vector, the energy along k_b has a parabolic shape and give 4 gaps by fitting by $\pm q_o$ vectors, states of degenerate energy, but by changing k_a the locus of these points is not a plane, but in the preceeding figure for example the whole Fermi lines. In this transformation the material has to become an insulator since a gap appears at each point of the Fermi-surface, but this is due to the oversimplifications of the

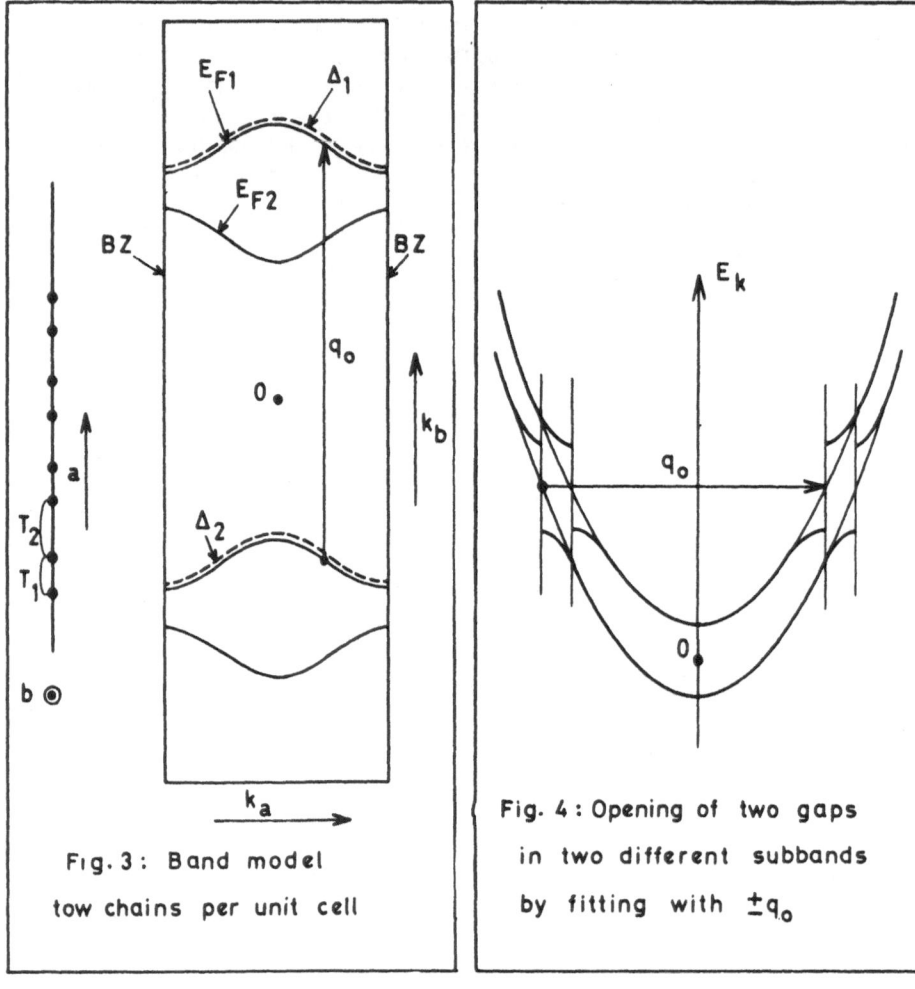

Fig. 3: Band model
tow chains per unit cell

Fig. 4: Opening of two gaps
in two different subbands
by fitting with $\pm q_o$

model : Neglecting second neighbours interactions, and curvature of energy in k_B space for example. In more realistic model the resulting pockets of electrons and holes, have higher symmetries, and at low T the material has lost any tendancy to show one dimensional properties.

TRANSPORT PHENOMENAS

The resistivity of $NbSe_3$ reflects the occurence of the two transitions at 145 and 59 K by a sharp increase. Well below the transition it keeps a metal like behaviour, showing the presence of free carriers, confirmed by experiments on de Haas-Shubnikov effect at low temperatures by Briggs and Monceau.

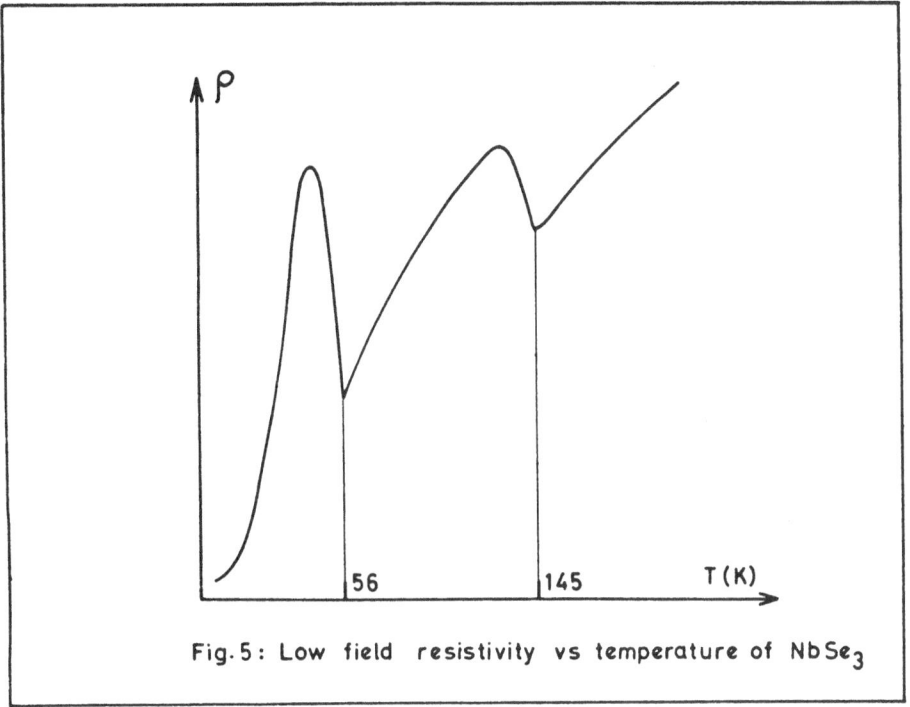

Fig. 5: Low field resistivity vs temperature of $NbSe_3$

The more striking feature is the non linearity of conductive properties in "high" electrical field. In the first papers, the conductivity at a given temperature below the transitions, was given as :

$$\sigma = \sigma_o + \sigma_1 \exp\left[-\frac{E_o(T)}{E}\right]$$

but it appears now that there is a threshold field E_c and that :

$$\sigma = \sigma_o \quad \text{if} \quad E < E_c$$

$$\sigma = \sigma_o + f\left[E-E_c\right] \quad \text{if} \quad E > E_c.$$

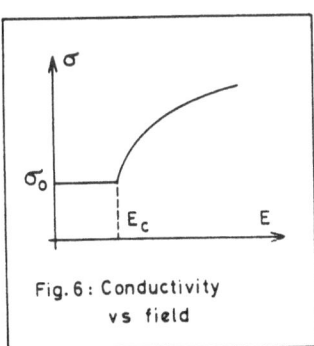

Fig. 6: Conductivity vs field

The extra conductivity is assigned to a depinning of the charge density wave which can carry current as Fröhlich has suggested, 25 years ago. Depinning fields depends of the sample and of the temperature but are in the range of 20 to 100 millivolts/cm, and so can be considered as very low, compared to other energies appearing in the problem.

Another striking feature, discovered by Ong, is that the electrical conductivity at frequency of a few GHertz, does not

show clearly the transitions, but looks like the "high" field resistivity, even for very weak currents. Here also this frequency
is difficult to compare to any other properties. Fleming and Grimes
have shown that just beyond the threshold field the electrical
noise increases very rapidly, and that it's power spectrum shows
rather narrow bands and some of their harmonics. The fundamental
frequency begins to be very low and then increases nearly linearly with current. Other frequencies appear at higher currents.

Other transport phenomenas like Hall effect, thermoelectric
effect, ... have been investigated but are beyond the scope of
this lecture.

THEORIES OF CDW TRANSPORTED CURRENT

Among many semi-phenomenological approaches, we have to notice an important theoretical work achieved by Overhauser and Boriak,
on microscopic basis. They have shown how moving CDW can carry
current, but on the basis of a jellium model for the ions. Since
for a uniform motion of the jellium the potential seen by electrons is in uniform translation, they can choose as galilean frame a set of axis linked to the CDW phase. So that in the absence
of scattering for free carriers, they have a time independent
hamiltonian and the current density is just :

$$j = nev_{CDW}$$

But the question is open, to know if for a(d) band which is more
realistically described by tight binding, than by nearly free
electrons, the result would be the same.

Assuming that the bottom of the (d) band can be described by
an effective mass, and that the ionic displacement is a sine wave,
one can solve the time dependent problem of the moving CDW in a
one dimensional model in the laboratory frame.

If ϕ_k is the state associated with $\vec{k}$ in the unperturbed (d)
band (without CDW) the classical result for a standing CDW is, for
the new state $\vec{k}$:

$$\begin{cases} \psi_k = e^{-\frac{it}{\hbar}\varepsilon_k} \cdot \left[A\phi_k + B\phi_{k-q} + C\phi_{k+q} \right] \\ \varepsilon_k = E_k^\circ - x(k) \end{cases}$$

where E_k° is the unperturbed energy of the d band (parabolic in k),
x the result of the perturbation due to CDW, A, B, C constants depending of the values of k and q_o. x is only important if $k \neq \pm q/2$.
If CDW is in motion at a speed v_{CDW}, the perturbation due to CDW

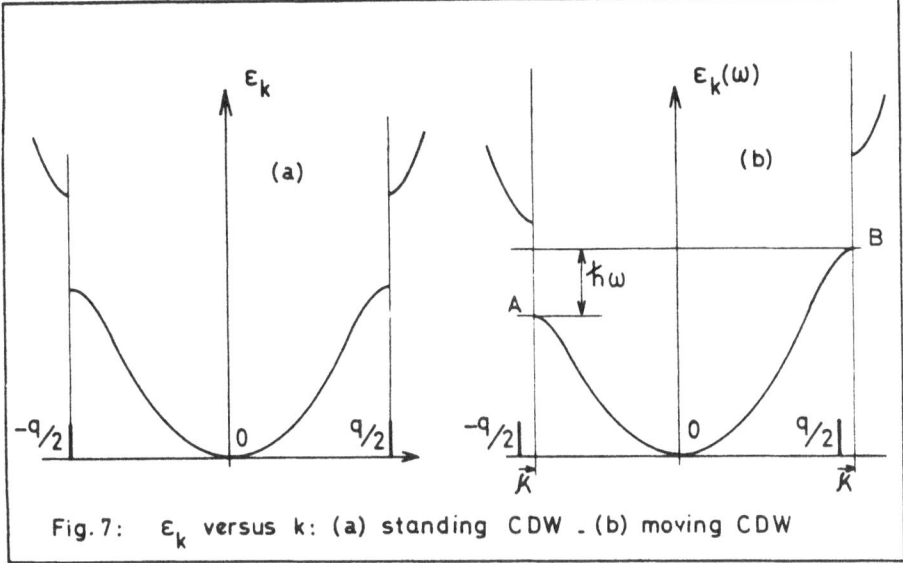

Fig. 7: ε_k versus k: (a) standing CDW . (b) moving CDW

have two components, which can be written :

$$\begin{cases} V_q\, e^{i(qr-\omega t)} \\ V_q^*\, e^{-i(qr-\omega t)} \end{cases}$$

where $\omega = v_{CDW} \cdot q$.

The solution can be found for the time dependent Schrödinger equation as :

$$\psi_{k,\omega} = e^{-\frac{it}{\hbar}\varepsilon_{k\omega}} \left[A'\phi_k + B'e^{+\frac{it}{\hbar}\hbar\omega} \cdot \phi_{k-q} + C'e^{-\frac{it}{\hbar}\hbar\omega} \cdot \phi_{k+q} \right]$$

So that the solution has three components with three different energies : $\varepsilon_{k,\omega}$ associated with ϕ_k, $(\varepsilon_{k,\omega}-\hbar\omega)$ associated with ϕ_{k-q} and $(\varepsilon_{k\omega}+\hbar\omega)$ associated with ϕ_{k+q}.

The eigenvalue problem is a third degree equation and by the transformation

$$\varepsilon_k = E_k^\circ - x\big[k,\omega\big].$$

One can show easily that :

$$x\big[k,\omega\big] = x\big[k+K,\ \omega = 0\big]$$

where $K = \dfrac{\omega m^*}{\hbar q} = \dfrac{m^* v_{CDW}}{\hbar}$.

By exagerating v values, the band structure for $\varepsilon_{k\omega}$ has the shape given fig. 7b.

The $\varepsilon_{k\omega}$ difference between points A and B is only due to E_k° values :

$$\varepsilon_A - \varepsilon_B = \frac{\hbar^2}{2m^*}\left[(\frac{q}{2} + K)^2 - (\frac{q}{2} - K^2\right] = \frac{\hbar^2}{2m^*} 2qK$$

$$= \frac{\hbar^2}{2m^*} 2q . \frac{\omega m^*}{\hbar q} = \hbar\omega .$$

So that, if the band is filled until the gap, and if : $\hbar\omega \ll \Delta$, by applying a continuous acceleration of (CDW) until it's v value, all states between A and B are filled and the continuous current will be ΣJ_k

$$J = \Sigma J_k = \Sigma\left[A'^2 j_k + B'^2 j_{k-q} + C'^2 j_{k+q}\right]$$

where j_k, j_{k-q}, j_{k+q} are the currents associated with k, k-q, k+q in the unperturbed lattice. Since $A'^2 + B'^2 + C'^2 = 1$

$$J = \Sigma_k\left[j_k + B'^2(j_{k-q}-j_k)+C'^2(j_{k+q}-j_q)\right]$$

If we assume that j_k is proportional to k the last two terms will give :

$$\propto \Sigma q (C'^2 - B'^2) = 0$$

since C' interchange with B' in the transformation $k+K \to -(k+K)$. The net result is

$$J = \sum_{(-q/2)+K}^{(q/2)+K}\left(j_k\right) = nev_{CDW}$$

But the problem is much more difficult if there are free electrons or hole pockets. For example impurities, can connect three states in energy : One conserving energy, one increasing the energy by $\hbar\omega$, the last one decreasing energy by $\hbar\omega$, due to the three components of each wave function.

The problem of thermal equilibrium itself is not clear since the main energy is $\varepsilon_{k\omega}$, but the mean value of energy is :

$$\bar{\varepsilon}_k = <\psi\left[i\hbar \frac{\partial}{\partial t}\right]\psi> = \varepsilon_{k\omega} + (C'^2-B'^2)\hbar\omega$$

SEMIPHENOMENOLOGICAL THEORIES

Since undoubdetly a moving CDW can carry current, many calculations have been done by assuming an effective charge per wave length to the CDW. One can develop Ginzburg-Landau approach of the problem near the second order transition for example (Mc. Millan, Maki), or consider a two-fluid model as in liquid helium (Bardeen).

An important problem is associated with the depinning of the CDW.
The charge density wave can be pinned by many defects and especially impurities, so that we can understand the existence of a
threshold field. But the intermediate regime is probably not a
motion of the whole CDW in the entire crystal. Maki has considered
soliton like solutions in motion in a one dimensional model. A
more realistic approach has been given by Lee and Rice who consider dislocation loops in the periodic lattice of the CDW deformation. Just as in crystals the motion of a dislocation throughout
the sample is equivalent to an increase of the length of one atomic unit, here of one wavelength, carrying the charge associated
with this wavelength.

Since thermal or electrical activation of dislocations is
highly improbable, they imagine the equivalent of Frank-Read sources for dislocations. When a critical stress (here the electrical
field) is reached, the source is activated, and give new dislocation loops.

OPEN QUESTIONS

The theory of Lee and Rice explains the existence of a threshold field, but this field for a given sample seems to have a
divergence at T_c. It may be that the
effective charge goes to zero at T_c
more rapidly that the energy of the
dislocation. But no calculation has
been done to explain this fact.

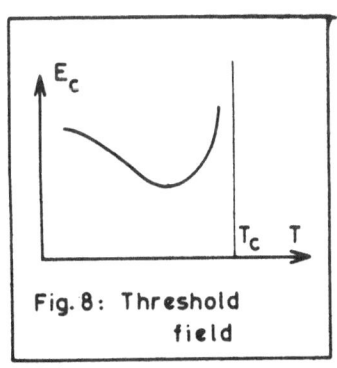

Fig. 8: Threshold field

The results of Ong and Monceau
on the high frequency resistive behaviour can probably be explain by
the high electrical polarizability
of the CDW, first pointed out by
Overhauser. In the presence of an
electric field, ionic and electronic
CDW are no longer in phase, just as
in atomic polarization, ion and electrons, are no longer centered at the
same point. But the energies invol-
ved in sharing ions and electrons are much lower in the case of
CDW which is just a slight modulation of the atomic potentials.
The net result has to be a high ε value, and possibly the measured
current was a displacement current, rather than a real one. But
here also some calculations are needed to acertain this hypothesis.

Another important question is the very existence of Fröhlich
superconductivity. Due to the existence of a gap, the motion of a

CDW is collective, and in principle insensitive to individual scattering. If the critical field is exceeded, an increase in current
may in principle, only be due to an acceleration of the CDW, without any further increase in electrical field ! Experimentally
however, the increase of the current seems to rub out the resistive anomaly, and conductivity seems to reach the limit of the
normal phase, but never exceeds it. Of course, there is an interaction between electron pockets and CDW, giving rise to mutual
friction with some higher value of conductivity, but linear in
field increment, which is the contrary of the Fröhlich assumption.

So if we look for Fröhlich free gliding mode, we have to try
with other materials which become insulating at the CDW transition,
but with wave vectors along the b axis different from (1/2), since
for this vector (Peierls pairing) there is an important intrinsic
pinning, due to the fact that in this case translation of CDW, is
linked with alternative destruction and creation of the CDW itself !

BIBLIOGRAPHY

Since reference numbers have not been given throughout the
text, we give here the chapters to which references are associated.

STRUCTURE, BAND THEORIES

- A. Meerschaut, J. Rouxel, J. Less Common Metals 39, 197 (1975).
- E. Bjerkelund, J.H. Fermor, A. Kjekshus, Acta Chemica Scand. 20,
  1836 (1966).
- J.L. Hodeau, M. Marezio, C. Roucau, R. Ayrolles, A. Meerschaut,
  J. Rouxel, P. Monceau, J. Phys. C : Solid State Physics 11, 4117
  (1978).
- J.A. Wilson, Phys. Rev. B6, 2866 (1978).
- R.M. Fleming, D.E. Moncton, D.B. Mc Whan, Phys. Rev. B18, 5560
  (1978).
- K. Tsutsumi, T. Tagagaki, M. Yamamoto, Y. Shiozaki, M. Ido,
  T. Sambongi, K. Yamaya, Y. Abe, Phys. Rev. Lett. 39, 1675 (1977).

TRANSPORT PHENOMENAS

- J. Chaussy, P. Haen, J.C. Lasjaunias, P. Monceau, G. Waysand,
  A. Waintal, A. Meerschaut, P. Molinié, J. Rouxel, Solid State
  Comm. 20, 759 (1976).
- P. Haen, P. Monceau, B. Tissier, G. Waysand, A. Meerschaut,
  P. Molinié, J. Rouxel (1975), Proceed. of Fourteenth Internatio-

nal Conference on Low Temperature Physics 5, 445.
- P. Monceau, J. Peyrard, J. Richard, P. Molinié, Phys. Rev. Lett.
  39, 161 (1977).
- P. Monceau, N.P. Ong, A.M. Portis, A. Meerschaut, J. Rouxel,
  Lett. 37, 602 (1976).
- N.P. Ong, P. Monceau, Phys. Rev. B16, 8, 3443 (October 1977).
- R.M. Fleming, C.C. Grimes, Phys. Rev. Lett. 42, 1423 (1979).
- N.P. Ong, Phys. Rev. B18, 5272 (1978).
- N.P. Ong, J.W. Brill., J.C. Eckert, J.W. Savage, S.R. Khanna,
  R.B. Somoano, Phys. Rev. Lett. 42, 811 (1979).

THEORIES OF TRANSPORTED CURRENT

- H. Fröhlich, Proc. Roy. Soc. Ser A 223, 296 (1954).
- A.W. Gverhauser, Advances in Physics 27, 3, 343 (1978).
- M.L. Boriak, A. W. Gverhauser, Phys. Rev. B :
    16, 12, 5206 (1977)
    16, 12, 5256 (1977)
    17, 6, 2395 (1978)
    17, 12, 4549 (1978).

SEMIPHENOMENOLOGICAL THEORIES

- R. Maki, Phys. Rev. Lett. 39, 46 (1977).
- R. Maki, Phys. Lett. 70A, 5, 449 (1979).
- J. Bardeen, to be published.
- J.S. Zmindzinas, Phys. Rev. B17, 10, 3919 (1978).
- P.A. Lee, T.M. Rice, Phys. Rev. B19, 3970 (1979).

SURVEY OF METAL CHAIN COMPOUNDS

P. Day

Oxford University, Inorganic Chemistry
Laboratory, South Parks Road,
Oxford, OX1 3QR, England

This review is a descriptive account of metal chain
compounds from a chemical point of view.   The most
famous one-dimensional molecular metals are the
mixed valence Pt cyanides and oxalates, but very
many transition metal complexes form crystals containing chains of metal atoms.   Some are conducting,
others not;   some show cooperative magnetism, others
not;   some have unusual optical properties and others
not.   Examples are used to identify those structural
and electronic features which lead to the observed
properties.

INTRODUCTION

Chains of closely spaced metal atoms are quite
a common feature of the crystal structures or inorganic compounds.   Do the metal atoms in these chains
interact with one another sufficiently to give the
crystals properties unusual enough to be interesting
to a physicist?   Clearly the answer to such a
question must be 'some do but some don't'.   The real
question, though, is 'which ones'?   The aim of this
chapter is to summarize and systematize structures of
a wide variety of inorganic and metal-organic chain
compounds, so as to draw out those salient features
of electronic structure and molecular or ionic packing
which might reasonably lead one to expect unusual
optical, magnetic or electron transport behaviour.

L. Alcácer (ed.), The Physics and Chemistry of Low Dimensional Solids, 305–320.

Since, apart from the elements, metal atoms
do not appear in crystal lattices uncoordinated to
other ions or molecules, we have to ask straightaway
whether the metal atoms in the chain compound in
question form a simple linear lattice such as
Figure 1(a), with other groups coordinated only in
equatorial positions, or whether the chain consists
of metal atoms bridged by one or more intervening
ligands, e.g. Figure 1(b).   I believe that this
distinction is fundamental, as shown later.

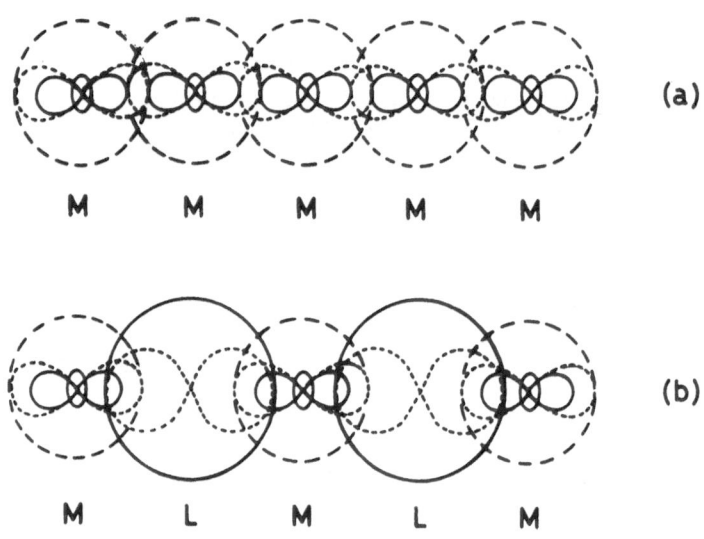

Figure 1.   Orbitals in metal chains (a) Directly
interacting, (b) Bridged.

LOCALIZED OR COLLECTIVE ELECTRONS?

Crucial to any description of electronic states
in a lattice is the decision whether to treat the
electrons in the valence shell as localized into
discrete orbitals, e.g. d- or f-shells, or delocalized
over all the centres.   Were the valence shells to be
only partially filled the former would lead to a Mott
magnetic insulator and the latter to a metal.   In
Figure 1 we show the atomic orbitals relevant to the
cases of direct metal-metal (M-M) overlap in the chain
and metal-ligand-metal (M-L-M-) overlap.   In either

case it is important to notice that (n+1)s and (n+1)p orbitals are involved in bonding, an addition to the nd orbitals which are partly occupied. Since the s and p orbitals are undoubtedly more spatially extended, one may plausibly assume that the equilibrium internuclear distance, M-M or M-L, is achieved by maximizing their overlap, leaving the d-orbitals to overlap to a greater or lesser degree. Assuming hydrogenic wavefunctions, the average radius of an orbital with quantum numbers n,l is

$$\langle r_{nl} \rangle = r_{nl}^{B} (3n-1)(3n-l+1)/4n^2 \tag{1}$$

where r_{nl}^{B} is the Bohr radius. Further, the one-electron energy of the n-shell is

$$E_n = -(2me^4/h^2)(Z/n)^2. \tag{2}$$

Thus if the equilibrium interatomic spacing is determined by maximizing the (n+1)s overlap, for example, this spacing is easily expressed as a multiple of the spacing which would apply if the bonding were entirely due to nd:

$$R_o^{(n+1)s} = R_o^{nd} (E_{nd}/E_{(n+1)s})^{\frac{1}{2}} (3n+3)(3n+4)/ \tag{3}$$
$$(3n-2)(3n-1)$$

Most of the compounds we are interested in contain a metal in a positive oxidation state. Under these circumstances the binding energy of the nd shell is always much greater than that of (n+1)s. Since many metal chain compounds contain Group VIII metals, consider the following pairs of ionization potentials as illustrative:

$$
\begin{array}{lllll}
Pt^o & 5d^9 6s^1 & \longrightarrow & Pt^I \; 5d^9 & 8.96 \text{ eV} \\
Pt^I & 5d^9 & \longrightarrow & Pt^{II} \; 5d^8 & 18.56 \\
Ni^I & 3d^8 4s^1 & \longrightarrow & Ni^{II} \; 3d^8 & 18.15 \\
Ni^{II} & 3d^8 & \longrightarrow & Ni^{III} 3d^7 & 35.16
\end{array}
\tag{4}
$$

In both cases $E_{nd}/E_{(n+1)s}$ is about 2 so for the 3d and 5d orbitals respectively we have

$$R_o^{4s} \sim 2.9 \; R_o^{3d}$$

$$\hspace{8cm} (5)$$

$$R_o^{6s} \sim 2.2 \; R_o^{5d}$$

Mott concluded some years ago (1) that there was a critical interatomic spacing R_c^{nl} beyond which the electrons in a partially occupied nl shell would be localized rather than collective, and Goodenough argued (2) that R_c^{nl} might be about 2.5 R_o^{nl}. Clearly this is only a rough estimate, but comparing it with eq. (5) we see that in 5d compounds collective behaviour is most probable, while the 3d compounds are more likely to be localized magnetic insulators. Furthermore, an intervening bridging group attenuates the M-M overlap in the chain by a factor S_{ML}^2, where S is an overlap integral between the M(nd) orbital and a ligand σ or π orbital. For this reason it is very unlikely that metal chain compounds containing bridging groups will show substantial collective electron behaviour.

Even if the d-orbitals of the metal chain overlap enough to make a description in terms of molecular orbital or band wavefunctions appropriate, we still do not have any electron density at the Fermi surface unless the band is partly occupied. Most transition metal chain compounds are formed by elements in positive oxidation states (though there are a few important exceptions, such as Chini's Pt carbonyls (3)). Since the nd-shell in the free ion then lies substantially below (n+1)s,p the nd-band in the chain probably will not overlap the s- and p-bands, even if it is very wide. Add to this the extreme directionality both of the d-orbitals and of the chain, and we see that partial occupancy of the d-shell as a whole is not enough, and that partial occupancy of a d-orbital of specific symmetry is required, for example the z^2 or xz orbital as shown in Figure 1. The d-shell as a whole can be partly occupied in any single valence compound (e.g. d^7 in CoO) but we can only have a partly occupied orbital (e.g. $(z^2)^x$ where $1 < x < 2$) in a compound with mixed valence.

Two major chemical factors have therefore been identified as crucial in relating structure to gross electronic (optical, magnetic and transport) behaviour in metal atom chain compounds. First is the presence

or absence of bridging groups and second is the
question of single or mixed oxidation states.   With
these in mind we now examine some examples.

SINGLE VALENCE CHAINS

     Single valence metal chains exist with numbers
of bridging groups from zero to three and with total
metal coordination numbers from two to six.   A list
of representative examples was given earlier (4) so
here we shall only summarize some of the more unusual
and interesting ones.   Schematic structures are shown
in Figure 2.

Singly Bridged Chains

     Chains consisting of alternating cations and
anions in simple strings, with no further groups
attached, are confined to nd^{10} configurations, in
Group IB, such as AgNCS and AuI.   Three coordinate
metal ions are often found in B-subgroup ions with
the ns^2 electron configuration distorted by the so-
called 'inert pair effect', and sometimes form chains
with rather complicated connectivity, for instance
the double chain in Sb_2O_3.   Four coordinate metal
ions are square planar or tetrahedral: chains
constructed by sharing opposite corners of squares
occur in the $5d^8$ compound AuF_3 while tetrahedra
sharing vertices are the dominant structural feature
in the many silicates and phosphates.   Five coordinate molecules stacked in chains are much rarer: the
only example known to me is $WOCl_4$, where the square
pyramidal molecules share apical oxygen atoms to form
a chain W-O ...W-O ... with alternating W-O distances.

     If the distances -X-M-X-M- along a chain were
all equal, while four more X groups completed an
octahedron around M, the stoichiometry of the
resulting chain would be MX_5.   Binary examples are
BiF_5 and UF_5, but ternary compounds exist with this
structure, such as Na_2FeF_5, $Sr(PbF_5)F$ and others with
mixed equatorial coordination about M, like $FeF_3.3H_2O$
and $CsMnCl_3.2H_2O$ (CMC).   All the 3d compounds of
this kind behave as one-dimensional antiferromagnets,
with near neighbour exchange constants of a few cm^{-1},
but often very low three-dimensional ordering
temperatures (e.g. CMC 4.9K) (5).

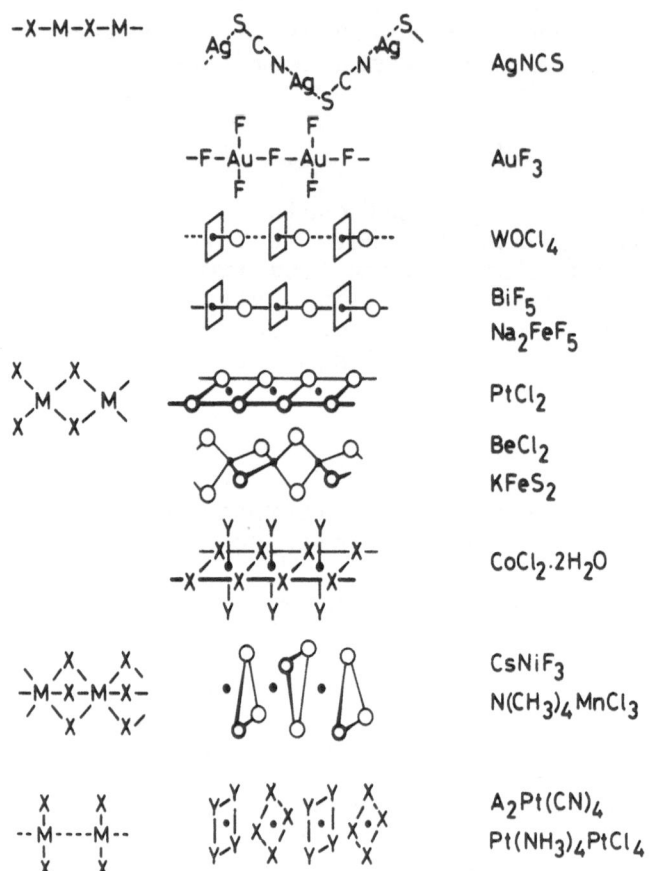

Figure 2.   Types of single valence metal chain.

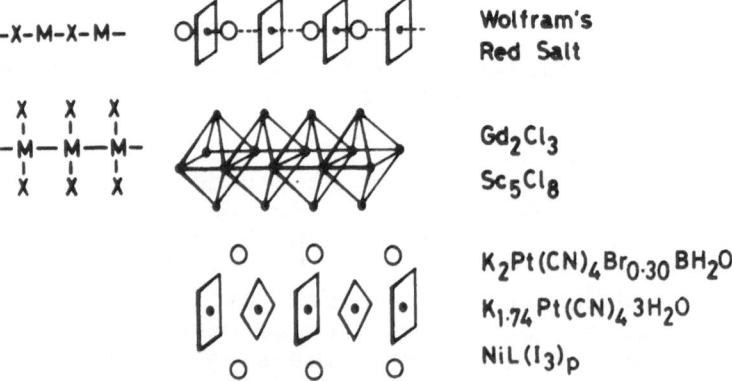

Figure 3.   Types of mixed valence metal chain.

Bifunctional aromatic molecules like pyrazine
(pz) and quinoxaline can also form linear bridges
in chain salts such as $Cu(pz)(NO_3)_2$, with an antiferromagnetic exchange constant of 8.5 cm^{-1} (6).

Doubly Bridged Chains

In a singly bridged chain the M-M interaction
has to go through the bridging group, but when two or
more bridging groups are involved the metal atoms may
be forced closely enough together for direct M-M
interaction to begin to play a role.   Such an effect
can be seen clearly (and was recognized many years ago)
in the MO_2 oxides (M = V, Nb, Ta, Cr, Mo, W) with
rutile structures.   These nd^1 and nd^2 compounds have
xy electrons which form electron pair bonds at low
temperatures, causing a lattice distortion with dimerization along the direction of the chains.   At higher
temperatures in VO_2 the electrons unpair and metallic
behaviour results, but this is quite exceptional, for
all the other compounds we are considering in this
section are insulators.   For instance, taking the
edge-sharing octahedral chains such as appear in the
rutile structure and spacing them out by substituting
more bulky apical groups, one arrives at the very
extensive series of compounds with stoichiometry
$MX_2.2Y$ (M = Mn, Fe, Co, Ni, Cu, Cd; X = halide or
thiourea; Y = H_2O, NH_3, pyridine etc.).   All of them
are magnetic insulators with localized ligand field
excited states in the visible.   They have been widely
studied as examples of one-dimensional magnetic correlations, including a number in which the near-neighbour
exchange along the chains is ferromagnetic (7).

Double bridges of a rather unusual kind are found
in the metal complexes of the planar aromatic macrocyclic ligand phthalocyanine.   The flat molecules
form stacks in such a way that nitrogen atoms from the
next molecules in the stack lie directly above and
below the metal ions, completing a tetragonally distorted octahedral coordination.   Overlap between the π-
electron clouds in adjacent molecules is no greater
than in most other aromatic molecular crystals such as
anthracene, so although phthalocyanine crystals have
a metallic lustre due to low energy π-π^* transitions
the single valence metal complexes are high resistance
semiconductors, though their photoconductivity was
widely studied some years ago (8).

Triply Bridged Chains

Where three anions form a triangular array
between neighbouring metal atoms in a chain the metal
d-shells will be brought together more closely than
in singly or doubly bridged compounds with comparable
metal-ligand distances.   Nevertheless only one compound of this kind, $BaVS_3$ (9), has a conductivity high
enough for it to be considered as containing band
rather than localized d-electrons.   A chain of octa-
hedra sharing opposite faces has stoichiometry MX_3,
a structural element found in binary trihalides of Ti,
Zr, Hf, Mo and Ru, and in ternary compounds with hexagonal perovskite structure.[1] All the trihalides are
insulators, and in $TiCl_3$ (d^1) and $MoBr_3$ (d^3), at least,
alternating long and short M-M spacings are found along
the chains, in much the same way as in the insulating
phase of VO_2.

The hexagonal perovskites AMX_3 contain many of
the classical examples of one-dimensional magnetism,
such as $N(CH_3)_4MnCl_3$ or TMMC (antiferromagnetic) and
$CsNiF_3$ and $RbFeCl_3$ (ferromagnetic) which have been
studied by optical, magnetic and neutron techniques.
A recent addition to the set, $NaScCl_2$ (10) is of
interest as illustrating how predictions of physical
behaviour based on chemical extrapolations can sometimes go wrong.   The best opportunity to realize the
requirements for collective electron behaviour in
this type of compound is found at the beginning of the
transition series where the radial extension of the
d-orbitals is greatest.   Nevertheless $NaScCl_3$ is
insulating, presumably because the trigonal distortion
of the octahedral crystal field round each d^1 ion is
such that the single valence shell electron occupies
the xy orbital, with its plane perpendicular to the
chain.

Directly Interacting Metal Ions

Metal-containing molecular complexes which are
planar, as found among the low-spin nd^8 elements,
provide a clear opportunity for the partly occupied
d-shells of adjacent metal atoms to be brought into
really close contact by forming directly interacting
stacks, with no intervening bridging ligands at all.
Among this class of compound are many examples in
which the optical properties in particular are substantially different from those of the individual molecular
units.   Nevertheless, even here the valence shell

electrons are near the localized limit, and only one
case of high conductivity, $Ir(CO)_3Cl$, is known (11).
Even in the latter the exact formula remains controversial, however, and there may be a small departure
from stoichiometry, leading to mixed valence.

The molecular stacks in this type of compound
may be formed either by neutral molecules, anions or
alternating anions and cations.   Apparently there are
no examples of cationic stacks.   Examples of neutral
stacks are the dimethylglyoximates (DMG) and Pb
phthalocyanine (PbPc).   The deep red colour of solid
$Ni(DMG)_2$ was one of the earliest instances to be
recognized of the optical consequences of stacking
planar complexes (12).   In solution the colour is
yellow, but in the crystal there is an extra absorption band at low energy with transition moment
directed along the stacks.   Nevertheless the crystals
are completely insulating.   The case of PbPc is
interesting because there have been reports (13) that
as a thin sublimed film it shows metallic conductivity.
On the other hand single crystal measurements made
many years ago (14) indicated a very high bulk
resistivity.

The most famous cases of 'extra' absorption bands
in planar molecular stacks are the tetracyanoplatinite
salts where the stacking unit is $Pt(CN)_4^{2-}$.   More than
a dozen such salts are known with Group IA and IIA
cations and various numbers of water molecules.   Their
colours range from red to green to yellow to colourless,
a remarkable variation considering that in solution
the isolated anions are colourless, with their lowest
absorption bands about 35 000 cm^{-1}.   The colour
variation comes about because changing the cation
varies the Pt-Pt spacing in the stacks from about 3.6
to 3.15 Å and the 'extra' absorption band shifts from
30 000 to 18 000 cm^{-1} (15).   In spite of the spectacular difference in optical properties between the
stacks and their constituent units we believe, nevertheless, that the lowest excited states in the crystal
can be described quite accurately as neutral Frenkel
excitons, i.e. that there is no electron exchange
between the d-shells of neighbouring molecules.   Our
evidence for this belief is that the frequency of the
'extra' absorption band is exactly proportionate to
R^{-3}, where R is the Pt-Pt spacing along the stack, as
required for interaction between molecular transition
dipoles (16,17).   Furthermore, extrapolation to R $= \infty$
predicts an excitation frequency very close to that of

the first fully allowed transition of $Pt(CN)_4^{2-}$
observed in solution.

Apart from Pt-Pt spacing, other parameters which
can be varied within the tetracyanoplatinite stacks
by changing the cation are the tilt and twist angles
of the molecular planes with respect to the stacking
axis and also, in one fascinating instance,
$Cs_2Pt(CN)_4H_2O$, formation of a spiral of Pt atoms (18).
The optical consequences are then quite bizarre, since
the exciton band now takes on rotational as well as
dipole strength (19).

MIXED VALENCE CHAINS

The main role of mixed valency in promoting
conductivity in metal chain compounds is to reduce the
on-site Coulomb repulsion which puts up a large energy
barrier to creating charge fluctuations along the
chain.   In a mixed valency compound the charge fluctuation is already built in, and only lattice polarisation (electron-phonon coupling) energy is required to
move it.   Of course, this remark is not valid for one-
dimensional materials alone, but where the path of
electron migration is substantially confined to a chain
the structural criteria for conductivity become very
much simplified.   On the other hand a variety of
phenomena connected with the electron-phonon interaction, such as Kohn anomalies, Peierls distortions and
charge density waves take on quite special importance
in one-dimensional compounds.

As with the single valence substances we find
profound differences in physical properties of mixed
valence metal chain compounds depending on whether the
valence shells of the metal atoms make contact by
direct overlap, or only through bridging groups.   In
the mixed valence case though, this is not just a
question of satisfying an overlap criteria for collective versus localized electronic behaviour.   Bridging
groups in the chain also introduce new longitudinal
phonon modes which can trap the charge fluctuation so
that, for example, one finds oxidation states which
alternate along the chain.   Some years ago we classified the physical properties of mixed valency compounds
in general by referring to the similarity or difference
in ligand field potential around the sites of differing
oxidation state (20).   In the context of mixed valency
chains we shall see that it is quite simply the

bridging groups which introduce this difference, so
the compounds become class II, or semiconducting,
rather than class III, metallic.   Schematic structures are given in Figure 3.

Singly Bridged Chains

The most famous and widespread singly bridged
mixed valency chains are those based on Pd and Pt
alternating with halide ions.   The classic example is
Wolfram's Red Salt $[Pt(C_2H_5NH_2)_4][Pt(C_2H_5NH_2)_4Cl_2]Cl_4$.
In this structure the ethylamine ligands form square
planar arrays round each metal atom, but the Cl atoms
are not equidistant from the Pt, but dimerized as
follows:

....Pt....Cl-Pt-Cl.....Pt....Cl-Pt-Cl.....

To a first approximation we have alternating Pt(II),
with square planar and Pt(IV) with tetragonally distorted octahedral coordination.   All the compounds of
this kind are insulators but, as the name implies,
they have intense absorption bands in the visible which
are not found in the monomer units alone (21).   They
are assigned to intervalence transitions, Pt(II) →
Pt(IV).   Similar dimerized diatomic chains are found
in Au(I,III) compounds like $CsAuCl_3$ where the alternation is between linear and square planar coordination.

Given that the electron-phonon interaction
trapping the charge fluctuation along these chains is
strong, their insulating properties are no surprise.
If the halide bridging atoms could be displaced so
they were equidistant from the metal atoms, the chain
would no longer be dimerized, and conductivity might
rise.   Under high pressure the Pt(II,IV) and Au(I,III)
compounds do indeed become conducting (22), the latter
even metallic (23)!   X-ray (24) and neutron (25)
diffraction has shown that as the pressure increases
the Cl atoms in $CsAuCl_3$ are forced towards the centre
of the Au-Au pairs.

Doubly Bridged Chains

The only example of a doubly bridged mixed valence
chain which I know is TlS, better formulated as
$Tl^I(Tl^{III}S_2)$, with a similar structure to $KFeS_2$.   The
$(TlS_2)_n$ is a chain of TlS_4 tetrahedra sharing opposite
edges, and the Tl(I) ions occupy sites between the
chains.   TlS is an insulator (mixed-valence class I).

Directly Interacting Metal Ions

Apart from $Ir(Co)_3Cl$ (11), metallic conducting
metal chain compounds are confined exclusively to this
category. We can distinguish three types of substance: (1) oxides of Pt containing interlocking
chains of planar PtO_4 units, (2) halides with a high
ratio of metal atoms to anions ("metal-rich" compounds)
and (3) the Pt complexes with molecular ligands such
as cyanide and oxalate.

The ternary Pt oxides all have formulae $M_xPt_3O_4$,
where M is Na, Mg, Cd or Ni. In every case there
are chains of closely spaced Pt atoms along each of
the three orthogonal crystal directions, and while
the intrachain spacing is very short (2.80-2.85 $\overset{\circ}{A}$) the
interchain spacing is only 3.43 $\overset{\circ}{A}$, so interaction
between chains is probably strong (26). Conductivities are 10-1000 $ohm^{-1}cm^{-1}$ for pellet compactions at
room temperature, but little single crystal work is
available.

Metal rich halides are a relatively new source
of metal chain compounds, but examples containing 4f,
3d and B-subgroup elements are now known, largely
thanks to preparative work from Simon (27) and
Corbett (28). When normal valence metal halides are
melted with excess of the metallic element the resulting phases contain metal clusters which then condense
into chains or sheets. Unfortunately, because they
are so sensitive to air and moisture, and hence
difficult to contact, little is known about their
conductivity. In general, though, we expect them to
be metallic (mixed valence class IIIB).

The most famous conducting metal chain compound
with mixed valency is $K_2Pt(CN)_4Br_{0.30}3H_2O$, KCP, whose
properties have been very thoroughly investigated and
reviewed (29). For the purpose of the present survey,
the salient structural feature of KCP is the stack of
closely spaced planar $Pt(CN)_4$ units (Pt-Pt 2.89 $\overset{\circ}{A}$).
Partial oxidation of the Pt chain is achieved by adding
anions in sites on either side (Figure 3). Since the
counter-ions lie in channels quite separate from the
conducting stacks the number of such ions can in
principle be varied to suit the electronic and steric
requirements of the Pt chain and, in particular,
generate a level of partial oxidation which is quite
incommensurate with the underlying lattice. Thus
charge fluctuations cannot be trapped by lattice

Table. Formulae of conducting metal chain compounds (for notation see text)

Non-stoichiometry of anions

L = CN

M:	K	K	NH$_4$	Rb	Rb	Cs	Cs	C(NH$_2$)$_3$
X:	Br$_{0.30}$	Cl$_{0.32}$	Cl$_{0.30}$	Cl$_{0.30}$	(HF$_2$)$_{0.26-0.4}$	F$_{0.19}$	(N$_3$)$_{0.25}$	Cl$_{0.25}$

L = planar macrocycle; M' = Ni

L':	OMTBP	TBP	OMTBP	DPG
p :	0.33	0.35	0.97	0.33

Non-stoichiometry of cations

L = CN (p = 0)

M:	K	Rb	Cs
m:	1.75	1.73	1.72

L = oxalate (p = 0)

M:	K	Mg	Co
m:	1.6	0.82	0.83

distortions periodic in a small and integral number
of repeat units along the chain as happens, for example, in Wolfram's Red Salt.   Electron-phonon coupling
occurs only at a much more subtle level, in the form
of anomalies in the acoustic phonon density-of-states,
and in Peierls instabilities (see chapter by Schultz).

Most generally, the formulae of the KCP salts can
be written as $M_mPtL_nX_p(H_2O)_q$.   Examples are known
with L = CN^-, oxalate and the PtL_n chains become partly
oxidized either by adding anions X or subtracting
cations M.   Table 1 shows some examples, mostly
prepared electrochemically (30).

A final set of partially oxidized compounds are
those derived from complexes with planar macrocyclic
ligands like Pc and porphins (tetrabenzoporphin, TBP;
octamethyltetrabenzoporphin, OMTBP) and diphenyl-
glyoximate (DPG).   Here, mixed valency is achieved by
adding iodine in the form of I_3^-.   Conductivities are
high (31   ), but it is quite likely that electron
transport takes place primarily through overlap between
p -orbitals on adjacent molecules, as in TTF-TCNQ, with
the metal playing only a minor role.

CONCLUSION

The aim of this chapter was to introduce some of
the immense variety of inorganic and metal-organic
compounds with structures containing chains of closely
spaced metal atoms, and to identify some of the arch-
itectural features which help to give them interesting
properties.   In assessing the likely extent of electron
delocalization we have emphasised two requirements:
the presence of anionic or molecular bridging groups
and of single or mixed valency.   Most important,
however, is to see that inorganic and organometallic
chemistry provide the physicist with an astonishing
wealth of materials with interesting and unusual
properties: continued preparative efforts will
certainly keep it that way.

REFERENCES

1.   N.F. Mott, Can J. Phys. 34, 1356 (1956).

2.   J.B. Goodenough, Magnetism and the Chemical Bond,
      New York, Interscience, 1963, p. 26.

3. J.C. Calabrese, L.F. Dahl, P. Chini, G. Longoni
    and S. Martinengo, J. Amer. Chem. Soc. 96, 2614
    (1974).

4. P. Day, Ann. N.Y. Acad. Sci. 313, 9 (1978).

5. For excellent reviews of one-dimensional magnetism
    see L.J. de Jongh and A.R. Miedema, Adv. Phys. 23,
    1 (1974) and M. Steiner, J. Villain and C.G.
    Windsor, ibid. 25, 87 (1976).

6. J.F. Villa and W.E. Hatfield, J. Amer. Chem. Soc.
    93, 4081 (1971).

7. S. Foner, R.B. Frankel, W.M. Reiff, B.F. Little
    and G.J. Long, Solid State Commun. 16, 159 (1975).

8. e.g. P. Day and R.J.P. Williams, J. Chem. Phys.
    37, 567 (1962).

9. G.D. Stucky, A.J. Schultz and J.M. Williams,
    Ann. Rev. Mater. Sci. 7, 301 (1977).

10. J.D. Corbett, personal communication.

11. A.H. Reis and S.W. Peterson, Ann. N.Y. Acad. Sci.
    313, 560 (1978).

12. For a review see P. Day, Inorg. Chim. Acta Rev.
    3, 81 (1969).

13. K. Ukei, J. Phys. Soc. Japan 40, 140 (1976).

14. G. Scregg, Chemistry Part II Thesis, Oxford,
    1962 (unpublished).

15. M.L. Moreau-Colin, Struct. Bonding (Berlin) 10,
    167 (1972).

16. P. Day, J. Amer. Chem. Soc. 97, 1588 (1975).

17. H. Yersin and G. Gliemann, Ber. Bunsenges.
    Phys. Chem. 79, 1050 (1975).

18. P.L. Johnson, T.R. Koch and J.M. Williams,
    Acta Cryst. B33, 1293 (1977); H.H. Otto,
    H. Schulz, K.H. Thiemann, H. Yersin and
    G. Gliemann, Z. Naturforsch. 326, 127 (1977).

19. F.D. Saeva, G.R. Olin, R.J. Ziolo and P. Day, J. Amer. Chem. Soc., in press.

20. M.B. Robin and P. Day, Adv. Inorg. Chem. and Radiochem. 10, 247 (1967).

21. P. Day, Low-dimensional Cooperative Phenomena, ed. H.J. Keller, New York, Plenum, 1975, p. 191.

22. L.V. Interrante, K.W. Browall and F.B. Bundy, Inorg. Chem. 13, 1158 (1974).

23. R. Keller, J. Fenner and W.B. Holzapfel, Mater. Res. Bull. 9, 1363 (1974).

24. W. Denner, Doctoral Dissertation, Karlsruhe (TH), 1977.

25. P. Day, C. Vettier and G. Parisot, Inorg. Chem. 17, 2319 (1978).

26. D. Cahen, J.A. Ibers and M.H. Mueller, Inorg. Chem. 13, 110 (1974).

27. A. Simon, Chem. Unserer Ziet 10, 1 (1976).

28. e.g. K.R. Poppelmeier and J.D. Corbett, J. Amer. Chem. Soc. 100, 5039 (1978).

29. Various chapters in Low-dimensional Cooperative Phenomena, ed. H.J. Keller, New York, Plenum, 1975.

30. J.M. Williams and A.J. Schultz, Molecular Metals, ed. W.E. Hatfield, New York, Plenum, 1979, p. 337.

31. B.M. Hoffman, T.E. Phillips, C.J. Schramm and S.K. Wright, Molecular Metals, ed. W.E. Hatfield, New York, Plenum, 1979, p. 393.

STRUCTURAL CHEMISTRY OF LINEAR CHAIN TRANSITION METAL COMPLEXES

Heimo J. Keller

Anorganisch-Chemisches Institut der Universität Heidelberg, Im Neuenheimer Feld 270, D-6900 Heidelberg 1/GFR

1. INTRODUCTION

Transition metal ions normally occur in very distinct geometric arrangements of their ligands. Very often certain metal ions in a distinct oxidation state prefer only one particular geometry of the nearest environment. The versatility of coordination chemistry is furthermore based on the fact, that there are ligands which "offer" donor atoms only in special spatial arrangements (like the "tripod" ligands or phthalocyanine). This opens the possibility of "forcing" metal ions in special, preselected ligand fields. Inter-ionic interactions in solid transition metal complexes are a prerequisite for collective physical properties. These interactions strongly depend on the sterical arrangement of the ligands. It follows then, that distinct cooperative properties can be produced "at will" by selecting appropriate metal ions and ligands. Since the main emphasis of this meeting is on very anisotropic collective behaviour, this paper concentrates on those species which allow "linear" and infinite interactions. There are quite a number of recent reviews in which different aspects of this problem have been discussed in detail (1-9) and two other lectures (J.S.Miller and P.Day) related to this problem have been presented at this meeting.

In principal two different types of lattices could be discerned:

321

L. Alcácer (ed.), The Physics and Chemistry of Low Dimensional Solids, 321–331.
Copyright © 1980 by D. Reidel Publishing Company.

(i)   Solids which are built up of individual <u>molecular</u>
        species. The collective properties are a result
        of interactions between the molecules and appear
        <u>additionally</u> to the slightly disturbed molecular
        properties ("molecular solids").

(ii)   Infinite, closed packed structures of different
        ions. The building blocks of the lattice have
        lost their individual properties after entering
        the lattice.

The second class of materials (transition metal oxides,
halides, tungsten and platinum bronzes, e.g.) are discussed in the paper by Prof. P. Day (Oxford) and extensive literature is available for these materials.

    Structural investigations on "molecular" transi-
tion metal complexes are certainly routine in inorganic
or metalorganic chemistry but detailed investigations
of the <u>relation</u> between the <u>molecular</u> structure, the
<u>crystal</u> structure <u>and</u> the <u>physical</u> properties of the
bulk material had not been studied until the sixties.
The knowledge of such a relation could lead to the
planned syntheses of new interesting materials. It is
the aim of this paper to discuss very briefly the present status in these efforts. The problem could be
summarized in the following two questions:

        (i)   What are the promising molecules and
        (ii) what are the promising conditions to
              prepare one-dimensional "molecular"
              transition metal compounds?

2.   PROMISING MOLECULES AND THEIR CLASSIFICATION

    The interactions in molecular transition metal
solids can be divided into two classes:

        (i)   Solids with directly interacting metal ions.
        (ii) Compounds in which the interaction occurs
              through ligand atoms (indirectly interacting,
              "bridged" metal ions).

2.1 Coordination geometries for directly interacting
        species

    Only metal ions with additional "free" coordina-
tion sites are able to form directly interacting systems. In principle the "unsaturated" ligand arrangement
in mononuclear complexes could be <u>linear</u>, <u>trigonal</u> pla-

nar or <u>tetragonal</u> planar. The association of all of
these types could lead to linear, directly interacting
chains of metal ions (figure 1).

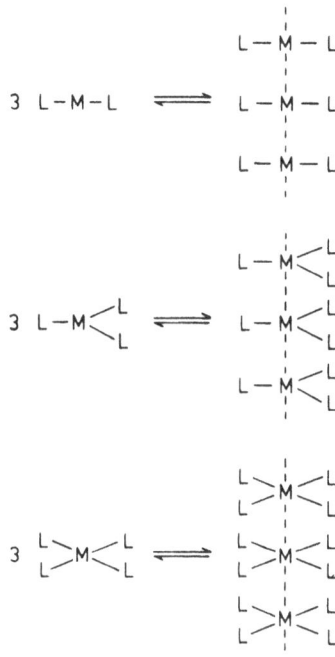

Figure 1: Schematic drawing of stacks built by aggre-
          gation of bi-, tri- and tetracoordinated
          metal complexes.

     With a few exceptions (e.g. bis(2.3-butanedione-
dioximato)gold(III)-dichloroaurat(I) with linear (Cl-
Au-Cl)⁻ units (10)) only planar four-coordinated species have so far been found as building blocks in directly interacting chains. The following part concentrates, therefore, mainly on these molecules. There are
several reasons to classify the so far investigated
molecules into two categories:

(i)   Planar compounds in which the geometry is "forced"
      by the ligand. Typical examples: metal complexes
      of phthalocyanine, porphyrine and similar macro-
      cycles.

(ii) Planar transition metal complexes, in which the
      electronic configuration of the central metal ions
      determines the molecular structure.

There are quite a number of compounds which cannot
be grouped unambigously in either (i) or (ii). These
are mainly the bis(α,ß-dionedioximato)metal compounds
of nickel, palladium and platinum. (The general molecular structure of these compounds is shown in figure 3,
lower line, central formula). Some physical data indicate that there are interactions between the ligands
of the complex molecules along the stacks which are
typical for the partially oxidized macrocyclic materials (11).

2.1.1 Macrocyclic ligands

The chemistry and physics of these very interesting compounds and especially of their partially oxidized derivatives have been reported recently by the
group at Northwestern University. The physical properties of several solid phthalocyaninato and porphyrina-
to complexes containing additional triiodide chains in
the channels of the stacked macrocycles have been discussed in detail. (12-14). The highly conducting compounds which are one-dimensional metals interact mainly
through their ligand orbitals the metal being a minor
perturbation in some cases. These materials can be regarded as a "link" between the highly conducting purely
organic charge transfer compounds and the one-dimensional transition metal complexes.

2.1.2 Metal ions with preferred planar four-coordina-
ted environment

Of all the elements in the Periodic Table only
metal ions of the group VIII and Ib occur frequently
in planar four-coordinated complexes (figure 2).

VIII			Ib
^{26}Fe	^{27}Co	^{28}Ni	^{29}Cu
^{44}Ru	^{45}Rh	^{46}Pd	^{47}Ag
^{76}Os	^{77}Ir	^{78}Pt	^{79}Au

Figure 2: Part of the Periodic Table showing the elements which
occur frequently in planar or linear geometries.

The promising molecules are, therefore, restricted to ions of these elements especially to the d^8-electron configuration (Co(I), Rh(I), Ir(I), Ni(II), Pd(II), Pt(II), Au(III)). An exception is the d^9 ion Cu(II) for which many examples of planar coordination are known. Since Cu(I), Ag(I) and Au(I) additionally occur in linear two coordinated complexes all the promising metal ions are found in this narrow region of the Periodic Table. Nevertheless an enormous variety of stacked materials with these central metal ions is known (15), ranging from integral oxidation state (IOS), mixed stacked solids with very different (linear and planar four-coordinated gold in bis(2,3-butanedionedioximato)-gold(III)dichloroaurate(I) (10)) or similar (Magnus' Green Salt structures (16)) site geometries to stronger interacting segregated stacked IOS crystals (tetra-cyanoplatinates(II) (17)) and from weakly interacting non integral oxidation state (NIOS) compounds (partially oxidized bis(dioximato)-nickel(II,IV) and palladium(II,IV)iodides (18-20)) to the partially oxidized one-dimensional metals of the tetracyanoplatinates(II, IV) ("KCP") (21-23). A few representative examples of well-known and thoroughly investigated molecules are shown in figure 3. Combination of the different complex ions contained in this scheme leads to numerous additional linear chain species.

PLANAR TRANSITION METAL COMPLEXES

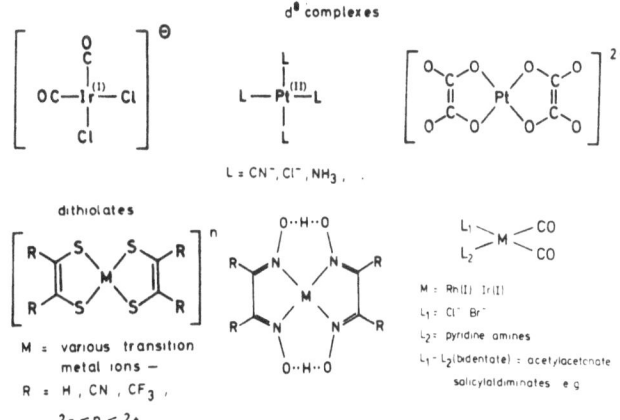

Figure 3: Representative examples of four-coordinated planar transition metal complexes with d^8 central ions.

In general one can state that the metal-metal interactions along the chains are stronger for the 5d
metals (especially Ir and Pt) compared to the solids
containing the lighter homologues. Most promising are
therefore compounds of the expensive metals iridium
and platinum.

If two different oxidation states of the· same metal complex are stable <u>in solution</u> mixed valence solids can be crystallized. The most famous examples being the broad class of tetracyanoplatinates(II,IV),
the bis(oxalato)platinates(II,IV) (24,25) (both from
aqueous solutions), the dicarbonyldihaloiridates(I,III)
(26) and the bis (α,ß-dionedioximato)metal(II,IV) complexes of nickel and palladium (18-20) (from organic
solvents).

2.2   Bridged linear chain compounds

The bridged linear chain metal complex solids formally can be constructed by mixing one of the above
mentioned geometries for directly interacting systems
and higher coordinated (octahedral e.g.) species (figure 4). A typical example being the Wolffram's Salt

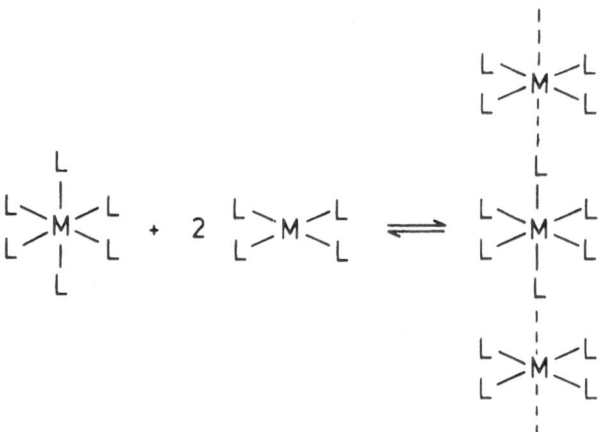

Figure 4: Reaction of octahedral and planar species to
          linearly bridged chain solids.

Analogues which consist of halide bridged stacks of
planar platinum(II) and octahedral platinum(IV) species
(figure 5) (27-30). In this way any of the above mentioned planar species capable of building linear arrays

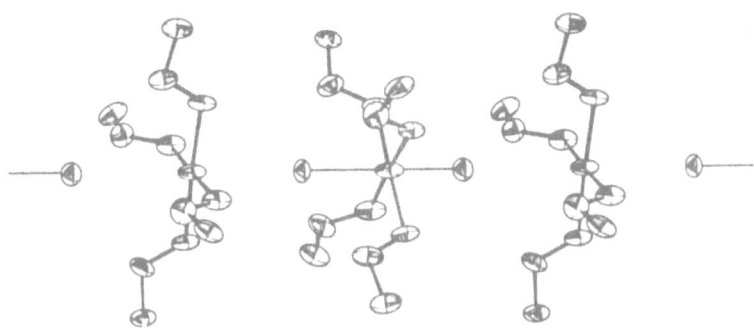

Figure 5: Structure of one Pt^{IV}──Br···Pt^{II} chain in
Tetrakisethylamineplatinum(II)-dibromotetra-
kisethylamineplatinum(IV)-tetrabromide (31).

could be combined with any other octahedral complex to
linearly bridged metal chains. Since the octahedral
ligand arrangement is very common for all metal ions
(transition and heavier main group elements) an "infinite" number of linear bridged systems could be prepared in this way in a very directed manner and numerous compounds of this type are known in the literature.

3.   PROMISING CONDITIONS

     In order to build up linearly interacting chains,
the reactive "unsaturated" coordination sites should
not be blocked by any other molecule. If partially oxidized complexes are crystallized from solution one
starts with the solvated molecules in their two different oxidation states. The piling up reaction of a partially oxidized species can be schematically represented as in figure 6 and could be regarded as a substitution reaction, because the active donor and acceptor
sites are coordinated to solvent molecules.

     Any donor-acceptor reaction in the solution will
lead to a blocking of the "open" coordination sites of
the complexes. Neither donors nor acceptors should be
used as solvents, therefore. If on the other hand the
solvent does not interact with the complex molecules
at all the compounds cannot be obtained from solution
(insolubility of starting materials). Though this aspect has been discussed in detail recently (32,33) the
importance of the reaction medium has not been recogni-

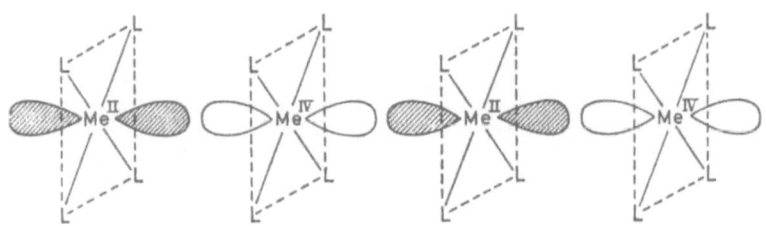

Figure 6: Schematic drawing of interacting molecules
          with d^8 central metal ions (filled d_z2 orbi-
          tals shown as shaded d_z2 lobes) and d^6 cen-
          tral metals ("empty" d_z2 orbital).

zed so far. Typical examples for preparations in "non-
interacting" solvents are the different partially oxidized macrocyclic and dioximato compounds. They can be
dissolved in weakly interacting <u>organic</u> solvents (because of their <u>organic</u> ligands). The subtle donor-
acceptor balance of the unoxidized and oxidized complex molecules in solution is not disturbed by the solvent. The so far known highly conducting mixed valence
materials like KCP and bis(oxalato)platinates(II,IV)
just "happen" to crystallize from <u>water</u> but many other
materials should be found after optimizing the problem
of reaction conditions especially finding suitable
solvents (SO_2 e.g.).

4. CONDUCTIVITY OF LINEAR CHAIN COMPOUNDS

     Since the highly conducting <u>organic</u> compounds were
the main emphasis of this meeting some remarks on the
electrical conductivities in linear chain metal complexes from a chemist's point of view might be appropriate. Again it is reasonable to divide the known materials in directly interacting and linearly bridged
systems.

4.1 Directly interacting systems

     As a summary of the known experimental data one
could state that directly interacting 1-dimensional metal chains are good conductors (1d metals), if the metal-metal distances are below 2.9 Å. This statement is
valid only for complexes interacting through their <u>me-
tal centers.</u> In the partially oxidized phthalocyaninato and other macrocyclic compounds, the distances bet-

ween the planes of these molecules are in the order of 3.2 Å as typical for organic π-systems) the conduction occurs not via the metal ions but via overlapping ligand π-orbitals. Some of these materials again behave like one-dimensional metals.

If the ligands do not interact electronically M-M distances above 2.9 Å will not allow metallic conductivities. The relative large intermolecular spacings in the α,β-dionedioximates ($\gtrsim$ 3.2 Å) make any metallic conductivity very improbable even in the partially oxidized compounds. The same arguements hold for other NIOS (KCP, partially oxidized bisoxalatoplatinates(II,IV)) as well as for IOS materials (Ir(CO)$_3$Cl) for which M-M distances below 2.9 Å lead to semimetals and high conductivities as well. Since the M-M distances can be reduced considerably by applying pressure, the conductivities can be increased in this way, but so far a conductivity drop has been observed in all cases after releasing the pressure.

4.2 Bridged systems

To my best knowledge metallic conductivity has not been found so far in one-dimensional linearly bridged systems whether partially oxidized or not. Evidently the transfer integrals between the metal ions and the bridging atoms are too small to allow a reasonable bandwidth. Using heavier elements as bridging atoms increases the interactions in many cases but a considerable increase in conductivity (metallic) could not be observed.

LITERATURE

(1)   Thomas, T.W., and Underhill, A.E.: 1972, Chem. Soc.Rev. 1, pp. 99 ff.
(2)   "Extended Interactions between Metal Ions in Transition Metal Complexes", ACS Symposium Series, Vol. 5, 1974, ed. L.V.Interrante, Washington, D.C.
(3)   "Low-Dimensional Cooperative Phenomena", NATO-ASI Series B, Vol. 7, 1975, ed. H. J. Keller, Plenum Press, N. Y.
(4)   "One-dimensional Conductors", 1975, Lecture Notes in Physics, Vol. 34, Springer Verlag, ed. H. G. Schuster.
(5)   Miller, J.S., and Epstein, A.J.: 1976, Prog.Inorg. Chem. 20, pp. 1 ff.

(6)   "Inorganic Compounds with Unusual Properties",
       1976, Adv. in Chem. Series 150, ed. R. B. King.
(7)   "Chemistry and Physics of One-dimensional Metals",
       1977, NATO-ASI Series B, Vol. 25, Plenum Press,
       N. Y., ed. H. J. Keller.
(8)   "Synthesis and Properties of Low-dimensional
       Materials", 1978, Ann. N. Y. Acad. Sci. 313, ed.
       J. S. Miller and A. J. Epstein.
(9)   "Molecular Metals", NATO-Conference Series VI,
       Vol. 1, Plenum Press, N.Y., 1979, ed. W.A.Hatfield.
(10)   Rundle, R.E.: 1954, J.Amer.Chem.Soc., 76,pp.3101.
(11)   Schramm, C.J., Stojakovic, D.R., Hoffman, B.M.,
       and Marks, T.J.: 1978, Science 200, pp. 47.
(12)   Phillips, T.E., and Hoffman, B.M.: 1977, J.Amer.
       Chem.Soc., 99, pp. 7734.
(13)   a) Petersen, J.L., Schramm, C.S., Stojakovic, D.
          R., Hoffman, B.M., and Marks, T.J.: 1977, J.
          Amer.Chem.Soc., 99, pp. 286.
       b) Marks, T.J., in ref. (8), pp. 609-614.
(14)   Hoffman, B.M., Phillips, T.E., Schramm, C.J., and
       Wright, S.K., in ref. (9), pp. 393 and references
       cited therein.
(15)   Keller, H.J. in "Mixed-Valence Compounds in Che-
       mistry, Physics and Biology", ed. Brown, D.B.,
       Reidel Publish. Comp., in press.
(16)   Miller, J.R.: 1965, J.Chem.Soc., pp. 713.
(17)   Yersin, H., Gliemann, G., Rössler, U.: 1977, So-
       lid State Comm. 21, pp. 915 and ref.cited therein.
(18)   Brown, L.D., Webster Kalina, D., McClure, M.S.,
       Schultz, S., Ruby, S.L., Ibers, J.A., Kannewurf,
       C.R., and Marks, T.J.: 1979, J.Amer.Chem.Soc.,
       101, pp. 2937.
(19)   Cowie, M., Gleizes, A., Grynkewich, G.W., Webster
       Kalina, D., McClure, M.S., Scaringe, R.P., Tei-
       telbaum, R.C., Ruby, S.L., Ibers, J.A., Kanne-
       wurf, C.R., and Marks, T.J.: 1979, J.Amer.Chem.
       Soc. 101, pp. 2921 and references cited therein.
(20)   Endres, H., Keller, H.J., Lehmann, R., and Weiß,
       J.: 1976, Acta Cryst., B32, pp. 627.
(21)   Williams, J.M., and Schultz, A.J., in ref. (9),
       pp. 337-368 and references cited therein.
(22)   Zeller, H.R.: 1973, Adv.Solid Stage Phys., 13,
       pp. 31.
(23)   Brown, R.K., and Williams, J.M.: 1979, Inorg.
       Chem., 18, pp. 1922 and references cited therein.
(24)   Krogmann, K., and Dodel, P.: 1966, Chem. Ber.,
       99, pp. 3402, pp. 3408.
(25)   Schultz, A.J., Underhill, A.E., and Williams,
       J.M.: 1978, Inorg.Chem. 17, pp. 1313.

(26) Rosencwaig, A., Ginsberg, A.P., and Koepke, J.W.:
      1976, Inorg. Chem. 15, pp. 2540.

(27) Matsumoto, N., Yamashita, M., Kida, S.: 1978,
      Bull. Chem. Soc. Japan 51, pp. 3514.

(28) Papavassiliou, G.C.: 1979, J. Phys. Solid State,
      12, L 297.

(29) Keller, H.J., Martin, R., Traeger, U.: 1978, Z.
      Naturforsch. 33b, pp. 1263.

(30) Campbell, J.R., Clark, R.J.H., Turtle, P.C.:1978,
      Inorg. Chem. 17, pp. 3622.

(31) Endres, H., Keller, H.J., Keppler, B., Martin,
      R., Steiger, W., and Traeger, U.: Acta Cryst.,
      submitted.

(32) Buse, K.D., Keller, H.J., and Nöthe, D.: 1976,
      Z.Naturforsch., 31b, pp. 194.

(33) Gitzel, W., Keller, H.J., Rupp, H.H., Seibold,
      K.: 1972, Z.Naturforsch. 27b, pp. 365.

(34) a) Reis, A.H., jr., Hagley, V.S., Peterson, S.W.:
         1977, J.Amer.Chem.Soc. 99, pp. 4184.
      b) Reis, A.H., jr., and Peterson, S.W., in ref.
         (8), pp. 571-578.

DESIGN AND SYNTHESIS OF HIGHLY CONDUCTING ONE-DIMENSIONAL MATERIALS

Joel S. Miller

Occidental Research Corporation
Irvine, California 92713

Unidimensionality (1-D) is ascribed to many organic and in-organic materials which possess an extended columnar structure in the solid state. Such materials possess a variety of physical properties which due to the unidimensionality are frequently significantly anisotropic. The most studied physical property is electrical conductivity. This arises from the substantial interest initially generated by W. A. Little in his proposal of 1-D high temperature superconductors and the subsequent observation of conductivities of 1-D materials that vary over an enormous range, nominally fifteen orders of magnitude, i.e., 10^{-12}-10^3 ohm^{-1}cm^{-1}. Most interestingly, some unidimensional substances possess a temperature range where the temperature dependence of the conductivity, $\sigma(T)$ is metal-like, i.e., $d\sigma^2/dT^2$, is positive. This and additional electrical and optical properties coupled with the materials low density and susceptibility for chemical modifications lend themselves to the technological applications in this electronic age.

To explore the structure function relationship and potential technological applications of inorganic, organic, and covalent polymers many subtances have been synthesized and additional ones are sought. In the past several years many investigators have conveyed their thoughts on the material design to achieve highly conducting materials. Table I lists review articles where the topic has been discussed in detail. I have written one of the articles and my thoughts have not really changed, thus, I believe it best for the reader to consult the articles listed in Table I for a prescription to designing new highly conducting unidimensional materials.

L. Alcácer (ed.), The Physics and Chemistry of Low Dimensional Solids, 333–337.
Copyright © 1980 by D. Reidel Publishing Company.

The key ingredient for the successful synthesis of new highly conducting materials is creativity.   Good fortune, however, should not be neglected.   One can easily summarize the criteria necessary for achieving a highly conducting 1-D chain.   For example, a planar molecule with an extended π system normal to the molecular plane that is furthermore capable of forming a solid with a fractional degree of charge transfer, is sought. Many real examples exist and some materials which fit the criteria do not form a highly conducting 1-D chain.

The creative synthetic chemist must integrate their expertise and intuition with the results of a decade of experience and fortune from scientists in the field as distilled in the Table I list of review articles. Imaginative ideas coupled with per-sistent diligence will lead to new materials.   The detailed exhaustive study of these new materials will give deeper, more penetrating insight into the chemical and physical properties of unidimensional systems and in some cases may reveal totally unexpected results and the observation of new phenomenon.   With these challenges the quest and study of new materials should be actively supported.

TABLE I

SELECTED ARTICLES DISCUSSING DESIGN AND SYNTHESIS OF HIGHLY CONDUCTING 1-D SUBSTANCE

Author	Title	Reference
Bloch, A.N.; D.O. Cowan, D.O.; Poehler, T.O.	"Design and Study of 1-Dimensional Organic Conductors"	in "Charge Transfer in Organic Semiconductors", Masuda, K. and Silver, M., Eds., pp. 167-176. Plenum Press. New York, N.Y.
Cowan, D.; Shu, P.; Hu, C.; Krug, W.; Carruthers, T.; Poehler, T.; Bloch, A.	"The Organic Metallic State: Some Chemical Aspects"	NATO Advance Studies Institute, 1977, 25B, 25-45.
Engler, E.M.	"Organic Metals"	Chemtech, 1976, 6, 276-279.
Garito, A.F.; Heeger, A.J.	"The Design and Synthesis of Organic Metals"	Acct. Chem. Res., 1974, 7, 232-240.
Haddon, R.C.	"Design of Organic Metals and Superconductors"	Nature, 1975, 256, 394-396.
Interrante, L.V.; Bray, J.W. Hart, H.R. Jr.; Kasper, J.S.; Piacente, P.A.; Watkins, G.D.	"Molecular Design of Solid-State Systems. Organic-Metal Complex π-Donor-Acceptor Compounds"	N.Y. Acad. Sci., 1978, 313, 407-416.

TABLE I - Contd

SELECTED ARTICLES DISCUSSING DESIGN AND SYNTHESIS OF HIGHLY CONDUCTING 1-D SUBSTANCE

Author	Title	Reference
Keller, H.J.	"Preparative Aspects of One-Dimensional Transition Metal Coordination Compounds"	NATO Advanced Studies Institute. 1975, 7B, 315-337.
Krogmann, K.	"Planar Complexes Containing Metal -Metal Bonds"	Angew. Chem. (Internat. Edit.), 1969, 8, 35-42.
Marks, T.J.	"Rational Synthesis of New Undimensional Solids: Chemical and Physical Studies of Mixed-Valence Polyiodides"	N.Y. Acad. Sci., 1978, 313, 594-616.
Miller, J.S.	"Design and Synthesis of Highly Conducting One-Dimensional Materials"	N.Y. Acad. Sci., 1978, 313, 25-60
Miller, J.S.; Epstein, A.J.	"One-Dimensional Inorganic Complexes"	Prog. Inorg. Chem., 1976, 20, 1-151.
Mueller-Westerhoff, U.T.; Heinrich, F.	"One-Dimensional and Pseudo-One-Dimensional Molecular Crystals"	A.C.S., Symp. Ser., 1974, 5, 392-401.

TABLE I - Contd

SELECTED ARTICLES DISCUSSING DESIGN AND SYNTHESIS OF HIGHLY CONDUCTING 1-D SUBSTANCE

Author	Title	Reference
Perlstein, J.H.	"Organic Metals-Intermolecular Migration of Aromaticity"	Angew. Chem. (Internat. Edit.), 1977, 16, 519-534.
Soos, Z.G.; Keller, H.J.	"Comparison of Columar Organic and Inorganic Solids"	NATO Advanced Studies Institute, 1977, 25B, 391-412.
Torrance, J.B.	"Preparation Of Organic Metals and Insulators: Some Ideas and	NATO Advanced Studies Institute, 1979, 1, 7-14.
Torrance, J.B.	"The Difference Between Metallic and Insulating Salts of Tetracy-anoquinodimethane (TCNQ): How to Design an Organic Metal"	Acct. Chem. Res., 1979, 12, 79-86.
Yagubskii, E.B.; Khidekel, M.L.	"High Temperature Excitonic Superconductivity: Synthetic Aspects'	Russ. Chem. Rev., 1972, 41, 1011-1026.

TWO BAND CONDUCTORS WITH VARIABLE BAND FILLING: PROBING THE PROPERTIES OF QUASI-ONE-DIMENSIONAL MATERIALS

Arthur J. Epstein[*] and Joel S. Miller[+]

Xerox Webster Research Center, Rochester, NY, 14644[*]
Occidential Research Corporation, Irvine, CA, 92713[+]

$(N\text{-methylphenazinium})_x(\text{phenazine})_{1-x}(\text{tetracyanoquinodimethanide})$, $(NMP)_x(Phen)_{1-x}(TCNQ)$, $0.5 \leq x \leq 1.0$, provides an unusually flexible system for probing the properties of highly conducting quasi-one-dimensional materials. The results of an extensive series of studies show that this material forms a two band system with delocalized electrons and only weak effects due to disorder. Though no anomalous behavior has yet been seen in this class for a commensurate value of charge density, a rapid change in properties is observed upon emptying of one of the two bands of carriers. This rapid change may reflect a crossover from large coulomb repulsion to small coulomb repulsion limits.

The conducting salts of the organic acceptor 7,7,8,8-tetracyano-p-quinodimethane (TCNQ) have been widely studied and are the basis for the examination of the relevance of a variety of models (1,2,3). The largest number of studies have been carried out on the one-to-one compound of TCNQ, 1, with the donor tetrathiafulvalene (TTF), (TTF)(TCNQ). The neutral TTF has two electrons in its highest occupied molecular orbital (HOMO) and the neutral TCNQ has an empty lowest unoccupied molecular orbital (LUMO). Upon formation of the (TTF)(TCNQ) compound, 0.59 electrons are transferred, on average, from each TTF to each TCNQ (4). Intensive study of this system and its closely related derivatives (5) has led to considerable insight into the physics and chemistry of highly conducting quasi-one-dimensional (1-D) molecular conductors.

Many issues remain in the field. Among these are questions concerning the magnitude and temperature dependence of the charge carrier mobility, the effects of disorder on the electronic structure of quasi-one-dimensional solids, the magnitude and

L. Alcácer (ed.), The Physics and Chemistry of Low Dimensional Solids, 339–351.

effects of the effective coulomb repulsion between two conduction
electrons at the same site (U), the role of commensurability in
determining electronic properties, and the role of fractional
charge transfer is determining electronic properties.

We have synthesized (6) and studied an isostructural series of
two band conductors. The number of electrons available to be
shared among the two types of chains is varied between 1.0 and
0.5 electrons per formula unit. This contrasts with (TTF)(TCNQ)
which has two electrons to be aportioned per formula unit. Study
of this system has provided detailed insight into the issues
noted above and has raised some questions concerning the understanding of electronic properties of quasi-one-dimensional
conductors.

NMP$^+$ TCNQ   Phen

1   2   3

The system studied is based upon (NMP)(TCNQ), which was among the
earliest known highly conducting organic 1-D conductors.
The HOMO of NMP$^\circ$ is half filled and the LUMO of TCNQ$^\circ$ is empty,
so that there is only one electron to be shared among the NMP, 1,
and the TCNQ, 2, by replacing some of the NMP molecules in (NMP)-
(TCNQ) with a neutral molecule whose HOMO is filled and is not
easily oxidized, the number of electrons available to be shared
among the donor and acceptor chains can be reduced. This was
successfully achieved through replacement of up to fifty percent
of the NMP moleculures by phenazine molecules, 3. The resulting
compounds, (NMP)$_x$(Phen)$_{1-x}$(TCNQ) with $0.5 \leq x \leq 1.0$ can be
synthesized through the four routes:

A) (NMP)(PF$_6$) + Li(TCNQ) + Phen + TCNQ
B) (NMP)(TCNQ) + (Phen)(TCNQ)
C) (NMP)(TCNQ)$_2$ + Phen
D) (NMP)(TCNQ)$_2$ + (Phen)(TCNQ) + TCNQ

$\longrightarrow$ (NMP)$_x$(Phen)$_{1-x}$(TCNQ)
$1.0 \geq x \geq 0.5$

X-ray determination of the unit cell parameters for varying
showed that the resulting (NMP)$_x$(Phen)$_{1-x}$(TCNQ) has similar ·but
not identical parameters to that of (NMP)(TCNQ) (6).

The detailed structure of the (NMP)(TCNQ) system itself has been controversial. The (NMP)(TCNQ) studied is the black highly conducting triclinic phase in which the donor and acceptor molecular form segregated stacks (7) with a = 3.8682Å, b = 7.7807Å, c = 15.735Å, α =91.67°, β = 92.71° and γ = 95.38°, see Figure 1.

The asymmetric location of the methyl groups of the NMP molecules introduce an intrinsic static disorder. The highly conducting forms of (NMP)(TCNQ) can exist with two different degrees of disorder of the methyl groups. The most studied of these forms, which we refer to as (NMP)(TCNQ),shows an increasing conductivity as the temperature is lowered from 400K to about 230K, at which temperature the conductivity presents a broad maximum, then decreases rapidly as the temperature is decreased further (8,9,10). On the basis of X-ray data this form, until a recent study by Pouget etal. (11), was assumed to correspond to a random disorder of methyl groups. Another less common and less well studied form, which we refer to as (NMP)(TCNQ)(IB), shows comparable values of room temperature conductivity, but the conductivity decreases very rapidly with decreasing temperatures, and consequently does not pass through a maximum value below room temperature (12). In this last form, diffuse X-ray scattering experiments showed that the methyl groups were ordered two dimensionally in the ab plans (a being the stacking direction) but without correlations between successive planes of this type (13). The Pouget et al. study (11) shows that for "disordered" (NMP)(TCNQ) the methyl groups are also ordered in the b direction with a coherence length (ξ) greater then 200Å (25 lattice constants) but that the alternation of the orientation of the methyl groups in the chain direction (a-axis) is only correlated over a few intermolecular spacings (the Pouget etal study gave ξ_a = 25Å or six lattice constants). Hence (NMP)(TCNQ) is not randomly disordered as earlier assumed (7) but is only less ordered than (NMP)(TCNQ)(IB).

Diffuse X-ray scattering studies (11) of (NMP)(TCNQ) shows two types of 1-D scattering bearing resemblance with the Kohn anomalies earlier observed in (TTF)(TCNQ). A first 1-D scattering is observed at room temperature at the wave vector 0.33a*; it couples three-dimensionally below approximately 200K. Further diffuse scattering is observed below 70K at half the previous wave vector (0.165a*). As expected in an intrinsically disordered system (as, for example, $K_2Pt(CN)_4Br_{0.3} \cdot 3H_2O$, KCP) no long range 3-D ordering is observed down to 20°K. These results cannot ascribe unambiguously the 0.33a* and 0.165a* scattering to $4K_F$ and $2K_F$ anomalies (as was done for (TTF)(TCNQ)), but strongly suggests that the charge transfer in (NMP)(TCNQ) is 2/3 electron. At low temperature, the distortions spread over a distance in the chain direction, ξ_{11} > 100Å, larger than that of the orientational disorder of the NMP molecules: $\xi_a \simeq$ 25Å. This indicates that the modulation in chain direction is not very sensitive to the

Table I
Charge Carrier Distribution, $(NMP)_x(Phen)_{1-x}(TCNQ)$

X	TCNQ	(NMP)(Phen)	Reference
1.00	0.67	0.33	(11)
0.80	0.60	0.20	(14)
0.50	0.50	0.00	(15,16,17)

Table II
Conductivity Parameters for $(NMP)_x(Phen)_{1-x}(TCNQ)$ (Ref. 19)

X	$\sigma(295K),ohm^{-1}cm^{-1}$	$\sigma_n(T_m)$	T_m,K	α	Δ,K	Δ_ℓ,K
1.00	200	1.17	220	4.1	900	500
0.94	100	1.27	205	3.9	800	400
0.81	100	1.85	155	3.7	575	275
0.63	70	1.26	175	2.2	400	200

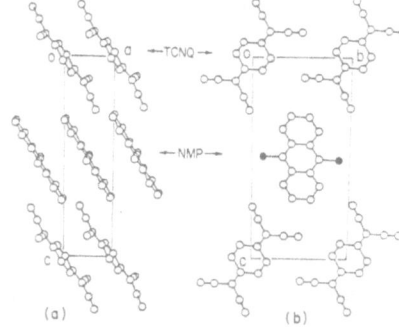

Fig. 1: (NMP)(TCNQ)(I): (a) View down the b-axis, (b) view down a-axis. Two methyl groups are shown for each NMP molecule, indicating the randomness in the methyl group location. (From Ref. 7).

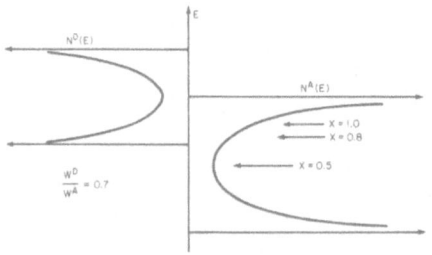

Fig. 2: Density of states vs. energy for the TCNQ band, $N^A(E)$, and the (NMP)-(Phen) band, $N^D(E)$ assuming large U.

NMP disorder and takes place on the TCNQ stacks and/or on the
averaged NMP chains. Temperature dependent DC conductivity
measurements after X-ray study unequivalocally associate the
highly conducting form with these properties.

Preliminary diffuse X-ray studies (14) of $(NMP)_{0.8}(Phen)_{0.2}(TCNQ)$
indicate only $2K_F$ scattering and an average of 0.6 charges per TCNQ
and 0.2 charges per donor stack repeat unit. Thermoelectric power
(15), electron spin resonance (16) and magnetic succeptibility
(17) experiments on $(NMP)_{0.51}(Phen)_{0.49}(TCNQ)$ show that for this
stoichiometry all of the charges are on the TCNQ stack and that
the mixed (NMP)(Phen) stack does not have any carriers. Table I
summarized the charge per chain as a function of x.

From the results of the experiments indicated above, it is clear
that the $(NMP)_x(Phen)_{1-x}(TCNQ)$ must be treated as a two chain
conductor, with the bandwidth along the TCNQ stacks denoted by W^Q,
and that along the mixed NMP-Phen stacks denoted by W^P. Assuming
1-D tight binding bands on acceptor (Q) and donor (P) stacks, the
density of states for the corresponding bands is given by (18)

$$\rho^{Q,P} = [(2,1)/\pi] \, N \, [(W^{Q,P}/2)^2 - (E^{Q,P})^2]^{-\frac{1}{2}} \qquad (1)$$

where $E^{Q,P}$ is the energy measured from the center of each band
and N is the number of sites in each chain. The (2,1) denotes
the large U, small U cases respectively. Such a two band system
is represented, for example, by two density of state functions of
widths W^Q and W^P and whose centers are separated by ΔE, as
illustrated in Figure 2. As the Fermi level is varied (variation
in x) in a rigid band model, both the total number of conduction
clcctrons per unit cell and their distribution between the donor
and acceptor chains varies.

Extensive studies have been performed of the temperature dependent
transport properties of $(NMP)_x(Phen)_{1-x}(TCNQ)$. The temperature
dependent DC conductivity (19), σ, is shown in Figures 3 and 4.
The samples studied all have similar behavior with a broad maximum
in $\sigma(T)$. The $\sigma(T)$ data for 60K < T < 400K is well fit by

$$\sigma_n(T) = AT^{-\alpha} \exp(-\Delta(x)/T) \qquad (2)$$

with $\Delta(x)$ constant for all samples of the same phenazine content and α a sample dependent constant in the range of 2 to 4,
Table II. The constant A is fixed by $\sigma_n(295K) = 1$. The solid
lines in Figures 3a and 4 show the fits obtained with these parameters in Eq. (2).

Eq. (2) suggests a model of a band semiconductor with two tight
binding bands each of width W separated by a gap $2\Delta(x)$. We have
previously shown (10,20,21) that for Maxwell-Boltzmann statistics

Table III
Thermoelectric Power Parameters for $(NMP)_x(Phen)_{1-x}(TCNQ)$
$S = - \mid k_B/e \mid \{\ell n2 + A/T + \epsilon\}$ (Ref. 15)

x	A,°K	Δ,°K*	b	ε
1.00	-126	900	0.75	.14
.95	- 92	812	0.80	.047
.86	- 72	666	0.81	.012
.63	0	357	1.00	.035
.54	0	262	1.00	.093

* From $\Delta = (900K)x^2$, see Ref. 19.

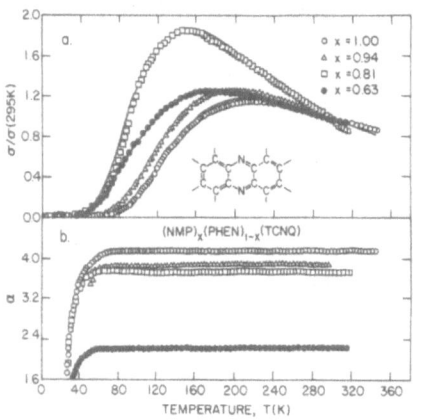

Fig. 3: (a) Normalized a-axis
conductivity for $(NMP)_x(Phen)_{1-x}$-
$(TCNQ)$. The solid lines are
computer fits to Eq. (2) with
values given in Table II.
(b) $\alpha(T)$ calculated from Eq. (2)
with experimental $\sigma(T)$ and $\Delta(x)$
from Table II. (Ref. 19).

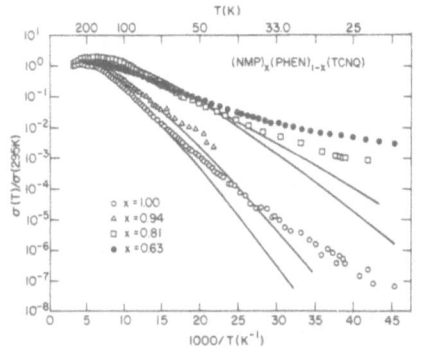

Fig. 4: Experimental
$log[\sigma(T)/\sigma(295K)]$ versus T^{-1}
for samples of Fig. 2, and
computer fits from Eq. (2) with
parameters given in Table II.
(Ref. 19)

this leads to an activated carrier concentration $n(T) \propto T^{\frac{1}{2}} \cdot$
$\exp(-\Delta/T)$ and mobility $\mu \approx 10(295/T)^{\alpha+0.5}$ cm^2/volt-sec. Similar
values of μ are obtained for the phenazine substituted samples.
Such $\mu(T)$ values have been calculated to TCNQ salts using single
phonon scattering of delocalized electrons. Examination of
Table II shows that $\Delta(x) = (900K)x^2$. The variation of Δ with x
can be examined as a variation with the charge density on the
TCNQ stack, N_Q, assuming that the TCNQ stack dominates the transport. There is no dramatic difference between the commensurate
case, $\Delta(N_Q = .67)$ and the incommensurate cases $\Delta(N_Q < 0.67)$, in
contrast with the behavior of (TTF)(TCNQ) (22). Further, the
variation of Δ with x or N_Q is not well modeled by mean field
behavior, $\Delta = 2E_F \exp[-\omega/g^2 \rho (E_F)]$ (23). Here E_F is the Fermi
energy, ω is the unperturbed $2K_F$ phonon frequency, and g is
dimensionless electron-phonon coupling constant.

Assuming a mean free path limited by impurities for T < 65K, $\sigma(T)$
$\propto \exp(-\Delta_\ell/T)$, Δ_ℓ is determined for each x, with $\Delta_\ell \approx \Delta/2$. This
suggests a large number of localized states near the center of
the gap. For T < 33K, $\sigma(T)$ becomes less T-dependent, as well as
increases with increasing Phen°, suggesting that hopping among
the increasing number of localized states in the gap dominates.

The thermolectric power (15), S(T), for (NMP)$_{0.54}$(Phen)$_{0.46}$(TCNQ)
is nearly identical to that of Qn(TCNQ)$_2$ (Qn = quinolinium) (24)
as well as that of Acridinium-TCNQ$_2$ (24) and (NMP)(TCNQ)$_2$ (25),
see Figure 5. Assuming that there is one electron transferred
from each NMP$^+$ to the TCNQ stacks in (NMP)$_{0.5}$(Phen)$_{0.5}$(TCNQ),
this system has the same number of charges per TCNQ as Qn(TCNQ)$_2$
(one-quarter filled band in a U=0 picture). The Qn(TCNQ)$_2$
(26) and Adn(TCNQ)$_2$ (27) materials have two acceptor stacks for
each donor stack. The (NMP)$_{0.5}$(Phen)$_{0.5}$(TCNQ) has one acceptor
stack for each donor stack as well as increased disorder. These
results demonstrate that for a large family of one-quarter filled
band systems the thermoelectric properties are nearly identical
despite very different crystal structures, donors, and degree of
disorder. S(T) depends only upon the number of electrons per TCNQ
at this band filling. The value $S \approx -60\mu V/°K$ [= - | k_B/e | $\ell n2$]
for a large number of quarter-filled band materials has been
understood as the effect of spin entropy (statistics), reflecting
the role of strong coulomb correlations (large U) in these materials (28-30). Recent experimental measurement of the magneto-
thermopower of Qn(TCNQ)$_2$ has verified this assignment (31). This
large U model for S(T) has been explicitly demonstrated theoretically for semiconductors with 0.5 electrons per site and finite
bandwidths (32).

S(T) for 0.5 < x < 1.0 is intermediate between that of (NMP)(TCNQ)
and Qn(TCNQ)$_2$. The plot of -S(T) versus T^{-1} reveals a systematic
behavior, Figure 5. In the high temperature regime (T > 100K),

S is fit by a linear function with the slope monotonically decreasing with decreasing x. In addition, all samples studied
have an intercept (constant term b) of S $\approx$ -60μV/°K. Hence

$$S = - \mid k_B/e \mid \{\ln 2 + A/T + \epsilon \} \tag{3}$$

Such a simple parametization of S(T) is unusual. The presence of
a constant contribution of $\sim$ -60μV/°K for the (NMP)$_x$(Phen)$_{1-x}$ ·
(TCNQ) family supports the important role of strong coulomb correlations in these materials. The second and third terms in
Eq. (2) are analyzed in terms of a simplified qualitative semiconductor model wherein the semiconducting gap occurs in the
lower Hubbard band. (Such separate spin and orbital contributions
will occur rigorously only in a strong coupling Hubbard model.)

Using results for three-dimensional intrinsic semiconductors
ignoring spin effects (33), A = [(b-1)/(b+1)]Δ, and b = μ_e/μ_h,
the ratio of electron to hole mobilities. The constant ϵ is
related to the electron and hole masses and the heats of carrier
transport (33). For high levels of phenazine doping, the
electron and hole subbands introduced by the gap into the lower
Hubbard band are nearly symmetric. For this case, $\mu_e \approx \mu_h$ and
b $\sim$ 1 leading to a small coefficient A. For low phenazine levels
μ_e and μ_h differ and A is expected to increase, as seen in
Figure 5. Table III summarizes the fit of Eq. (3) to the data.

The measurement of the frequency ω, dependence of the room temperature polarized reflectivity both parallel and perpendicular to
the highly conducting a-axis in (NMP)$_x$(Phen)$_{1-x}$(TCNQ) (34) is the
 first unequivocal demonstration that the minimum in the reflectivity spectrum in the region of 1 ev, often observed for linear
chain TCNQ salts, is indeed a collective (plasma) response due
to the number of charge carriers per unit cell in these compounds.
The Drude formulation of the plasma frequency:

$$\omega_p{}^2 = 4\pi \, Ne^2/\epsilon m^* \tag{4}$$

can be used to analyze the data provided the effective mass m* is
appropriately evaluated for a tight binding "metallic" band of
width W and proper account is taken of U. The background dielectric
constant, ϵ, is assumed to be constant in this series and e is
the electronic charge. Averaging the effective mass in Eq. 4
over the entire tight binding band leads to

$$\omega_p{}^2 = [(2,1)e^2 S_F aW/4\pi^2 \epsilon \hbar^2] \sin (aq_F) \tag{5}$$

Here (2,1) refers to either the small or large U case respectively,
a is the lattice constant is the highly conducting direction, S$_F$
is the Fermi surface area normal to the a direction, q$_F$ is the
Fermi wave vector, and $\hbar$ is Planck's constant divided by 2π. For

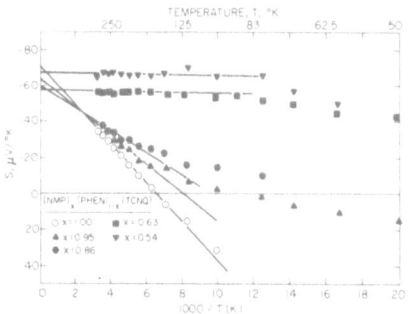

Fig. 5: Thermoelectric power vs T^{-1} for $(NMP)_x(Phen)_{1-x}$-(TCNQ). The straight lines are a fit of Eq. (3) to the data points using the parameters of Table III. (Ref. 15).

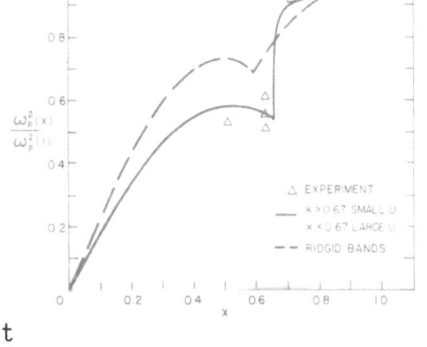

Fig. 6: An anomalous red shift of ω_p occurs at $0.6<x<0.7$. The dash dot line is a fit obtained by lowering the Fermi level of two rigid large U bands as a function of x, with $W^P=0.7W^Q$ and $\Delta E=0.5W^Q$. The solid curve is a fit to the data assuming $W^P/W^Q<0.1$ and a crossover from small U/W to large U/W as x decreases below 2/3 (see Ref. 34).

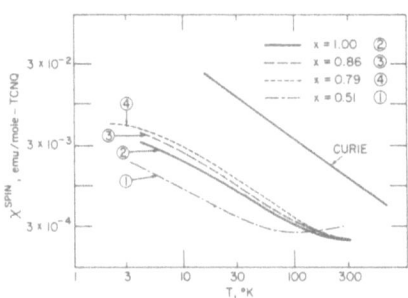

Fig. 7: Molar spin susceptibility versus T for $(NMP)_x(Phen)_{1-x}(TCNQ)$ (see Ref. 17).

a small band gap at the Fermi energy, Eq. 5 should remain approximately valid. To sum the contributions from the two types of conducting chains, the approximation $\omega_p^2 = \omega_p^{Q2} + \omega_p^{P2}$ is used.

The rapid red shift of ω_p for x < 2/3, Figure 6, can be analyzed via two different mechanisms. Assuming large U and a density of states as in Figure 2, a rapid emptying of the (NMP)(Phen) band occurs for a fixed value of x, leading to the behavior illustrated by the dashed line in Figure 6. Here the ratio of the bandwidths was taken as $W^P/W^Q = 0.7$ and the shift in the band centers, ΔE, as $0.5 \ W^Q$. In the second case, smooth variation in the charge transfer with x is assumed [number of electron per TCNQ taken as $(x + 1)/3$] with a change in coulomb repulsion from small U for x > .7 to large U for x < .7 and $0 \le W^P/W^Q \le .1$.

The sharpness of the transition leads one to suspect that both mechanisms occur simultaneously. That is, the conductive electrons appear as if U/W is small when the two chains are occupied. When the donor chain is empty the remaining electrons, now only in acceptor chains, feature large U/W. This change in the magnitude of the effective coulomb repulsion upon emptying of the donor band would accentuate the effect of either mechanism alone. The simultaneous change could be related to variation in onsite coulomb repulsion with charge density within each stack. Alternatively, the presence of a partly filled donor chain with delocalized carriers may determine the effective screening of the onsite coulomb repulsion (35). Upon emptying half the chains of delocalized electron, the screening may become less effective and the system appear as having large U/W. Comparisons with magnetic (16,17) and diffuse X-ray studies (11,14) are consistent with the large U/W small U/W crossover picture, though thermoelectric power results (15) suggest that electrons of high q (near q_F) retain a significant coulomb repulsion for all x.

The electron spin resonance studied (16) have clearly shown that as x decreases from 1.0 to 0.5 an increasing fraction of spins is on the TCNQ stacks. For the x = 0.5 case, the ESR g tensor and its temperature dependence is nearly indistinguishable from that of $Qn(TCNQ)_2$ (37) showing that for this case, all spins are on the TCNQ stack. Static susceptibility, χ, measurements (17), Figure 7, show that for x = 0.5, the magnitude and temperature dependence of $\chi(T)$ are nearly identical to that of $Qn(TCNQ)_2$, supporting that all spins are on the TCNQ stack for x = 0.5 and that this system has large U/W.

For all x studied, $\chi(T) = AT^{-\beta}$ with $\beta \sim .7 - .8$ for low temperatures (T < 40K) (16,17,36). This is associated with random exchange among a small number of localized spins (37). However x < 2/3, there is a rapid decrease in the coefficient A. This is consistant with the depopulation of the phenazine stack for this x so that for x < 2/3, only unpaired localized spins on the TCNQ stack contribute. For x > 2/3 localized spins on both stacks contribute.

For T > 100K, there is a difference in $\chi(T)$ for x greater or less then 2/3. For x < 2/3 $\chi(T)$ appears much like that of $Qn(TCNQ)_2$ with large U and an exchange of J $\sim$ 300K. For x > 2/3, $\chi(T)$ decreases, consistent with either an increasing gap is the spin excitation spectrum [similar to increasing $\Delta(x)$ in Table II] or the presence of small U/W on both stacks for x > 2/3.

The observation of $2K_F$ scattering in $(NMP)_x(Phen)_{1-x}(TCNQ)$ for x > .8 supports the concept of small U/W in this region. A change over to large U for x < 2/3 would show that U is a property of the solid and not dependent on the isolated molecule alone (38). Experimental detection of diffuse X-ray scattering for x < 2/3 would help clarify these issues.

In sum the $(NMP)_x(Phen)_{1-x}(TCNQ)$ system provides the synthetic flexibility that enables one to probe the physics of the 1-D solid state. Results of extensive studies demonstrate that the effects of disorder are weak leading only to low temperature effects, that a two band system is formed with delocalized electrons on both stacks, and there is no indication of any anomalous behavior for values of x corresponding to commensurate charge densities. The system has been analyzed as being a semi-conductor with a large strongly temperature dependent mobility. The variation of the band gap with x does not follow a mean field behavior. Finally, detailed studies of the optical and magnetic behavior suggest a possible crossover from large U/W for x > 2/3 to small U/W for x < 2/3 coincident with the emptying of alternate planes of delocalized electrons.

ACKNOWLEDGMENT

We are indebted to numerous individuals for the discussions, cooperation and collaborations which enabled these studies to occur. In particular, we acknowledge J. Bardeen, P. M. Chaikin, W. G. Clark, R. Comès, E. M. Conwell, C. B. Duke, J. Hammann, S. Megtert, J.-P. Pouget, H. Rommelmann, M. J. Rice, D. J. Sandman, M. L. Slade, A. P. Troup, and B. A. Weinstein for their contributions and critical comments.

REFERENCES

(1)   See, for example, 1979, Quasi-One-Dimensional Conductors, Lecture Notes in Physics 95 and 96.

(2)   Highly Conducting One-Dimensional Solids, ed. by Devreese,
        J.T., etal. (Plenum Press, NY) (1979).
(3)   Synthesis and Properties of Low Dimensional Materials, ed.
        by Miller, J.S. and Epstein, A.J., 1978, Ann. N.Y. Acad.
        Sci. 313.
(4)   Comès, R., Shapiro, S.M., Shirane, G., Garito, A.F., and
        Heeger, A.J.: 1976, Phys. Rev. B14, pp. 2376-2383.
(5)   Schultz, T.D., and Craven, R.A.: in Ref. (2), pp. 147-
        225.
(6)   Miller, J.S. and Epstein, A.J.: 1978, J. Am. Chem. Soc.
        100, pp. 1639-1641.
(7)   Fritchie, C.J.: 1966, Acta Cryst. 20, pp. 892-898;
        Morosin, B.: 1975, Phys. Lett. 53A, pp. 455-456.
(8)   Epstein, A.J., Etemad, S., Garito, A.F., and Heeger, A.J.:
        1971, Solid State Comm. 9, pp. 1803-1808; 1972, Phys.
        Rev. B5, pp. 952-977.
(9)   Coleman, L.B., Cohen, J.A., Garito, A.F., and Heeger, A.J.:
        1973, Phys. Rev. B7, pp. 2122-2128.
(10)   Epstein, A.J., Conwell, E.M., Sandman, D.J., and Miller, J.S.:
        1977, Solid State Comm. 23, pp. 355-358.
(11)   Pouget, J.P., Megtert, S., Comès, R., and Epstein, A.J.:
        1979, Phys. Rev. B, in press.
(12)   Fujii, G., Shirotani, I, and Nagano, H.: 1977, Bull Chem.
        Soc. Jpn. 50, pp. 1726-1730.
(13)   Kobayashi, H.: 1975, Bull. Chem. Soc. Jpn. 48, pp. 1373-
        1377.
(14)   Pouget, J.-P., Comès, R.: private communication.
(15)   Epstein, A.J., Miller, J.S., and Chaikin, P.M.: 1979,
        Phys. Rev. Lett. 43, pp. 1178-1181.
(16)   Troup, A.P., Epstein, A.J. and Miller, J.S.: to be
        published.
(17)   Epstein, A.J. and Miller, J.S.: to be published.
(18)   Miller, J.S., and Epstein, A.J.: 1976, Prog. Inorg. Chem.
        20, pp. 1 - 151.
(19)   Epstein, A.J., and Miller, J.S.: 1978, Solid State Comm.
        27, pp. 325-329.
(20)   Epstein, A.J., and Conwell, E.M.: 1977, Solid State Comm.
        24, pp. 627-630.
(21)   Epstein, A.J., Conwell, E.M, and Miller, J.S.: 1978,
        Ann. N.Y. Acad. Sci. 313, pp. 183-209.
(22)   Andrieux, A., Schulz, H.J., Jérome, D., and Bechgaard, K.:
        1979, J. de Physique - Lett. 40, pp. L-385-L-389.
(23)   Sham, L.J.: In Reference (2), pp. 227-245.
(24)   Kwak, J.F., Beni, G., and Chaikin, P.M.: 1976, Phys.
        Rev. B 13, pp. 641-646; Buravov, L.I., Fedutin, D.N., and
        Shchegolev, I.F.: 1971, Soviet Physics JETP 32, pp. 612-
        616.
(25)   Epstein, A.J., Miller, J.S., and Chaikin, P.M.: to be
        published.

(26)   Kobayashi, H., Marumo, F. and Saito, Y.: 1971, Acta
         Cryst. B27, pp. 373-378.
(27)   Kobayashi, H.: 1974, Bull. Chem. Soc. Jpn. 47, pp. 1346-
         1352.
(28)   Beni, G.: 1974, Phys. Rev. B10, pp. 2186-2189.
(29)   Kwak, J.F., and Beni, G.: 1976, Phys. Rev. B13, pp. 652-
         657.
(30)   Chaikin, P.M.: 1978, Proc. Int. Conf. on Thermoelectricity
         in Metallic Conductors, Aug. 1977 (Plenum Press, N.Y.),
         p. 359.
(31)   Chaikin, P.M., Kwak, J.F., and Epstein, A.J.: 1979, Phys.
         Rev. Lett. 42, pp. 1178-1182.
(32)   Conwell, E.M.: 1978, Phys. Rev. B18, pp. 1818-1823.
(33)   Tauc, J.: 1962, Photo and Thermoelectric Effects in
         Semiconductors (Pergamon Press, N.Y.).
(34)   Weinstein, B.A., Slade, M.L., Epstein, A.J. and Miller, J.S.:
         to be published.
(35)   Similar concepts are discussed in Gutfreund, H., and
         Little, W.A.: in Ref. (2), pp. 305-372; Bloch, A.N.:
         1979, Abstracts of Papers PHYS 47 (Joint Am. Chem. Soc./
         Chem. Soc. Jpn. Chemical Congress, Honolulu, Hawaii,
         April 1-6, 1979).
(36)   Hammann, J., Clark, W.G., Epstein, A.J., and Miller, J.S.:
         to be published.
(37)   Clark, W.G., Hammann, J., Sanny, J., and Tippie, L.C.:
         1979, Lecture Notes in Physics 96, pp. 255-264.
(38)   Torrance, J.B.: 1979, Acct. Chem. Res. 12, pp. 79-86.

CONDUCTING ORGANIC POLYMERS: DOPED POLYACETYLENE

A. J. Heeger[†] and A. G. MacDiarmid[‡]

Laboratory for Research on the Structure of Matter
University of Pennsylvania, Philadelphia, PA 19104

Contents

---

† Dept. of Physics, U. of Pennsylvania
‡ Dept. of Chemistry, U. of Pennsylvania

L. Alcácer (ed.), The Physics and Chemistry of Low Dimensional Solids, 353–391.
Copyright © 1980 by D. Reidel Publishing Company.

I. INTRODUCTION

The electronic structure of polyenes has been a subject of
interest for many years. The unsaturated bonds which characterize the polyenes have an important effect on their electronic
structure and the corresponding electronic properties. In a
polyene (see Fig. 1) three of the four carbon valence electrons
are in sp^2 hybridized orbitals; two of the σ-type bonds are links
in the backbone chain while the third forms a bond with some side
group (e.g., H in Fig. 1). The remaining valence electron has
the symmetry of the $2p_z$ orbital and forms a π bond in which the
charge density is perpendicular to the plane of the molecule. In
terms of an energy-band description, the σ bonds form low-lying
completely filled bands, while the π bond would correspond to a
half-filled band. The π bond could be metallic provided there is
negligible distortion of the chain, and an independent-particle
model proved to be satisfactory.

The possibility of a distortion of the molecular structure
intrigued the early investigators. J. E. Lennard-Jones,[1] in an
early application of Huckel theory of molecular orbitals, investigated the electronic structure of polyenes. His conclusion, that
in the limit of an infinite chain the bonds tended to a constant
value of 1.38 A, was later supported by studies of Coulson.[2] However, the theoretical conclusions seemed less than satisfactory
in view of the experimental observations.[3] Using either molecular-
orbital (MO) theory or a free-electron model in which the π electrons are considered free to delocalize along the chain, the
energy for a transition to the lowest excited state decreases
linearly with the reciprocal of the chain length. This rule was
observed to work rather well for the short polyenes. However,
experiments indicated that for very long chains the transition
energy reached a limiting value of about 2 eV.[3] Later Kuhn[4]
demonstrated that bond alternation could serve as a possible
explanation of the energy gap in long polyenes.

The first convincing analysis was that given by Lounget-
Higgins and Salem;[5] by assuming a well-defined model they were
able to carry out the calculation without recourse to estimates
of physical parameters. They found the uniform infinite chain
unstable with respect to bond alternation. Hence for an infinite
polyene the stable configuration is one of unequal bond lengths.
This result is no more than a restatement of the one-dimensional
(1-D) Peierls instability[6,7] for a special case. For long polyenes
where bond alternation becomes important the electron is subject
to a periodic potential with a period of twice the original undistorted chain. Such a potential introduces a gap at $2k_F$ causing a
change of character in the polyene from a metal to a semiconductor.

The importance of correlation in long chain polyenes has
been stressed by Ovchinnikov et al.[8] and others.[9] Ovchinnikov

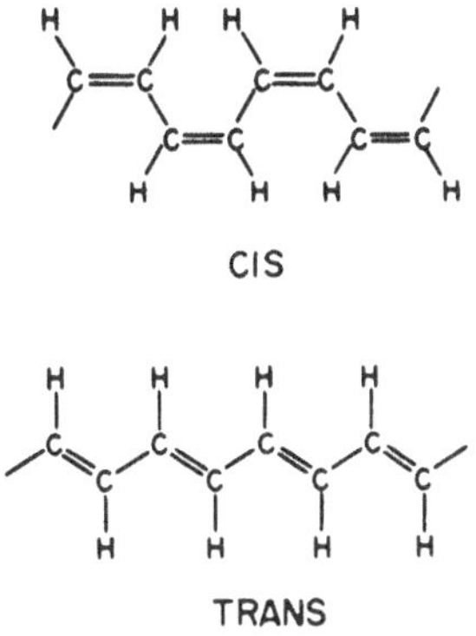

Fig. 1.   Molecular structure of cis and trans isomers
of polyacetylene, $(CH)_x$.

et al.[8] argued that the electronic gap of ~ 2 eV seen in the
long chain polyenes is due almost entirely to correlation with
the Peierls effect playing little or no part. However, there is
not general agreement on this point. Grant and Batra[10] estimated
a somewhat larger single particle energy gap, whereas Duke et al.[11]
showed that a calculation including a combination of bond alternation and Coulomb interaction yields results in agreement with
photoemission and optical absorption data.

Linear polyacetylene, $(CH)_x$, is the simplest conjugated
organic polymer (Fig. 1) and is therefore of special fundamental
interest in the context of the above discussion. From theoretical
and spectroscopic studies of short chain polyenes, the π-system
transfer integral can be estimated as $|t| \simeq 2$-2.5 eV. Such an
estimate implies that the overall bandwidth would be of order
8-10 eV; $W_{||} = 2Z|t|$. As a result of this large overall band
width and unsaturated π-system $(CH)_x$ is fundamentally different
from either the traditional organic semiconductors made up of
weakly interacting molecules (e.g., anthracene, etc.), or from
other saturated polymers with monomeric units of the form $\binom{R}{C}\binom{R}{C}$
where there are no π-electrons (e.g., polyethylene, etc.). Polyacetylene is more nearly analogous to the traditional inorganic
semiconductors. However, the transverse bandwidth due to interchain coupling is much less. The nearest-neighbor interchain
spacing[12] of 4.39 Å implies a transverse bandwidth $(W_\perp)$ comparable
to the bandwidth (longitudinal) in molecular crystals like tetra-
thiafulvalenium-tetracyano-p-quinodimethanide (TTF-TCNQ);[7] thus
$W_\perp \sim 0.1$ eV. Weak interchain coupling is therefore implied, and
the system may be regarded as quasi-one-dimensional. Consequently,
although the 1-D Peierls instability is not a full explanation of
the electronic structure of polyacetylene, it provides a useful
starting point of discussion for many of the electronic properties.

Interest in this semiconducting polymer has been stimulated by the successful demonstration of doping with associated control of electrical properties over a wide range;[13],[14] the electrical conductivity of films of $(CH)_x$ can be varied over 12 orders of magnitude from that of an insulator ($\sigma \sim 10^{-9}\,\Omega^{-1}\,cm^{-1}$) through semiconductor to a metal ($\sigma \sim 10^3\,\Omega^{-1}\,cm^{-1}$).[15],[16] Various electron donating or accepting molecules can be used to yield n-type or p-type material,[17] and compensation and junction formation have been demonstrated.[17],[18] Optical-absorption studies[19] indicate a direct band-gap semiconductor with a peak absorption coefficient of about $3\times10^5\,cm^{-1}$ at 1.9 eV. Partial orientation[20] of the polymer fibrils by stretch elongation of the $(CH)_x$ films results in anisotropic electrical[21] and optical properties[19] suggestive of a highly anisotropic band structure. The electrical conductivity of partially oriented metallic $[CH(AsF_5)_{0.1}]_x$ is in excess of 2000 $\Omega^{-1}\,cm^{-1}$.[21] The qualitative change in electrical and optical properties at dopant concentrations above a few percent have been interpreted as a semiconductor-metal transition[16] by analogy to that observed in studies of heavily doped silicon. However, the anomalously small Curie-law susceptibility[22] components in the lightly doped semiconductor regime indicate that the localized states induced by doping below the semiconductor-metal transition are nonmagnetic. These observations coupled with electron spin resonance studies of neutral defects[23],[24] in the undoped polymer have resulted in the concept of soliton doping; i.e. localized domain-wall-like charged donor-acceptor states induced through charge transfer doping.[25],[26]

The initial results obtained on conducting polymers have generated considerable interest from the point of view of potentially low cost solar energy conversion. Experiments utilizing polyacetylene, $(CH)_x$, successfully demonstrated rectifying junction formation.[17],[18] In particular a p-$(CH)_x$:n-ZnS heterojunction solar cell has been fabricated with open circuit photovoltage of 0.8 Volts.[18] In related experiments, a photoelectrochemical photovoltaic cell was fabricated using $(CH)_x$ as the active photoelectrode.[27]

In this review, we will concentrate primarily on aspects of the work carried out at the University of Pennsylvania, bringing in contributions of colleagues at other institutions where appropriate. The focus will be on the fundamental physics and physical mechanisms. Detailed discussion of the more chemical aspects can be found elsewhere.[13],[14]

II. SEMICONDUCTING $(CH)_x$

IIA. Band Structure[10],[25]

As an initial approximation we consider a model in which the

π-electrons of <u>trans</u>-$(CH)_x$ are treated in a tight binding approximation and the σ electrons are assumed to move adiabatically with the nuclei. Let u_n be a configuration coordinate for displacement of the $n\underline{th}$ CH group along the molecular symmetry axis (x), where $u_n=0$ for the undimerized chain. The Hamiltonian is

$$H = -\sum_{hs} (t_{n+1,n} c^+_{n+1,s}\ c_{n,s} + h.c.) + \sum_n \frac{K}{2}(u_{n+1}-u_n)^2 + \sum_n \frac{M}{2} u_n^2 \qquad (1)$$

where to first order in the u's,

$$t_{n+1,n} = t_o - \alpha(u_{n+1}-u_n). \qquad (2)$$

M is the mass of the CH unit, K is the spring constant for the σ energy when expanded to second order about the equilibrium undimerized systems, and c^+_{ns} (c_{ns}) creates (annihilates) a π electron of spin s on the $n\underline{th}$ CH group. The band structure of the perfect infinite dimerized <u>trans</u> structure is shown in Fig. 2. In this perfect structure, the displacements are of the form

$$u_{no} = \pm (-1)^n u_o, \qquad (3)$$

where $\pm$ corresponds to the two possible degenerate structures (double bonds pointing "up" and double bonds pointing "down"). The transfer integrals for the perfect chain are

$$t_{n+1,n} = \begin{cases} t_o - t_1 & \text{"single" bond} \\ t_o + t_1 & \text{"double" bond} \end{cases} \qquad (4)$$

The overall bandwidth is $4t_o \approx 8\text{-}10$ eV; whereas the energy gap $Eg = 4t_1$ depends on the magnitude of the distortion. For example if $Eg = 4t_1 \approx 1.4$ eV, the value of u_o which minimizes the ground state energy (assuming $K = 10.5$ eV/Å^2)[28] is $u_o = 0.042$ Å in the direction of the CH displacement while the bond length change due

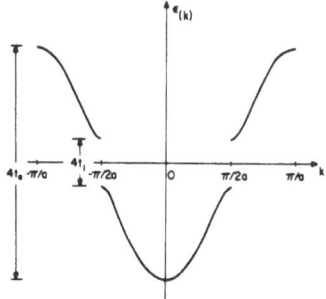

Fig. 2. Band structure of perfectly dimerized $(CH)_x$.

to bond alternation is $\sqrt{3}\ u_0 = 0.073$ Å. Larger values of the
energy gap require larger distortions; e.g. $Eg = 4t_1 = 2$ eV corresponds to $u_0 = .056$ Å. In the presence of weak bond alternation
with static distortion u_0

$$\mathcal{E}(k) = \pm 2 \sqrt{t_0^2 \cos^2 ka + t_1^2 \sin^2 ka} \tag{5}$$

where a is the c-c distance along the chain in the uniform structure. More detailed band calculations have been carried out by
Grant and Batra[10] including consideration of the cis-structure and
the effects of interchain coupling. The interchain coupling is
expected to give transverse bandwidth of the order of a few tenths
of an eV. Because of the crystal potential associated with the
doubled unit cell of the cis-$(CH)_x$ structure, there is an energy
gap even for uniform bonds. Consequently we expect the experimental energy gap for cis-$(CH)_x$ to be greater than that for
trans-$(CH)_x$.

IIB. Optical Absorption[19,29,30]

Optical absorption data for trans-$(CH)_x$ and cis-$(CH)_x$ are
shown in Figures 3a and 3b. The absorption coefficient for trans-
$(CH)_x$ begins a slow increase around 1.0 eV rising sharply at 1.4
eV to a peak at about 1.9 eV. The magnitude of the absorption
coefficient (3×10^5 cm^{-1}) at the peak is comparable with the peak
value in typical direct gap semiconductors of about 10^6 cm^{-1}.
For the cis isomer the absorption maximum is blue shifted by about
0.3 eV consistent with a somewhat larger gap.

A detailed analysis[29] of the absorption spectrum has been
carried out in terms of the joint density of states for the optical
interband transition. Using the tight-binding results discussed
above and assuming weak interchain coupling the optical joint
density of states takes the form sketched in Fig. 3c. The dashed
curve represents the 1-D limit where the $(\varepsilon - \varepsilon_g)^{-\frac{1}{2}}$ singularity
occurs at the band edge. Interchain coupling removes the singu-
larity as described above shifting the maximum away from E_g^{direct}
by an amount of order $W_\perp$. The possibility of an indirect gap,
which may arise from interchain coupling, is indicated by extending
the curve below E_g^{direct}.

Collecting these results we can make a qualitative picture of
the optical absorption for a highly anisotropic (quasi-1-D) semiconductor. The optical absorption will begin once the photon
energy is larger than the indirect gap and then increase rapidly
to the quenched singularity, decreasing for larger photon energies.
Such a picture agrees with the data for polyacetylene shown in
Fig. 3. Taking the absorption peak to be 1.9 eV, one estimates
the transverse bandwidth to be ≤ 0.5 eV. This value for $W_\perp \sim 2zt$
is consistent with the intermolecular transfer integrals in other

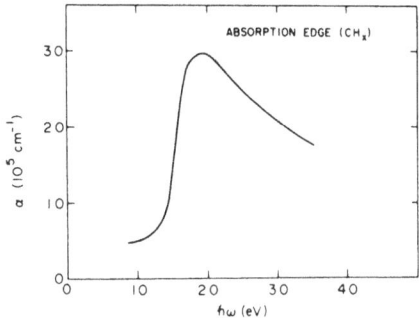

Fig. 3a. Absorption coefficient as a function of frequency; <u>trans</u>-(CH)$_x$ (ref. 29).

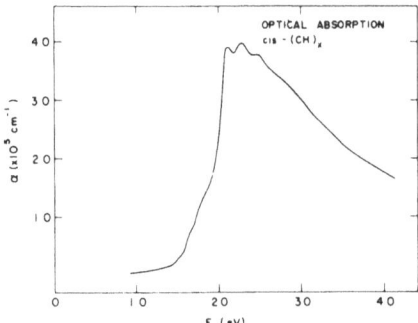

Fig. 3b. Absorption coefficient as a function of frequency; <u>cis</u>-(CH)$_x$ (ref. 29).

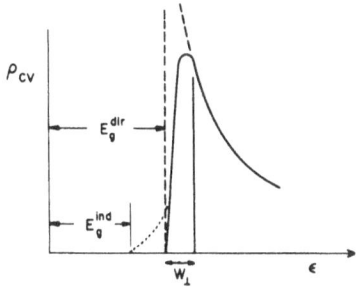

Fig. 3c. The joint optical density of states corresponding to the band structure of Fig.2 The rounding of the square-root singularity arises from interchain coupling ($W_\perp$) (ref. 29).

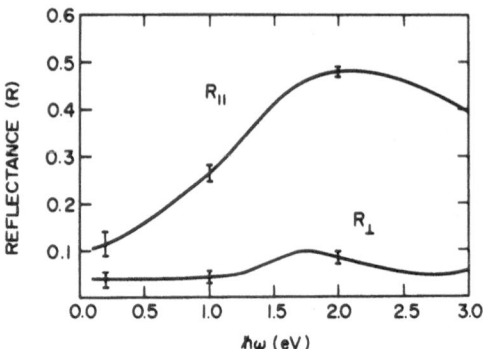

Fig. 4.   Polarized reflectance as a function of fre-
          quency from partially oriented $(CH)_x$ $(1/1_o=2.5)$.
          $R_{||}$ and $R_\perp$ refer to light polarization parallel
          and perpendicular to the orientation direction
          (ref. 29).

molecular crystals such as TTF-TCNQ.[7] This interpretation of the
optical absorption appears to be applicable to $(CH)_x$ as well as
many other low-dimensional solids. Using the above analysis as a
guide one would estimate from the data a direct gap of ~1.4 eV.

IIC.   Visible and Infrared Reflectance[21,29]

     Figure 4 shows the polarized reflectance data from oriented
films $(1/1_o \sim 2.5\text{-}3)$ of pure trans-$(CH)_x$ where 1_o and 1 are the
unstretched and stretched lengths, respectively. The large optical
anisotropy induced by orientation is clearly evident. The reflection data are consistent with a semiconductor model for pure
$(CH)_x$ in qualitative agreement with the absorption data described
above. The interband transition is apparent in $R_{||}$ in the region
of 2 eV with the reflectance decreasing to a low frequency value
of $R_{||} \approx 0.1$, implying a dielectric constant of $\epsilon_{||} \approx 3\text{-}4$, in agree-
ment with the value measured[31] at microwave frequencies (note that
the low density of fibrils in the $(CH)_x$ films implies that the
true value of $\epsilon_{||}$ within a given fibril is $\epsilon \sim 10\text{-}12$). The perpen-
dicular reflectance $R_\perp$ is small throughout the measured range
indicative of quasi-one-dimensional behavior and relatively weak
interchain coupling. The low frequency limiting value, $R_\perp \approx 4\%$,
implies $\epsilon_1 \approx 1.5$.

     The Kramers-Kronig analysis was used to analyze the reflec-
tance data for partially oriented pure trans-$(CH)_x$. The results
for $\sigma_1(\omega)$ and $\epsilon_1(\omega)$ are presented in Figs. 5 and 6. The results
are precisely as expected for a semiconductor; $\sigma(\omega)$ rises from
zero at low frequencies to a peak value of $4\text{x}10^3\,\Omega^{-1}\text{cm}^{-1}$ at about
16,000 cm^{-1}. $\epsilon(\omega)$ is negative at high frequencies crossing zero

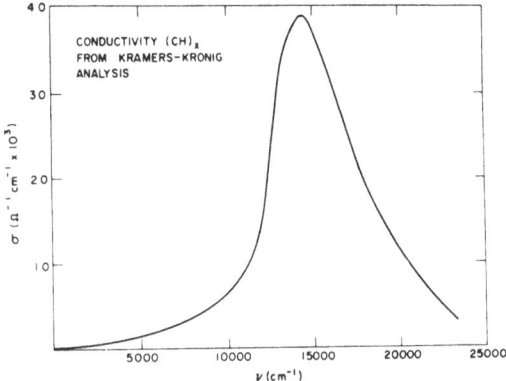

Fig. 5. $\sigma(\omega)$ for <u>trans</u>-$(CH)_x$ as obtained from Kramers-
Kronig analysis of the $R_{\parallel}$ reflection spectrum
(ref. 29).

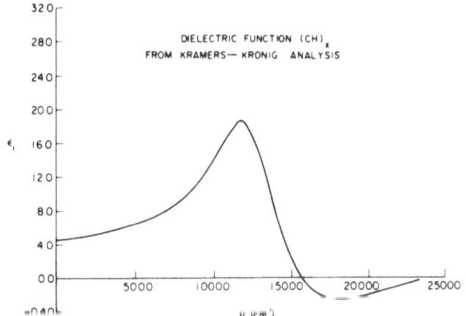

Fig. 6. $\epsilon(\omega)$ for <u>trans</u>-$(CH)_x$ as obtained from Kramers-
Kronig analysis of the $R_{\parallel}$ reflection spectrum
(ref. 29).

at about 16,000 cm^{-1}. At low frequencies ϵ approaches a limiting
value $\epsilon(o) \simeq 4$ consistent with the values determined directly
from $R(\omega \rightarrow 0)$ and the microwave data.

The absorption and reflection data are consistent and imply
a semiconductor band structure with an energy gap arising from
bond alternation of about 1.5 eV. However, as indicated in the
introduction, this viewpoint is not universally accepted. Ovchinnikov et al.[8,9] have argued that the energy gap extrapolated to
infinite chain polyenes is too large to be accounted for by simple
band theory of the dimerized chain, and concluded that Coulomb
correlations play an important role. It has been shown[32] that the
gap due simultaneously to correlation and a Peierls dimerization
is of the form $\Delta \approx (\Delta_{corr}^2 + \Delta_{alt}^2)^{\frac{1}{2}}$. Various other theoretical

studies have shown that Coulomb correlation may be important in this system. Duke et al.[11] use a CNDO-S2 calculation scheme on polyenes of varying lengths, $C_{4n+1}H_{4n+2}$ (n = 1,2,3,4) to find an approximation to the band structure, electron photoemission spectra, and the lowest electronic transition energy. Using a configuration interaction analysis, Duke et al.[11] also include the lowest B_uMO. The energy difference is then the lowest singlet molecular exciton. The exciton band is then extrapolated to n → ∞ in order to predict the transition energy in polyacetylene, obtaining $\Delta E \approx 2.0$ eV. The agreement between the extrapolated value and the experimental absorption edge is quite good.

The optical data appear consistent with either of two models for the electronic structure of polyacetylene. The first view is simple single-particle (band theory) point of view, while the second view models $(CH)_x$ as a strongly correlated system where single-particle ideas are invalid. Since excitons may be viewed as bound electron-hole pairs, whereas a direct π-π^* transition would create a free electron and hole, direct observation of electron-hole pair generation through photoconductivity and/or photovoltaic studies is of critical importance.

IID. Photovoltaic Effect[18,27,33]

Photovoltaic studies have been carried out using trans-$(CH)_x$: n-ZnS heterojunctions, photoelectrochemical cells using $(CH)_x$ as the active photoelectrode, and Schottky barriers formed with Indium metal on $(CH)_x$.[33] The photoresponse of the trans-$(CH)_x$: n-ZnS heterojunction diode is shown in Fig. 7. One sees two photoresponse edges, at approximately 1.4 eV and 2.6 eV. The magnitude of the peak at 0.9 eV was dependent upon previous illumination and thus probably arises from trapping states in the gap. The data therefore are consistent with optical studies of $(CH)_x$ which imply a gap energy of about 1.5 eV. However, the relatively low quantum efficiency below 2.5 eV may indicate primary exciton formation. Thus the actual single particle energy gap may be somewhat higher than 1.5 eV. Under illumination of approximately 1 sun, the $(CH)_x$:ZnS heterojunction gives an open circuit photovoltage of 0.8 V.

Photoelectrochemical photovoltaic cells have been fabricated[27] using polyacetylene, $(CH)_x$, as the active photoelectrode. Using a sodium polysulfide solution as electrolyte, $V_{oc} \approx 0.3$ Volts and $I_{sc} \approx 40$ μ amps/cm^2 were obtained under illumination of approximately 1 sun. With the initial configuration, the cell efficiency was limited by the series resistance, the small effective area of the electrode configuration, and the absorbance of the solution. The PEC photoresponse falls rapidly at photon energies below 1.4 eV, consistent with optical studies of $(CH)_x$ which suggest a band gap energy of about 1.5 eV. However, again, the relatively low

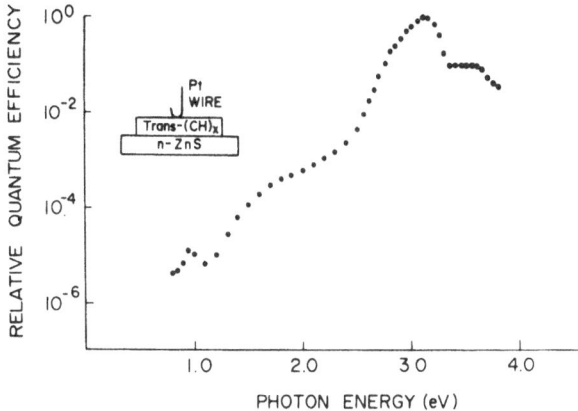

Fig. 7. Photoresponse of n-ZnS:undoped trans-$(CH)_x$ n-p
heterojunction. Quantum efficiency normalized
to 3.1 eV peak where the absolute quantum
efficiency is of order unity. (ref. 18).

quantum efficiency below 2.5 eV may indicate primary exciton
formation.

In summary, optical absorption and reflection data are consistent with the band structure of a direct gap semiconductor with
an energy gap of about 1.5 eV. Photovoltaic response is observed
with an onset near 1.5 eV. However, the relatively low quantum
efficiency strongly suggests that the true single particle band
gap is somewhat higher with the edge at 1.5 eV being due to a
weakly bound exciton.

III. DOPING OF $(CH)_x$ [13-17,34]

When pure polyacetylene is doped with a donor or an acceptor,
the electrical conductivity increases sharply over many orders of
magnitude at low concentration, then saturates at higher dopant
levels, above approximately 1%. The maximum conductivity for
nonaligned cis-$[CH(AsF_5)_{0.14}]_x$ is in excess of $10^3 \Omega^{-1}$-cm^{-1} with
similar values obtained for a variety of dopants. The typical
behavior for the conductivity as a function of dopant concentration
(y) is shown in Figure 8. The general features appear to be the
same for the various donor and acceptor dopants, but with detailed
differences in the saturation values and the critical concentration at the "knee" in the curve (above which σ is only weakly
dependent on y). These transport studies suggest a change in
behavior at a critical concentration (y_c); a semiconductor to
metal (SM) transition.

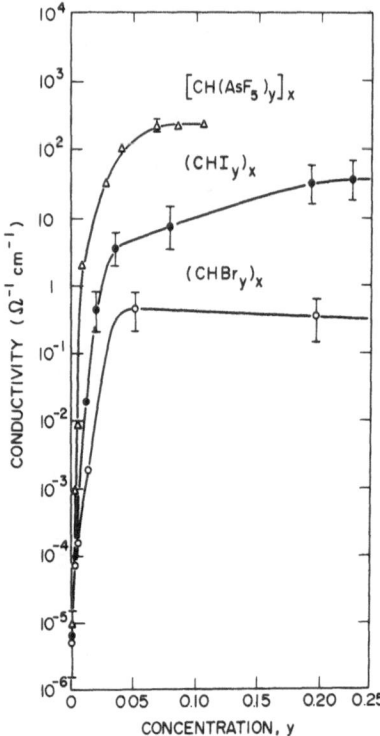

Fig. 8.   Electrical conductivity (room temperature) as
a function of dopant concentration.

IIIA.   Localized Donor/Acceptor States: Light Doping

The simplest model of the localized donor or acceptor state
at light doping $(y < y_c)$ follows the traditional semiconductor
approach and pictures the electron or hole, with effective mass
m^* determined by the band structure loosely bound to the charged
center by Coulomb forces in a dielectric medium. In lightly
doped $(CH)_x$, instead of substitutionally replacing the host as in
silicon, the impurity resides very close to the polymer chain, on
the surface of the 200 Å fibrils and/or between individual chains.
At light doping levels we assume isolated impurities interacting
with a single polymer chain. At heavy doping levels impurity
interactions will become important, however, well below the SM
transition this should not be a problem. The impurity could either
donate an electron to, or accept an electron from the chain. In
the acceptor case, the hole on the chain would be free to delocalize if it were not for the Coulomb binding to the impurity. The
resulting localized states are bound states of the hole on the
polymer chain in the vicinity of the charged donor or acceptor
ion. The binding energies of such quasi-one-dimensional hydrogenic

impurity states have been shown to be the same as for donors and acceptors in traditional semiconductors.[31] Note that these hydrogenic bound states will contain one electron or hole with spin $\frac{1}{2}$ and thus will give a Curie-law contribution to the magnetic susceptibility.

The assumption that charge transfer doping leads to electrons and holes assumes a rigid band structure along the lines of traditional semiconductor physics. However, as discussed in IIA, the bond-alternation energy gap in trans-$(CH)_x$ may be viewed as the result of the $2k_F$ instability of the coupled electron-lattice system in a quasi-1d metal. In this point of view, the effect of charge-transfer doping would be to change $k_F = (\pi/2)n$, where n is the number of carriers per unit length along the chain in the undistorted metallic state. In the pure polymer, there is one electron per carbon atom corresponding to a half-filled band so that $n_0 = a^{-1}$, where a is the uniform carbon-carbon bond length and $k_F = k_F^0 = \pi/2a$. The resulting charge-density-wave distortion occurs with a superlattice period $b_s^0 = 2\pi/2k_F = 2a$, implying a dimerized bond-alternating structure as observed in $(CH)_x$. Doping will change k_F; $k_F = (\pi/2)n_0(1 \pm c)$, where c is the impurity concentration, and the plus or minus sign is appropriate for donor or acceptor doping. We have assumed complete charge transfer. The resulting superlattice period will shift away from $b_s^0 = 2a$ to a new value $b_s = 2a/(1 \pm c)$, incommensurate with the carbon-chain polymer structure. However, the commensurability energy strongly favors locking $2k_F$ at the zone boundary; a commensurate bond alternation with period 2a places π-bond charge between carbon atoms [Fig. 10(a)], whereas in the incommensurate case, the charge-density maxima and minima would not be related to the atomic positions. As a compromise the doped structure breaks up into domains; i.e. regions of bond-alternation separated by charged domain walls.

The formation of domain-walls, or solitons, on long chain polyenes has been studied theoretically by Su, Schrieffer and Heeger,[25] and by Rice.[26] Following their work, we consider two domains with B phase to the left and A phase to the right of a soliton which is at rest at the origin, as illustrated in Fig. 9a. To determine the properties of the soliton (width, energy, mass, spin, etc.) the ground state energy of the system was determined for an arbitrary displacement pattern which reduces to the A and B phases as one moves to the far right or left. In practice, the displacements were varied in a segment containing N CH groups located symmetrically about n=0, and matched onto perfect A and B phases on either side. The ground state energy was then calculated for any set of displacements, u_n, in the segment. Defining the staggered order parameter $\Psi_n = (-1)^n u_n$ (which reduces to $\pm u_0$ for the A and B phases and is zero by symmetry at the center of the wall), and choosing as a trial function

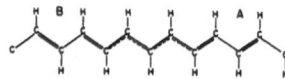

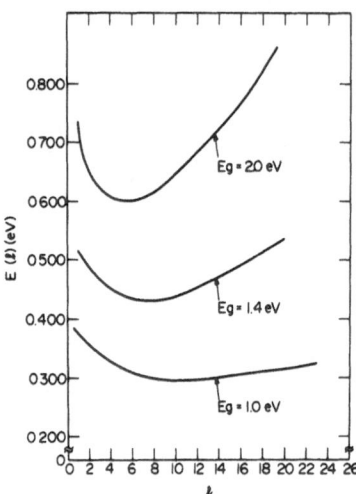

Fig. 9 (ref. 25).
        a. Schematic of soliton separating A and B
           phases.
        b. Soliton energy as a function of wall width
           for three values of Eg.

$$\Psi_n = u_0 \tanh\left(\frac{n}{\ell}\right) \tag{7}$$

the system energy $E(\ell)$ was determined as shown in Figure 9b for
$4\beta_1$ = 1.0, 1.4, and 2.0 eV. The minimum occurs at $\ell \simeq 7$ for $4\beta_1$ =
1.4 eV so that the wall is quite diffuse (for $4\beta_1$ = 2.0 eV, $\ell \simeq 5$
and for $4\beta_1$ = 1.0 eV, $\ell \simeq 9$). The energy to create the soliton at
rest is $E_s \simeq 0.4$ eV for the 1.4 eV gap energy (0.6 eV for $4t_1$ =
2.0 eV, 0.3 eV for $4t_1$ = 1.0 eV). Using the adiabatic approximation for the electronic motion, it was shown that the effective
mass of the soliton is related to the δu_n for a small change in
wall position δx_s by

$$M_s = M \sum_n \left(\frac{\delta u_n}{\delta x_s}\right)^2 \simeq \frac{4u_0^2 M}{3\ell a^2}, \tag{8}$$

where the second equality holds if we use (7) and replace n by
$n-x_s/a$ in the derivative. Using the values obtained above for
$4t_1$ = 1.4 eV, one finds $M_s \simeq 6\ m_e$, where m_e is the electron mass.

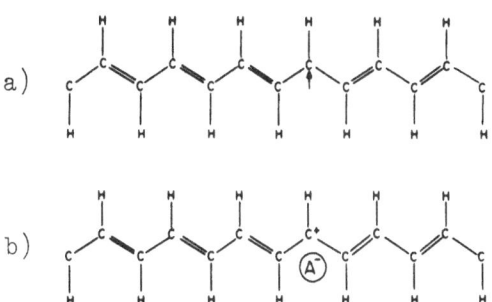

Fig. 10. a. Neutral soliton defect in trans-$(CH)_x$
(paramagnetic).
b. Soliton kink after charge transfer
(diamagnetic).

The electronic structure of the soliton exhibits a localized state ω_0 at the center of the gap, containing one electron for the neutral kink. While this localized state is spin unpaired, the distorted valence band continues to have spin zero. Thus, the neutral soliton has spin $\frac{1}{2}$. The static susceptibility therefore will contain a Curie law contribution and can be used to count the number of soliton defects present.

Since the localized state occurs at the gap center, i.e. the chemical potential, the relevance of the solitons to the doping of $(CH)_x$ depends on the energy for creation of a soliton, E_s, as compared with the energy required for making an electron or a hole, $\frac{1}{2} E_g = \Delta$. If $\Delta < E_s$, charge transfer doping would occur by creating free band excitations; if $\Delta > E_s$, soliton formation would be favored. For $E_g = 1.4$ considered above, $E_s \sim 0.4$ eV $< \Delta = 0.7$eV. Thus, the soliton bound hole leads to stabilization over the free hole by $(\Delta - E_s) \sim 0.3$ eV with correspondingly larger values if the gap is larger. The same stabilization energy holds for adding an electron, which would occupy $\varphi_0(x)$ of the soliton rather than being placed at the conduction band edge. Self consistent field effects arising from electron-electron interactions would split the ionization and affinity levels of ω_0 about the center of the gap and change these estimates accordingly. We note that the ionized solitons formed upon doping with acceptors (or donors) will be non-magnetic.

The origin of the localized state can be seen from elementary chemical arguments. Fig. 10a shows a schematic diagram of a neutral kink localized on a single lattice site. Satisfying the carbon valence at the center of the kink requires a non-bonded pi-electron $(S=\frac{1}{2})$ localized on the kink (Fig. 10a). Ionization by a nearby acceptor leads to a charged, non-magnetic, kink (Fig.

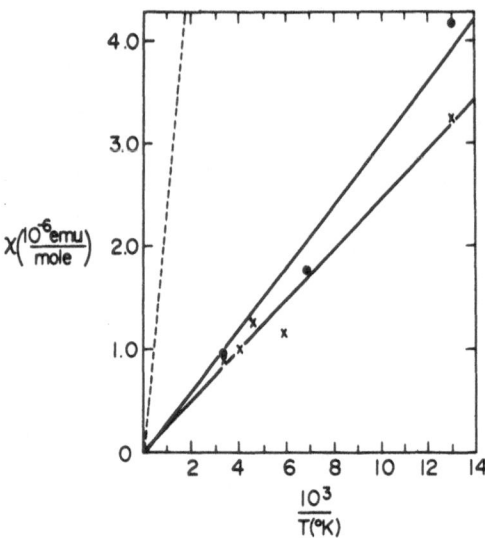

Fig. 11.   Spin susceptibility of nonmagnetic $(CH)_x$
ooo undoped sample. xxx $[CH(AsF_5)_{0.006}]_x$.
Both samples were from same batch of undoped
polymer. The dotted line represents the
Curie law expected for y=0.006. (ref. 22)

10b). For simplicity the kinks are shown localized on one lattice
site whereas minimization of the energy will spread the kink over
many sites as described above.

An important difference between the conventional semiconductor doping (electrons or holes bound in the Coulomb potential
of the ion) and soliton doping is the magnetic character of the
resulting localized states. Because of the spin degeneracy of a
single hole (or electron) in a localized state, the magnetic susceptibility would be expected to obey a Curie law. Double occupancy of such weakly bound states in the gap is inhibited by
electron-electron Coulomb interactions. On the other hand, the
charged ionized soliton localized state is diamagnetic.

Experimental studies[22] have shown that the localized states
induced by doping are non-magnetic in agreement with the soliton
theory outlined above. For example, a series of electron spin
resonance measurements (10 GHz) were carried out on a sample
lightly doped (y=0.006) with AsF_5 and an undoped trans-$(CH)_x$ as
shown in Fig. 11. Both samples exhibited Curie-like susceptibilities between 77 K and 295 K. Assuming $S = \frac{1}{2}$ for the observed
spins, the concentrations of unpaired local moments are 850 and
630 per 10^6 carbon atoms for y=0 and 0.006, respectively. The
local-moment concentration in undoped trans-$(CH)_x$ is roughly consistent with previous measurements by Shirakawa[23] and Goldberg

et al.[24] of $300/10^6$ carbon atoms. Differences are probably due
to minor variations in film preparation and isomerization.
Significantly, however, the trans-$(CH)_x$ showed no enhanced local-
moment concentration upon doping. Shown also in Fig. 11 (dashed
line) is the Curie-law susceptibility which would correspond to
one unpaired spin generated per AsF_5 for $y = 0.006$. From these
results one concludes that the localized states induced by light
doping below the SM transition are non-magnetic. This result was
confirmed by Faraday balance measurements. The residual weak
Curie law from defects in the undoped $(CH)_x$ decreases in magnitude
as y increases rather than increasing in proportion to the dopant
concentration.

Other evidence relevant to the soliton question is summarized
by Su, Schrieffer and Heeger[25] and by Fincher, Ozaki, Heeger and
MacDiarmid.[31] Although considerable work needs to be done in this
area, the experimental results are in qualitative agreement with
the soliton doping mechanism.

IIIB. Semiconductor-Metal Transition and the Metallic State

The semiconductor-metal transition in doped polyacetylene has
been studied by a variety of experimental techniques including
electrical conductivity, thermopower, magnetic susceptibility,
far-ir transmission, and ir reflectance; all as a function of
dopant concentration. In general qualitative changes are evident
in essentially all the physical properties, indicative of semiconductor behavior below y_c and metallic behavior above y_c where
y_c is about 0.03 (i.e. ~ 3 mole% dopant).

i) Transport: Conductivity[14,15] and Thermopower[35]. Typical
data for electrical conductivity (room temperature) as a function
of dopant concentration are shown in Fig. 8. The change in behavior above ~ 3% is evident. Similar changes are observed in
the temperature dependence; samples with $y < y_c$ show activated
behavior with the activation energy decreasing with increasing
dopant concentration. For $y > y_c$, the activation energy is sufficiently small that the interfibril contacts play a limiting role.

Since thermoelectric power is a zero-current transport coefficient, thermopower measurements can provide important information
on such a system. The interfibril contacts should be unimportant
allowing an evaluation of the intrinsic metallic properties.
Moreover, since the thermopower (S) can be viewed as a measure of
the entropy per carrier, measurements as a function of y can be
used to study the SM transition. One generally expects a large
thermopower for semiconductors (few carriers with many possible
states per carrier) whereas in the metallic state the entropy of
the degenerate electron gas is small ($\sim k_B(k_B T/E_F)$ per carrier).

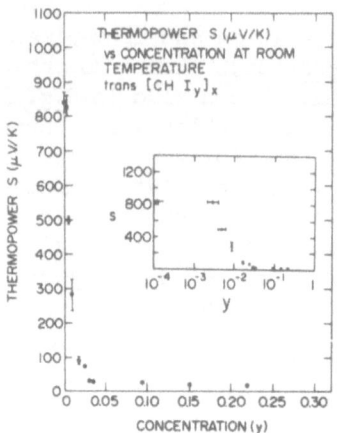

Fig. 12. Thermopower of (CHI$_y$) as a function of iodine
concentration y. The inset shows the data
with y on a log scale; the value for undoped
(CH)$_x$ is indicated with an arrow (ref. 35).

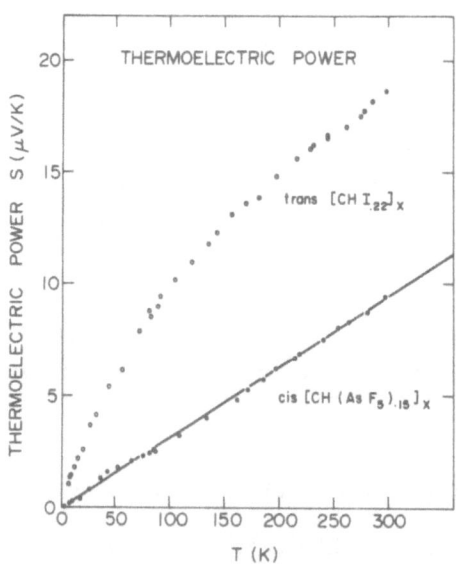

Fig. 13. Temperature dependence of the thermopower for
heavily doped metallic (CH)$_x$.

Thermopower studies were carried out by Park et al.[35] The
room temperature results, S vs y for trans-$(CHI_y)_x$ are shown in
Fig. 12. The sign of S is positive throughout the entire concentration range indicating p-type behavior consistent with charge
transfer doping to the iodine acceptor and the formation of I_3^-.
The metal insulator transition is clearly evident. Note that S
remains constant up to approximately y = 0.003 (0.3%) (see inset)
then falls steeply for 0.003 < y < 0.03. Above 3 mole %, S is
nearly independent of y decreasing to S = +18.5uV/K in the heavily
doped metallic limit. From this change in behavior, we infer a
critical concentration, $n_c \approx 3$ mole% (1 mole% I_3^-), for the semiconductor-metal transition in $(CHI_y)_x$. Below 0.3%, the thermopower
is insensitive to the dopant implying that the transport is dominated by defects and/or impurities in the synthesized trans-polymer.

The thermopower in the metallic high concentration limit is
shown in Figure 13 where S vs T is plotted on a linear scale.
Figure 13 includes the results for both trans-$(CHI_{0.22})_x$ and trans-
$[CH(AsF_5)_{0.16}]_x$. The results for the AsF_5 doped sample are in
excellent agreement with those obtained by Kwak et al.[36] S is a
linear function of T whereas for the iodine doped sample, there
is somewhat more curvature.

                                              In both cases, the
thermopower decreases smoothly toward zero as T → 0 in a manner
typical of metallic behavior. The magnitude and temperature
dependence of S (Fig. 13) are characteristic of a degenerate electron gas. Thus the TEP results independently verify the metallic
state above y_c and thereby confirm that the temperature dependent
conductivity (decreasing with decreasing temperature) is limited
by interfibril contacts; i.e. interrupted metallic strand behavior.

For a nearly filled band (i.e. p-type) metallic system, the
thermopower can be written as

$$S = + \frac{\pi^2}{3} \left(\frac{k_B}{|e|}\right) kT \left[\frac{d \ln \sigma(E)}{dE}\right]_{E_F} \qquad (9)$$

where $\sigma(E) = n(E)|e|\mu(E)$ and $n(E)$ is the number of carriers con-
tributing to $\sigma(E)$, $dn(E)/dE = g(E)$ is the density of states (both
signs of spin) and $\mu(E)$ is the energy dependent mobility. Assuming
energy independent scattering ($\mu(E)$ independent of E)

$$S = + \left(\frac{k_B}{|e|}\right) \frac{\pi^2}{3} k_B T_\eta (E_F) \qquad (10)$$

where $\eta(E_F) = g(E_F)/N$ is the density of states per carrier. The
experimental results (Fig. 13) are in good agreement with eq. 10;
the results imply $\eta(E_F) = 1.36$ states per eV per carrier. Since
there are 0.15 carriers per carbon atom in $[CH(AsF_5)_{0.15}]_x$

(assuming complete charge transfer), the thermopower data yield
for the density of states, $g(E_F) \approx 0.2$ states per eV per C atom.
The curvature and somewhat larger values found for the $(CHI_{0.22})_x$
data may imply an additional contribution arising from energy
dependent scattering.

For dilute concentrations well below the semiconductor-metal
transition the thermopower is large and essentially temperature
independent. Such behavior can be understood for a dilute concentration of carriers (holes) which hop among a set of localized
states. In this case, where the kinetic energy of the carriers is
negligible, the thermopower is given by the Heikes formula[37,38]

$$S = + \frac{k_B}{|e|} \ln[(1-\rho)/\rho] \tag{11}$$

where $\rho = n/N$ is the ratio of the number of holes (n) to the number of available sites (N) (eq. 11 assumes spinless Fermions; the
inclusion of spin degeneracy changes the expression to S =
$+(k_B/|e|)\ln(2-\rho)/\rho)$. Identification with the experimental data
for undoped trans-$(CH)_x$ requires that $\rho \sim 10^{-4}$ and temperature
independent. The results therefore suggest that in the undoped
polymer, the conductivity is due to a small number of residual
carriers ($\rho_0 \sim 10^{-4}$) provided by defects and/or impurities, and
that the mobility results from hopping. The insensitivity of the
thermopower to iodine concentrations less than 0.3 mole% (or 0.1%
I_3^-) is consistent with this interpretation and implies that $\rho_0 <$
10^{-3}. Moreover, the large TEP together with $\rho_0 \sim 10^{-4}$ implies
that the number N of available sites is comparable to the number
of carbon atoms per unit length in the polymer chain.

The localized state hopping transport inferred from the
thermopower measurements below y_c is consistent with the soliton
doping mechanism. Motion of the charged localized domain-walls
would be expected to be via diffusive hopping. Moreover, for a
fixed impurity concentration the number of charged kinks would be
independent of temperature in agreement with the data for undoped
$(CH)_x$ and for dopant concentration $y < y_c$. Finally, although the
domain wall would be distributed over a group of carbon atoms, in
the dilute limit the center of mass of the wall could take any
position along the chain so that the number of available sites
would be of order the number of carbon atoms.

ii) Magnetic Susceptibility.[22] The magnetic properties of
a material can provide important details of its electronic structure. In the heavily doped samples, a temperature-independent
paramagnetic contribution to the susceptibility suggests the
existence of a degenerate Fermi sea of metallic charge carriers.
From the Pauli formula the Fermi-surface density of states $N(E_F)$
may be obtained; $\chi_p = \mu_B^2 N(E_F)$, where μ_B is the Bohr magneton and

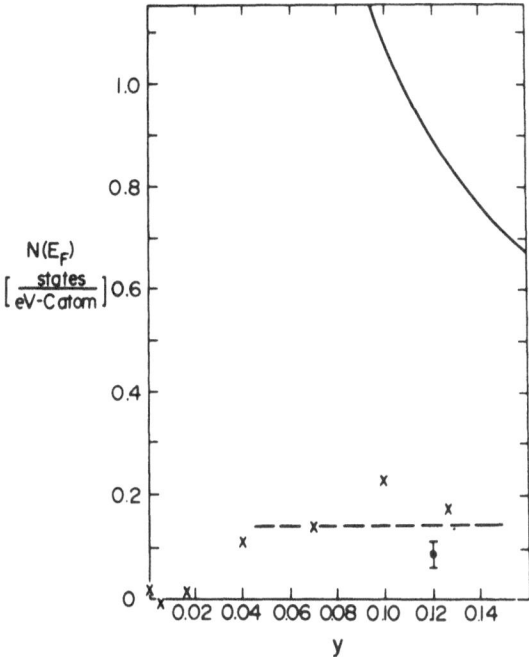

Fig. 14. Fermi-surface density of states vs AsF$_5$ concentration. The solid curve represents a
rigid band one-dimensional density of state
($m^*/m = 1$). The dashed curve would result
from a half-filled simple tight-binding band
with transfer integral $t = 2.25$ eV (bandwidth
of 9 eV). x Faraday balance; ● ESR (see ref.22).

$N(E_F)$ includes both spin directions. The onset of such a temperature-independent Pauli term in the susceptibility as a function
of dopant concentration indicates the transition from the semiconducting to the metallic state.

Magnetic susceptibility studies have been carried out as a
function of dopant concentration in $[CH(AsF_5)_y]_x$. The measured
total susceptibilities for all samples studied were diamagnetic
indicating the dominance of atomic core contributions.

Treating the paramagnetic deviation from the prediction of
Pascal's constants as a Pauli susceptibility, the Fermi-surface
density of states $N(E_F)$ is obtained

$$\chi_{meas} - \chi_{Pascal} = \mu_B^2 \, N(E_F) \tag{12}$$

These results are plotted in Fig. 14. The density of states in
the metallic regime, $N(o) \approx 0.15$ states/eV/carbon atom, is in

excellent agreement with that inferred from the thermopower.

Anisotropies found in the electrical[21] and optical[19] properties of partially oriented films suggest that heavily doped polyacetylene may be described as a quasi-one-dimensional metal. Theoretical one-dimensional band structures predict a divergence in the density of states at the band edge. Plotted in Fig. 14 for comparison is the Fermi-surface density of states for $[CH(AsF_5)_y]_x$ in a one-dimensional rigid-band model assuming $m^*/m = 1$ and one free hole generated per AsF_5. Clearly, the experimental results give no indication of such a one-dimensional band-edge anomaly.

A possible explanation of this result is that above the SM transition bond alternation and the associated energy gap no longer exist. That is, the band structure is that of a one-dimensional metal with a half-filled conduction band. The dashed line in Fig. 14 represents a calculation of $N(E_F)$ for such a half-filled band in a simple tight-binding model with transfer integral 2.25 eV. Raman spectra[39] of metallic iodine doped $(CH)_x$, however, continue to show the two absorption bands identified with C=C and C-C stretching modes with only minor frequency shifts compared with the pure polymer. Moreover, optical absorption and reflection data imply that the interband transition is present above the SM transition. Thus bond alternation persists into the metallic regime. On the other hand, x-ray photoemission studies indicate nonuniform doping at the highest concentrations.[40] An inhomogeneous metallic state with undistorted metallic domains coexist with regions of bond-alternated semiconductor is consistent with the data.

Alternatively, the small density of states may result from interchain coupling. Although the interchain distance is even greater than the intermolecular distances in molecular crystals such as TTF-TCNQ, it has been suggested that even this small amount of three-dimensional coupling is sufficient to quench the one-dimensional band-edge divergence in the density of states.[10]

The Pauli susceptibility of $[CH(AsF_5)_y]_x$ seems to turn on continuously at AsF_5 concentrations of a few atomic percent. Within the limits of experimental uncertainty the susceptibility of $[CH(AsF_5)_{0.04}]_x$ is temperature independent (77-295 K). Therefore, the results of this work provide no evidence of strong correlation enhancement of the spin susceptibility near the SM transition.

iii) <u>Infrared Absorption and Reflection.</u>[16,41,42] To verify the existence of the semiconductor-to-metal transition, far infrared transmission data were taken on samples of varying concentrations of iodine and AsF_5 (with qualitatively similar results).

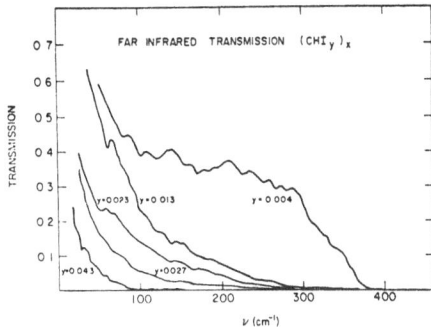

Fig. 15. Infrared transmission through $(CHI_y)_x$ films
(ref. 41).

The data[16,41] for a series of iodinated samples with $y \rightarrow y_c$ are
shown in Fig. 15. At all frequencies the transmission decreases
with increasing concentration. As $y \rightarrow y_c$, the far ir transmission
cuts off at progressively longer wavelengths consistent with an
approach to metallic behavior. Results for $y \rightarrow 0.05$ are not shown
because the transmission level is less than the amount of scattered light ($\sim 1\%$) in the apparatus.

Since the long wavelength reflectance is particularly sensitive to the free carrier density, the ir reflection spectra demonstrate, with special clarity, the occurrence of the semiconductor-
metal transition and to the formation of the metallic state. The
results indicate that the critical concentration range is 1-3% in
agreement with the values inferred from dc transport, far ir transmission, thermopower and magnetic studies. The anisotropy of the
polarized reflectance implies that the free carrier contribution
is polarized along the polymer chains.

The reflectance results obtained from oriented $(CH)_x$ films
doped with AsF_5 are shown in Figure 16. The error bars are indicative of the variations obtained from different samples separately prepared and mounted. Figure 16 contains data for
$[CH(AsF_5)_y]_x$ with y = 0, 0.006, 0.034, 0.093 and 0.134. Since the
transition from semiconductor to metal occurs at AsF_5 concentration
of about 1-3%, the data of Figure 16 span the full concentration
range. Metallic behavior is clearly indicated at the highest concentrations; the reflectance increases with decreasing frequency
with a value approaching 90% at the longest wavelengths. On the
other hand for light doping ($y < 0.01$) the reflectance remains
small even at the longest wavelength consistent with semiconducting
behavior. The sharp increase in reflectance below 0.2 eV in the
y = 0.034 sample is indicative of the presence of free carriers;
evidently this sample is close to the boundary between semiconductor and metal.

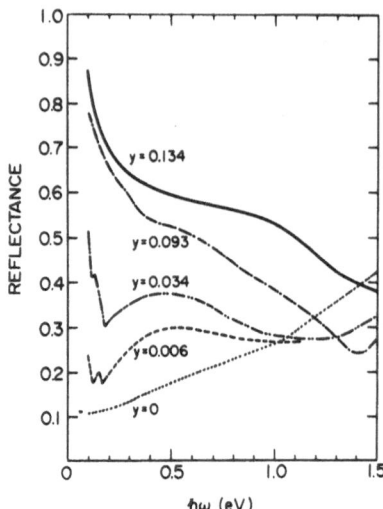

Fig. 16.   Reflectance $(R_{||})$ for oriented $[CH(AsF_5)_y]_x$ (ref. 42).

The broad reflectance peak centered near 0.5 eV is evident for both the AsF_5 doping (Fig. 16) and the iodine doping. For light doping a well-defined maximum is observed suggesting the introduction of states within the energy gap. At higher levels the 0.5 eV band grows in strength with an increasingly long, low frequency tail. Qualitatively the 0.5 eV maximum which shows up at the lightest doping levels appears to evolve continuously into the free carrier reflectance in the metallic regime. Evidently the states introduced into the gap at low doping grow in number (the magnitude of R) and in length (the low frequency cut-off) as the doping level increases. The series of spectra suggest an interpretation in terms of small metallic regions which gradually grow and coalesce into an extended metallic state at the highest doping concentrations.

iv) <u>Kramers-Kronig Analysis of Metallic $[CH(AsF_5)_{0.13}]_x$.</u>[29] A Kramers-Kronig transform has been carried out to analyze the reflectance data (0.1 eV-3.0 eV) of metallic $[CH(AsF_5)_{0.13}]_x$. At the lowest frequencies (Fig. 16), the data were extrapolated to approach unity as $\nu \rightarrow 0$ assuming a Hagen-Rubens behavior of the reflectance below 0.1 eV. Because the results of a Kramers-Kronig transform are very sensitive in the far infrared to the precise manner in which $R(\omega) \rightarrow 1$ as $\omega \rightarrow 0$, an accurate determination of $\sigma(\omega)$ in this frequency region is extremely difficult. However, it was found that in the middle ir (above 1500 cm^{-1}) $\sigma(\omega)$ is insensitive to the details of the low frequency extrapolation. The results for $\sigma(\omega)$ are shown in Fig. 17. The broad peak in $\sigma(\omega)$ centered around 16,000 cm^{-1} results from the interband

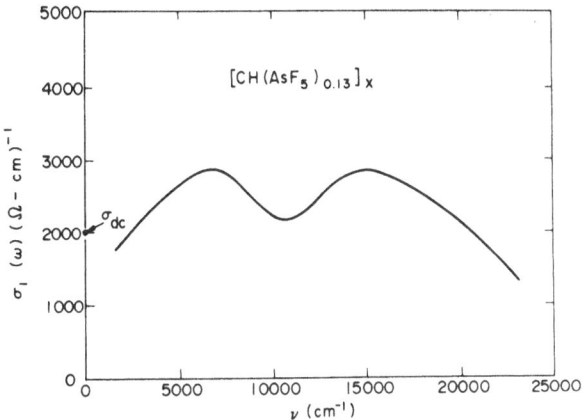

Fig. 17. $\sigma_{\|}(\omega)$ for heavily doped metallic $[CH(AsF_5)_{0.11}]_x$ as obtained from the $R_\|$ reflection spectrum (ref. 29).

transition. The conductivity then begins to fall at lower frequencies until 11,000 cm^{-1}, where it again rises as one might expect for a system with free carriers. However, a somewhat surprising result is that below 6000 cm^{-1} the conductivity again decreases. This behavior is markedly different from the simple Drude behavior where $\sigma(\omega)$ is a monotonically decreasing function of frequency $[\sigma_{Drude} = \sigma_0/(1+\omega^2\tau^2)]$. The slow drop in $\sigma(\omega)$ continues until 1500 cm^{-1}, which is the lower limit of the analysis. Between 1000 and 1500 cm^{-1} a weak dependence on the low frequency extrapolation was observed with the behavior below 1000 cm^{-1} strongly dependent upon the extrapolation. Above 1500 cm^{-1} the results are independent of the extrapolation except for minor adjustments of the values in the third or fourth significant digit, details beyond the resolution of this analysis. The value $\sigma_{dc} \simeq 2 \times 10^3 \, \Omega^{-1} \text{cm}^{-1}$ found through direct dc measurements on heavily doped (AsF_5) oriented samples of $(CH)_x$ is indicated in Fig. 17. Thus, the low frequency limit of this Kramers-Kronig analysis is quite consistent with other independent measurements. The decrease of σ below 6000 cm^{-1} is a real feature of $\sigma(\omega)$ and not an artifact of the extrapolation procedure.

Useful information can be extracted by applying the sum rule relations to the data from the metallic state (Fig. 17). The effective number of electrons per molecule participating in the free-carrier optical transitions for energies less than the interband transition is given by the oscillator strength

$$8 \int_0^{\omega_c} \sigma(\omega)d\omega = \frac{4\pi Ne^2}{m^*} \; \eta_{eff}(\omega_c), \tag{13}$$

where $\eta_{eff}(w_c)$ is the fractional number of carriers contributing to the metallic conductivity. Using $\nu_c = 11,000$ cm^{-1}, we find the oscillator strength ($\simeq 4 \times 10^{31}$) approximately equal to the total π-electron oscillator strength in the polymer. The large oscillator strength therefore suggests that heavy doping removes the bond-alternation leading to a uniform bond length polyene; a quasi-one-dimensional broad band metal with all π electrons contributing to the transport, i.e., $\eta_{eff} \simeq 1$. Such a picture is consistent with the small value of the density of states at E_F as inferred from magnetic susceptibility[22] and thermopower[35] studies. On the other hand, after doping, Raman data continue to show the two carbon-carbon stretch frequencies[39] (diminished in intensity), and optical studies[19,29] continue to show evidence of the unshifted interband transition (no Burstein shift), both characteristic of the bond alternated semiconductor. A possible explanation is an inhomogeneous metallic state (possibly resulting from inhomogeneous doping within the fibrils[40]) with undistorted metallic domains coexistent with regions of bond alternated semiconductor. Such a picture is consistent with all the data and explains the apparent absence of a Burstein shift upon doping, the residual Raman lines, and the observed sensitivity of the strength of the interband transition (after doping) to the different dopants.

The low frequency decrease in $\sigma(w)$ extrapolating toward the dc value cannot be understood in terms of a simple Drude-Lorentz model of the conductivity. However, it is intuitively clear that the fibril nature of the polymer will have a very strong effect on the dc transport properties. Moreover, if the metallic system is highly anisotropic on a microscopic scale as suggested by dc and optical studies of partially oriented films, impurities and defects will have an especially strong effect; in one dimension disorder leads to localization of states. Thus we anticipate that the low frequency transport will be limited by the imperfect polymer structure.

Electron microscopy[21,23,43] photographs of $(CH)_x$ films show a fibril structure with fibril diameter of about 200 A. The individual fibrils form a multiply coupled array through branching. Typical length to diameter ratio for the fibrils appear to be in the range of about 5-10. Therefore the polymer can be viewed as an effective medium made up of $(CH)_x$ fibrils at a volume filling factor of about $f = \frac{1}{3}$. For the as-grown films, the fibrils are randomly oriented in the plane; stretch orientation leads to partial alignment. From examination of the electron micrographs and related x-ray data[44] we estimate approximately 75% alignment, i.e., the alignment factor, $\alpha \simeq 0.75$.

Within the doped $(CH)_x$ metallic fibrils the intrinsic frequency-dependent conductivity and dielectric functions are denoted $\sigma_1(w)$ and $\epsilon_1(w)$. The bulk properties of the metallic polymer can be related to $\sigma_1(\omega)$ and $\epsilon_1(\omega)$ through effective-medium theory (see ref. 29). The average medium conductivity can be written

$$\langle \sigma_1(\omega) \rangle \simeq \frac{\alpha f \sigma_1(\omega)}{1+[4\pi\sigma(\omega)/\omega]^2 g^2 (1-\alpha f)^2} \qquad (14)$$

where f is the filling factor, α is the fractional alignment factor, and g is the typical depolarization factor ($g = (b/a)^2 [\ln(2a/b)-1]$) for an ellipsoid of revolution, where b/a is the ratio of minor to major semiaxes. In writing the above expression, we have assumed that, consistent with the metallic behavior, $4\pi\sigma_1/\omega > |\epsilon_1|$. Moreover, as a result of the observed dc and optical anisotropy, we consider only the response to components of the applied field parallel to the $(CH)_x$ fibrils.

The important feature of Eq. 14 is that at low enough frequencies, the electric field within an "interrupted strand" is screened by the depolarization field due to the charge buildup at the boundary. The characteristic cutoff frequency ω_c is given by

$$[4\pi\sigma_1(\omega_c)/\omega_c]g(1-\alpha f) = 1 \qquad (15)$$

and corresponds to the condition when the depolarization factor in the denominator of Eq. 14 exceeds unity. From Fig. 17, we estimate $\omega_c \sim 7000$ cm^{-1} at which point the (maximum) medium conductivity is approximately $3\times10^3 \Omega^{-1}$ cm^{-1}. Taking $\alpha f \sim 0.25$ as described above, Eq. 15 yields $g \sim 10^{-2}$, or $b/a \sim 10^{-1}$, in good agreement with the fibril morphology. The results therefore imply that the fibril structure of $(CH)_x$ leads to metallic polymers which may be viewed as interrupted metallic strands. From the magnitude of the cutoff frequency we infer strand dimensions consistent with fibril dimensions observed in electron microscopy studies. Therefore the interrupted strands are tentatively identified with the branched fibrils; localization due to microscopic disorder appears to be relatively unimportant in these crystalline polymers.

An estimate of the intrinsic conductivity within the individual metallic doped $(CH)_x$ fibrils can be obtained by inverting eq. 14. At ω_c, the denominator is ~ 2, so that assuming $\alpha f \simeq \frac{1}{4}$, we find $\sigma_1(\omega_c) \simeq 2.4\times10^4 \Omega^{-1}$ cm^{-1}. The intrinsic dc conductivity is undoubtedly higher. If we assume a Drude dependence, $\sigma_1(\omega) = \sigma_0/(1+\omega^2\tau_0^2)$, where τ_0 is the Drude scattering time for the metallic state. It is difficult to obtain a direct measurement of τ_0 from the available data. However, the decrease in $\langle \sigma_1(\omega) \rangle$ observed from 7000 to 11,000 cm^{-1} in Fig. 14 implies that $\omega_c\tau_0 > 1$. Thus from this analysis we are able to estimate the intrinsic dc conductivity within a single fibril of metallic $(CH)_x$; $\sigma_{intrinsic}^{dc} \gtrsim 2 \times 10^4 \Omega^{-1}$ cm^{-1}.

v) Mobility Changes at the SM Transition. Direct measurements of the mobility (e.g. time of flight, etc.) are not available

Hall effect data[45] have been reported in the metallic regime; however, the Hall coefficient appears to be dominated by the fibril structure of the composite medium. Some information on the transport mobility in the semiconductor and metal limits can be obtained directly from the conductivity, $\sigma = ne\mu$ where n is the density of carriers with charge e, and μ is the mobility.

In the undoped trans-$(CH)_x$, the thermopower results imply a carrier concentration of 10^{-4} resulting from residual impurities, defects, etc. Thus, since the number of carbon atoms per unit volume is $n_c \simeq 2\times10^{22} cm^{-3}$ (based on the measured density of 0.4 grams/cm^3), the residual carrier density in undoped trans-$(CH)_x$ is $\sim 10^{18} cm^{-3}$. With $\sigma \simeq 10^{-5} \Omega^{-1}\text{-}cm^{-1}$ (see Fig. 8), one finds μ(undoped) $\simeq 10^{-4} cm^2/V\text{-}sec$.

In the metallic regime, e.g. $[CH(AsF_5)_{0.1}]_x$ the numbers of carriers can be estimated from the oscillator strength (see subsection iv above). The results imply that in the heavily doped metallic state all the π-electrons contribute to the metallic transport; thus n(metal) $\simeq 2\times10^{22} cm^{-1}$. Taking the intrinsic conductivity to be $\sigma_{intrinsic}^{dc} \sim 2\times10^4 \Omega^{-1}\text{-}cm^{-1}$ as inferred above, one obtains μ(metal) $\simeq 10 cm^2/V\text{-}sec$. Note that assuming that all π-electrons contribute requires that heavy doping removes the bond-alternation (see subsection iv above) leading to a uniform bond length polyene. Taking a somewhat more conservative point of view one could estimate the number of carriers by assuming unit charge transfer with one carrier per dopant. For $[CH(AsF_5)_{0.1}]_x$, the corresponding value would be $n\sim2\times10^{21}$ with $\mu\sim10^2 cm^2/V\text{-}sec$!

The high mobility in the metallic state provides strong evidence of the validity of a band theory approach with delocalized states in this disordered metallic polymer. The inferred values of μ are comparable to or greater than the mobilities in the best metals (e.g. for copper $\mu \sim 50 cm^2/V\text{-}sec$ at room temperature).

The low mobility inferred for the undoped polymer and the remarkable increase on going through the SM transition are surprising. The low mobility in undoped $(CH)_x$ is, however, consistent with the thermopower results which imply hopping in the undoped polymer. Hopping transport is not expected for electrons or holes in a broad band ($w\sim 8\text{-}10$ eV) system unless strong electron-lattice coupling leads to large effective mass carriers. The soliton doping mechanism discussed above (Sec. IIIA) does lead to relatively large effective mass, e.g. $M_s \sim 6m_e$ for Eg = 1.4 eV, $M_s \sim 12 m_e$ for Eg = 2.0eV. Thus the low mobility is consistent with soliton formation induced by charge transfer doping. Of course, more conventional polaron effects and/or disorder localization could also play a role.

The low mobility inferred indirectly from the dc transport may not be appropriate to optically induced electron-hole pairs

particularly when generated in the high electric field of a junction. Such optically induced minority carriers would move rapidly
to the interface and out of the polymer before a kink could form.
Thus, the optically induced carrier mobilities may be comparable
to the band-like values found in the metallic regime ($\mu \sim 10$-100
cm^2/V-sec). This important question can only be resolved by
more direct measurements involving photogenerated carriers.

   vi) The Mechanism of the SM Transition. As an initial point
of view we treated this transition as similar to that seen in
heavily doped semiconductors. In this case, one expects the
halogen and AsF_5 dopants to act as acceptors with localized hole
states in the gap, with the hole bound to the acceptor in a
hydrogen-like fashion. For low concentrations, one expects the
combination of impurity ionization and variable range hopping to
lead to a combination of activated processes as observed experimentally. However, as extensively discussed by Mott[46] and others,
[97] if the concentration is increased to a critical level, then the
screening from carriers will destroy the bound states giving an
insulator-to-metal transition. This will occur when the screening
length becomes less than the radius of the most tightly bound
Bohr orbit of the hole and acceptor in the bulk dielectric;

$$n_c^{\frac{1}{3}} \simeq (4a_H)^{-1} \left(\frac{m^*}{m\varepsilon}\right) \tag{16}$$

where a_H is the Bohr radius, ε is the dielectric constant of the
medium and m^*/m is the ratio of the band mass to the free electron
mass. Assuming $m^*/m \simeq 1$ and using $\varepsilon \simeq 10$ from ir reflection measurements, we estimate $n_c \sim 10^{20}$-10^{21} cm^{-3}. Since the density of
carbon atoms is about $2 \times 10^{22} cm^{-3}$, n_c would be in the range of a
few percent assuming one carrier per dopant. The good agreement
with our experimental results is probably fortuitous in view of
the much over-simplified model. More importantly, the magnetic
properties of doped $(CH)_x$ are inconsistent with the SM transition
being a Mott transition in the conventional sense. There are two
important points of disagreement:
   1) The localized states at light doping are non-magnetic
       (see Sec. IIIA).
   2) There is no evidence of strong correlation enhancement of
       the Pauli susceptibility near the SM transition (see
       Sec. IIIBii and Fig. 14).
Thus the SM transition in doped polyacetylene is fundamentally
different from that observed in traditional semiconductors where
1) and 2) above play a central role.

   As shown above, the optical-ir reflectance data suggest an
interpretation in terms of small metallic regions which gradually
grow and coalesce into an extended metallic state at the highest
doping concentrations. In addition, the oscillator strength

analysis (Sec. IIIBiv) leads to the conclusion that all the π-
electrons contribute to the metallic conductivity suggesting that
heavy doping removes the bond alternation leading to uniform bond
length metallic trans-$(CH)_x$. This is consistent with the small
density of states inferred from χ and TEP studies. In the context of the soliton doping mechanism discussed above, this picture
of the SM transition could be at least qualitatively understood.
Isolated charged solitons would be non-magnetic, in agreement
with experiment. An attractive interaction between solitons
would lead to a coalescence into uniform bond-length metallic
regions. As the dopant density increases, the metallic regions
would begin to overlap giving, eventually, a metallic state. A
naive estimate of the critical concentration would be $y_c = 1/2N$
where N is the hald-width of the soliton; taking $N \approx 7$ one obtains
$y_c \approx 6\%$. Inclusion of interchain coupling could lead to a perco-
lation threshold at lower values. The interpretation of the SM
transition as arising from soliton-soliton interactions has been
considered in more detail recently by Rice,[48] Schrieffer[49] and
Horowitz.[50] This interesting possibility will undoubtedly
stimulate further theoretical and experimental study.

IV. SEMICONDUCTOR PHYSICS AND DEVICE APPLICATIONS[17,18,27]

A variety of rectifying junctions have been fabricated using
doped and undoped $(CH)_x$. Schottky diodes formed between metallic
AsF_5-doped $(CH)_x$ and n-type semiconductors indicate high
$[CH(AsF_5)_y]_x$ electronegativity. The p-type character of undoped
trans-$(CH)_x$ is confirmed by Schottky barrier formation with low
work function metals. An undoped p-$(CH)_x$:n-ZnS heterojunction
has been demonstrated with open circuit photovoltage 0.8 V. These
results point to the potential of $(CH)_x$ as a photosensitive
material for use in solar cell applications.

As an example, the I-V and C-V characteristics of a Schottky
junction formed with n-GaAs: (metallic)$[CH(AsF_5)_y]_x$ are shown in
Fig. 18. The interface was fabricated by direct polymerization
of the $(CH)_x$ film on the semiconductor surface. Several n-GaAs:
(metallic)$[CH(AsF_5)_y]_x$ diodes were produced. As indicated by C-V
and J-V characteristics, all had barrier heights in the range 0.8-
1.0 V consistent with results for most metal: n-GaAs interfaces.
Photosensitive diodes were fabricated using thin semi-transparent
$[CH(AsF_5)_y]_x$ films. Illumination through the polymer with a tungsten lamp produced open circuit voltages in the range 0.4-0.6 V.

The formation of rectifying Schottky junctions has been
demonstrated with undoped trans-$(CH)_x$ as the semiconductor. Metal
contacts to the $(CH)_x$ films were formed by pressing small area
metal tips onto polymer films in inert atmosphere. The undoped
polymer forms ohmic contacts with highly electronegative metals

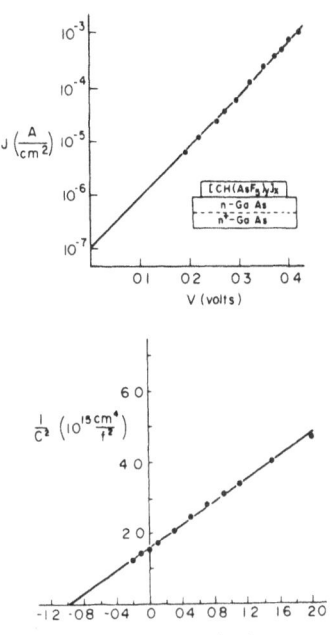

Fig. 18.

(Cu, Au, Pt) and rectifying junctions (Schottky diodes) with
relatively low electronegativity metals (Na, Ba, In). The systematically different behavior with metals of different electronegativity shows that undoped trans-$(CH)_x$ is a p-type semiconductor
consistent with thermopower data.

A p-n heterojunction was produced[18] from undoped trans-$(CH)_x$:
n-ZnS as shown in the inset of Fig. 7. Both free standing films
(pressure contact) and thin film polymerization of $(CH)_x$ directly
on the n-ZnS single crystal were employed to form the interface.
In each case rectifying junctions were obtained. The data (Fig. 7)
indicate significant photoresponse at energies well below the 3.7
eV ZnS band gap; the peak absolute efficiency occurring at 3.1 eV
is ~ 0.003. This result is clearly a severe underestimate of the
actual photogeneration efficiency due to inefficient collection
of photocarriers by the small area point contact. Capacitance
measurements imply that the active area of the $(CH)_x$:ZnS diode
was confined to a small region around the point contact and was
far smaller than the entire 0.15 cm^2 interface area. To quantify
this the following experiments were carried out: Junction capacitance was measured on the original cell and, subsequently, after
doping with AsF_5 to reduce the series resistance. The $(CH)_x$ film
was then removed, and replaced over the identical area by a
sputtered Au contact, and the capacitance was measured. The
junction capacitance of the $(CH)_x$:ZnS diode was < 20 pf, whereas

(a) (CH)$_x$ ELECTRODE CONFIGURATION

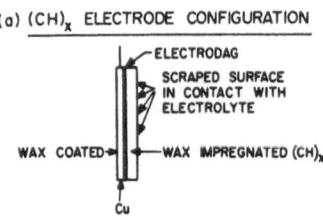

(b) PEC CELL CONFIGURATION

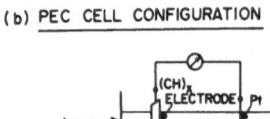

Fig. 19.

after doping with AsF$_5$, the capacitance increased to ~ 10 nf.
The junction capacitance of the Au:ZnS diode was also ~ 10 nf.
Therefore an area correction to the quantum efficiency of order
10^2-10^3 is implied leading to an actual quantum efficiency for the
(CH)$_x$:n-ZnS junction approaching unity at 3.1 eV. Evidently, this
represents photoactivation of carriers in (CH)$_x$ since such a
photoresponse from ZnS at a photon energy 0.6 eV below its band
gap with light illuminating the junction through ~1 mm of ZnS is
not reasonable.

   Photoelectrochemical photovoltaic cells[27] have been fabricated using polyacetylene as the active photoelectrode. Utilization of polyacetylene as a photoelectrode in a photoelectrochemical (PEC) cell offers attractive prospects. In general, semi-
conductor-electrolyte junctions are relatively insensitive to the
quality of the semiconductor and are favored over solid state
junctions with regard to trapping and surface recombination. The
porous fibrillar microstructure of semiconducting (CH)$_x$ should be
an advantage in PEC cells since the electrolyte can come into
effective contact with the large surface area (~ 40 m^2/g).
Since photoexcitation of polyacetylene involves only pi-electrons
with the backbone binding sigma-bond remaining intact, photo-
dissolution may not be a serious problem.

   The (CH)$_x$ electrode configuration utilized is shown in Fig.
19. To reduce series resistance, the (CH)$_x$ film is contacted to
a thin copper sheet using Electrodag. The entire structure is
then dipped into molten paraffin wax; the wax is subsequently
partially removed from the active surface by scraping with a
razor blade thus exposing only the outermost fibrils to the
electrolyte. The wax-filled electrode was utilized in these

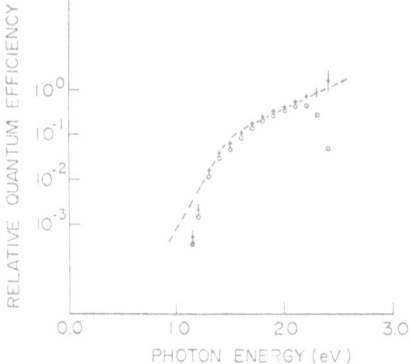

Fig. 20.

initial studies as a simple means of preventing permeation of the electrolyte through the porous film to the Electrodag backing.

The aqueous electrolyte consisted of a saturated Na_2S solution to which sulfur has been added to saturation. The alcohol/water electrolyte consisted of either a) 1 volume of saturated aqueous sodium polysulfide and two volumes of isopropyl alcohol to which sulfur has been added to saturation, or b) 1 volume of the saturated aqueous sodium polysulfide solution to which 1 volume of isopropyl alcohol has been added. The resulting sodium polysulfide solution acts as an effective redox couple

$$S_y^{-2} + 2e^- \rightleftarrows S_{y-n}^{-2} + S_n^{-2} \qquad (y \geq n+1)$$

Since $(CH)_x$ is p-type, photogenerated electron-hole pairs are separated at the $(CH)_x$-electrolyte interface with electrons injected into the electrolyte. The resulting $S^=$ ions migrate to the Pt electrode where they lose electrons to reform the $S_y^=$ species. The liberated electrons return to the $(CH)_x$ electrode via the external circuit. Since $(CH)_x$ is hydrophobic, the alcohol was added to the solution to achieve improved contact at the $(CH)_x$ surface.

The photoresponse spectrum of the $(CH)_x$ PEC cell is shown in Fig. 20 plotted as relative quantum efficiency vs photon energy. The open circles represent data obtained directly from the cell with light incident on the photoelectrode through the highly colored orange solution which rapidly becomes opaque above 2.2 eV. The crosses represent the quantum efficiency corrected for light absorption by the electrolyte. The absolute quantum efficiency is approximately 1% at 2.4 eV. The dashed curve on Fig. 20 represents[51] the photoconductivity response of $(CH)_x$ (normalized to

the PEC cell data at 2 eV) as measured at the University of
Pennsylvania using standard techniques. The nearly identical
shapes of the PEC photoresponse and the $(CH)_x$ photoconductivity
curves indicate that the former arises from electron-hole generation in the semiconducting polymer with subsequent minority
carrier (electron) transfer into the sodium polysulfide solution
at the interface.

In a general context, these junction formation experiments
demonstrate that doped $(CH)_x$ can be used in a manner similar to
conventional semiconductors. The Fermi level can be set (and
moved) by selected dopants. Photogeneration of carriers has been
demonstrated both in solid state p-n and Schottky junctions and
in the PEC configuration. Quantum efficiencies are low below
2 eV but increase to values approaching unity at higher energies.

V. PROGRESS ON NEW MATERIALS

Considering possible polyacetylene derivatives, replacement
of some or all of the hydrogen atoms in $(CH)_x$ with organic or
inorganic groups, copolymerization of acetylene with other acetylenes or olefins, and the use of completely different conjugated
polymers, a large new class of conducting organic polymers can be
anticipated with electrical properties that can be controlled over
the full range from insulator to semiconductor to metal. Work
along these lines has been initiated. Substituted polyacetylene
derivatives are being studied at a number of laboratories with
initial results that look promising. The work on poly(para-
phenylene)[52] and polypyrrole have demonstrated a second system
which can be doped (n- and p-type) etc. with resulting properties
similar to those of doped $(CH)_x$.

Continuing collaborative work[53] at the U. of Massachusetts
and the U. of Pennsylvania, and independent studies of Shirakawa[54]
in Tokyo have resulted in the synthesis of a new form of polyacetylene with variable density, through the isolation of a gel
as a synthetic intermediate step. Scanning electron micrographs
show the fibril structure characteristic of $(CH)_x$, but with a
fibril diameter of ~ 600 Å to 800 Å. Electrical conductivity
and thermopower measurements on samples of the undoped polymer
and on samples doped with iodine or AsF_5 imply properties nearly
identical with previously studied as-grown films of $(CH)_x$, but
with a lower density of $(CH)_x$ fibrils per unit volume.

The density of the "foam-like" materials was in the range
0.02 gm/cm^3 to 0.04 gm/cm^3 with some variation from sample to
sample. Pressed films were obtained with density 0.1 gm/cm^3 to
0.4 gm/cm^3 depending on the final thickness. These values are to
be compared with the 0.4 gm/cm^3 density of the as-grown $(CH)_x$

films. Since the flotation density of the as-grown $(CH)_x$ films is 1.2 gm/cm^3, the volume filling fraction, f, of fibrils is approximately 1/3. For the "foam-like" material, $f \sim 0.015$ to 0.03 whereas the variable thickness pressed films have intermediate values for f.

Electrical conductivity and thermopower data were obtained from the "foam-like" material and from pressed films. Measurements were carried out on undoped samples and on samples heavily doped with iodine or AsF_5. The thermopower of the undoped polymer is insensitive to the density; the "foam-like material and the pressed film yield thermopower values of approximately +900 $\mu V/K$. Any variations are comparable with the typical variations observed earlier in as-grown film samples from different synthetic preparations. Similarly, after heavy doping, the results are insensitive to the density with values for the "foam-like" material, pressed film and as-grown films in good agreement for each dopant. Comparison with the variation in S as a function of dopant concentration studied in detail earlier leads to the conclusion that the heavily doped samples are all metallic. The thermopower results imply that the various forms of the doped and undoped polymers are microscopically identical. The low density materials simply consist of fibrils at a smaller filling fraction, f.

The electrical conductivity data are consistent with this conclusion. The conductivity of the undoped "foam-like" material is nearly two orders of magnitude below that of the high density pressed film. Although there may be some increase in the interfibril contact resistance, the reduction in f is of major importance. This conclusion is strengthened by the observation that the conductivity activation energy (obtained from the temperature variation of the conductivity near room temperature) is 0.25 eV for both samples. This value is comparable to the 0.3 eV value typically obtained from as-grown films.

Similar results are obtained with the heavily doped polymers. The conductivity (176 $ohm^{-1}-cm^{-1}$) of the $\lceil CH(AsF_5)_{0.06} \rceil_x$ prepared from the low density pressed cis-$(CH)_x$ film ($\rho = 0.1$ gm/cm^3) is correspondingly lower than that of the conductivity (560 $ohm^{-1}-cm^{-1}$, 1200 $ohm^{-1}-cm^{-1}$) of $\lceil CH(AsF_5)_{0.14} \rceil_x$, $[CH(AsF_5)_{0.10}]_x$, respectively, prepared from as-grown cis-$(CH)_x$ film ($\rho = 0.4$ gm/cm^3). The room temperature conductivity of $(CHI_{0.06})_x$ increases by a factor of forty on going from the low density "foam-like" materials to the high density pressed film. In all cases, the resulting conductivity increases with the filling fraction of fibrils. Note that the increased fibril size in the "foam-like material and pressed films does not appear to significantly alter the resulting transport properties of the doped polymers. The resulting doped and undoped variable density $(CH)_x$ polymers can be viewed as effective media in which the dc electrical transport is determined by the volume filling fraction of conducting fibrils.

VI. CONCLUSION

The emergence of polyacetylene as a prototype conducting organic polymer can be viewed in the context of the recent developments in research on quasi-1d organic conductors. In only a few years, organic conductors have progressed from tiny, brittle, molecular crystals with single crystal electrical conductivity less than 10^2 $(\Omega\text{-cm})^{-1}$ to large area flexible $(CH)_x$ polymer films with conductivity systematically controllable over a remarkable range (less than $10^{-10}\Omega^{-1}\text{-cm}^{-1}$ to greater than $2\times10^3\Omega^{-1}\text{-cm}^{-1}$. The interest in the fundamental physics and chemistry of this novel system and the initial success in device fabrication have stimulated a broad based interdisciplinary study of doped conducting polymers in many laboratories throughout the world. New concepts, new materials, and new and unusual electronic properties have already been demonstrated. It appears safe to conclude that this field will continue to grow; and that by means of a three fold interdisciplinary effort involving chemistry, physics and polymer science, conducting polymers will develop into a broad class of materials of considerable scientific interest and potential technological importance.

REFERENCES

1. Lennard-Jones, J. E.: 1937, Proc. R. Soc. A 158, p. 280.
2. Coulson, C. A.: 1938, Proc. R. Soc. A 164, p. 383.
3. Brooker, L. G. S.: 1951, J. Am. Chem. Soc. 73, p. 1087; 73, p. 5332.
4. Kuhn, H.: 1948, Helv. Chim. Acta 31, p. 1441.
5. Lounget-Higgins, H. C. and Salem, L.: 1959, Prov. R. Soc. A 25, p. 172.
6. Peierls, R. E.: 1955, "Quantum Theory of Solids" (Oxford University, London), Chap. 5.
7. "Low Dimensional Cooperative Phenomena", edited by H. J. Keller (Plenum, New York, 1975); "Chemistry and Physics of One Dimensional Metals", edited by H. J. Keller (Plenum, New York, 1977); "One Dimensional Conductors", edited by J. Devreese (Plenum, New York, 1978).
8. Ovchinnikov, A. A., Ukrainski, I. I., and Kventsel, G. V.: 1978, Usp. Fiz. Nauk. 108, p. 81 [Sov. Phys. Usp. 15, p. 575 (1978)].
9. Szabo, A., Langlet, J., and Malriev, J.: 1976, Chem. Phys. 13, p. 173; Cojan, C., Agrawal, G. P. and Flytzanis, A.: 1977, Phys. Rev. B 15, p. 909; Kertez, M., Koller, J., and Azman, A.: 1977, J. Chem. Phys. 67, p. 1180.
10. Grant, P. M. and Batra, I. P.: 1979, Solid State Commun. 29, p. 225.
11. Duke, C. B., Paton, A., Salaneck, W. R., Thomas, H. R., Plummer, E. W., Heeger, A. J. and MacDiarmid, A. G.: 1978, Chem. Phys. Lett. 59, p. 146.

12. Hsu, S., Signorelli, A., Pez, G., and Baughman, R.: 1978, J. Chem. Phys. 68, p. 5105; 69, p. 106.

13. Heeger, A. J. and MacDiarmid, A. G., Proc. of the Dubrovnik Conf. on Quasi One-Dimensional Conductors, p. 361, Lecture Notes in Physics 96, Ed., S. Barisic (Springer-Verlag Berlin-Heidelberg New York, 1979); MacDiarmid, A. G. and Heeger, A. J., "Molecular Metals", Ed. W. Hatfield, NATO Conference Series VI: Materials Science, (Plenum Press, New York and London, 1979), p. 161. See also accompanying paper in this volume by A. G. MacDiarmid and A. J. Heeger for more details on chemical aspects of the problem.

14. Shirakawa, H., Louis, E. J., MacDiarmid, A. G., Chiang, C. K., and Heeger, A. J.: 1978, Chem. Commun. p. 578; Chiang, C. K., Druy, M. A., Gau, S. C., Heeger, A. J., Shirakawa, H., Louis, E. J., MacDiarmid, A. G. and Park, Y. W.: 1978, J. Am. Chem. Soc. 100, p. 1013.

15. Chiang, C. K., Park, Y. W., Heeger, A. J., Shirakawa, H., Louis, E. J., and MacDiarmid, A. G.: 1978, J. Chem. Phys. 69, p. 5098.

16. Chiang, C. K., Fincher, C. R., Jr., Park, Y. W., Heeger, A. J., Shirakawa, H., Louis, E. J., Gau, S. C., and MacDiarmid, A. G.: 1977, Phys. Rev. Lett. 39, p. 1098.

17. Chiang, C. K., Gau, S. C., Fincher, C. R., Jr., Park, Y. W., MacDiarmid, A. G., and Heeger, A. J.: 1978, Appl. Phys. Lett. 33, p. 181.

18. Ozaki, M., Peebles, D., Weinberger, B. R., Chiang, C. K., Gau, S. C., Heeger, A. J. and MacDiarmid, A. G.: 1979, Appl. Phys. Lett. 35, p. 83.

19. Fincher, C. R., Jr., Peebles, D. L., Heeger, A. J., Druy, M. A., Matsumura, Y., MacDiarmid, A. G., Shirakawa, H., and Ikeda, S.: 1978, Solid State Commun. 27, p. 489.

20. Shirakawa, H. and Ikeda, S. (to be published).

21. Park, Y. W., Druy, M. A., Chiang, C. K., Heeger, A. J., MacDiarmid, A. G., Shirakawa, H., and Ikeda, S.: 1979, Polymer Lett. 17, p. 195.

22. Weinberger, B. R., Kaufer, J., Heeger, A. J., Pron, A., and MacDiarmid, A. G.: 1979, Phys. Rev. B 20, p. 223.

23. Shirakawa, H., Ito, T., Ikeda, S.: 1978, Die Macromolecular Chemie 179, p. 1565.

24. Goldberg, I. B., Crowe, H. R., Newman, P. R., Heeger, A. J., and MacDiarmid, A. G.: 1979, J. Chem. Phys. 70, p. 1132.

25. Su, W. P., Schrieffer, J. R., and Heeger, A. J.: 1979, Phys. Rev. Lett. 42, p. 1698.

26. Rice, M. J.: 1979, Phys. Lett. 71A, p. 152.

27. Chien, S. N., Heeger, A. J., Kiss, Z., MacDiarmid, A. G., Gau, S. C., and Peebles, D. L.: Appl. Phys. Lett. (to be published)

28. Ooshika, Y.: 1957, J. Phys. Soc. Japan 12, pp. 1238, 1246.

29. Fincher, C. R., Jr., Ozaki, M., Tanaka, M., Peebles, D. L., Lauchlan, L., Heeger, A. J. and MacDiarmid, A. G.: Phys. Rev. B (Aug. 15, 1979).

30. Ito, T., Shirakawa, H., and Ikeda, S.: 1975, J. Polym. Sci.,
    Chem. Ed. 13, p. 1943.
31. Fincher, C. R., Jr., Ozaki, M., Heeger, A. J., and MacDiarmid,
    A. G.: 1979, Phys. Rev. B 19, p. 4140.
32. Bychkov, Y. A., Gorkov, L. P., and Dzyaloshinskii, I. E.:
    1966, Sov. Phys. JETP 23, p. 489.
33. Tani, T., Gill, W. D., Clarke, T. C., and Street, G. B.:
    Preprint, IBM Symposium on Conducting Polymers, March 29, 30,
    1979.
34. Nigrey, P. J., MacDiarmid, A. G., and Heeger, A. J.: (in
    press, 1979), Chem. Commun. This paper discusses electro-
    chemical doping techniques.
35. Park, Y. W., Denenstein, A., Chiang, C. K., Heeger, A. J.,
    and MacDiarmid, A. G.: 1979, Sol. State Commun. 29, p. 747.
36. Kwak, J. F., Clarke, T. C., Greene, R. L., and Street, G. B.:
    1978, Bull. Am. Phys. Soc. 23, p. 56.
37. Heikes, R.: "Buhl International Conference on Materials",
    Ed. by E. R. Shatz (Gordon and Breach, New York, 1974).
38. Chaikin, P. M. and Beni, G.: 1976, Phys. Rev. B 13, p. 647.
39. Harada, T., Shirakawa, H., and Ikeda, S.: Chem. Lett. (to be
    published; Street, G. B. (private commun.).
40. Salaneck, W. R., Thomas, H. R., Duke, C. B., Paton, A.,
    Plummer, E. W., Heeger, A. J., and MacDiarmid, A. G.:
    J. Chem. Phys. (in press).
41. Fincher, C. R., Jr., Ph.D. Thesis, U. of Pennsylvania (1979)
    (unpublished).
42. Tanaka, M., Fincher, C. R., Jr., Heeger, A. J., Druy, M. A.,
    and MacDiarmid, A. G.: 1979, Bull. Am. Phys. Soc. 24, p. 327.
43. Shirakawa, H. and Ikeda, S.: 1971, Polym. J. 2, p. 231;
    Shirakawa, H., Ito, T., and Ikeda, S.: 1973, Polym. J. 4,
    p. 460; Ito, T., Shirakawa, H., and Ikeda, S.: 1974, J.
    Polym. Sci. Polym. Chem. Ed. 12, p. 11; Ito, T., Shirakawa,
    H., and Ikeda, S.: 1975, J. Polym. Sci. Polym. Chem. Ed.
    13, p. 1943.
44. Shirakawa, H. (private communication).
45. Seeger, K., Gill, W. D., Clarke, T. C., and Street, G. B.:
    1978, Solid State Commun. 28, p. 873.
46. Mott, N. F.: 1972, Advances in Physics 21, p. 785.
47. See for example, Ziman, J. M., "Principles of the Theory of
    Solids", (Cambridge Univ. Press, 1972) pp. 168-170.
48. Rice, M. J. and Timonen, J. (preprint)
49. Schrieffer, J. R., (private communication).
50. Horovitz, B., (private communication).
51. Peebles, D. L., Tanaka, M., Heeger, A. J., and MacDiarmid,
    A. G., (to be published).
52. Baughman, R. H., Ivory, D. M., Miller, G. G., Shacklette,
    L. W., and Chance, R. R.: Preprint, IBM Symposium on Con-
    ducting Polymers March 29-30, 1979; Shacklette, L. W., Chance,
    R. R., Ivory, D. M., Miller, G. G., and Baughman, R. H.,
    "Synthetic Metals" (in press).

53. Kanezawa, K. K., Diaz, A. F., Gardini, G. P., Gill, W. D., Kwak, J. F., and Street, G. B.: 1979, Bull. Am. Phys. Soc. 24, p. 326; IBM Symposium on Conducting Polymers, March 29-30, 1979.
54. Wnek, G. E., Chien, J. C. W., Karasz, F. E., Druy, M. A., Park, Y. W., MacDiarmid, A. G., and Heeger, A. J., Polymer Lett. (in press).
55. Shirakawa, H. and Ikeda, S., IBM Symposium on Conducting Polymers, March 29-30, 1979 and ACS/AJS Chemical Congress, Honolulu, April 1-6, 1979.

ORGANIC METALS AND SEMICONDUCTORS: THE CHEMISTRY OF POLYACETY-
LENE, $(CH)_x$, AND ITS DERIVATIVES

Alan G. MacDiarmid and Alan J. Heeger

Departments of Chemistry and Physics
University of Pennsylvania
Philadelphia, Pa.   19104

ABSTRACT

Both cis- and trans-forms of polyacetylene, $(CH)_x$, may be prepared as silvery, flexible, polycrystalline, semiconducting films.
The cis-films can be stretched to over three times their original
length with partial alignment of the $(CH)_x$ fibrils. Through chemical or electrochemical doping, the electrical conductivity of the
films can be increased over twelve orders of magnitude with properties ranging from insulator ($\sigma < 10^{-10}$ ohm^{-1}cm^{-1}) to semiconductor to metal ($\sigma > 10^3$ ohm^{-1}cm^{-1}). By the use of donors or acceptors,
n-type or p-type polymer, respectively, is produced. Photoelectrochemical photovoltaic cells have been fabricated using $(CH)_x$ as
the active photoelectrode. For example, using a sodium polysulfide
solution as an electrolyte, V_{oc} ~0.3 volts and I_{sc} ~40 μ amps/cm^2
were obtained under an illumination of ca. 1 sun.

Polyacetylene, $(CH)_x$, is the simplest possible conjugated organic polymer and is therefore of special fundamental interest.
It can be prepared in the form of lustrous, silvery, flexible,
polycrystalline films having any desired cis/trans content by
catalytic polymerization of gaseous acetylene, C_2H_2 (1-4):

CIS   TRANS

393

L. Alcácer (ed.), The Physics and Chemistry of Low Dimensional Solids, 393–402.
Copyright © 1980 by D. Reidel Publishing Company.

The cis-rich films can be stretched easily at room temperature in
excess of three times their original length with concomitant partial alignment of the $(CH)_x$ fibrils (5,6). Dark red gels of toluene in $(CH)_x$ may be prepared using a lower catalyst concentration
(7). Highly porous, very low density, "foam-like" $(CH)_x$ can be
obtained from these gels (7). Both cis- and trans-$(CH)_x$ are p-
type semiconductors (8) which can be treated with a variety of p-
or n-type dopants with concomitant increase in conductivity to give
a series of semiconductors and ultimately, "organic metals." This
report will be directed primarily towards a description of the
more chemically oriented aspects of $(CH)_x$ and its derivatives.

1. DOPING OF $(CH)_x$ FILMS

The various types of dopants and doping procedures, the nature
of the $(CH)_x$ chain, and the nature of the dopant in the films will
be described below. The terms "cis" and "trans" used in conjunction
with a doped film will refer to the principal isomeric composition
before doping and does not imply that the isomeric composition
either remains constant or changes during the doping process.

1.1. P-type doping

1.1.1. Dopants and methods of doping. When either cis or trans
films are exposed to the vapor of electron-attracting substances
(p-type dopants) such as Br_2, I_2, AsF_5, H_2SO_4, $HClO_4$, etc. (9,10)
they become "doped" with the species and their electrical (11)
(Table I) and optical (12) properties change markedly. Dopant

TABLE I

DOPANTS FOR $(CH)_x$ [a,b]	Conductivity $(ohm^{-1}cm^{-1})$ 25°C
cis-$(CH)_x$	1.7×10^{-9}
trans-$(CH)_x$	4.4×10^{-5}
A. p-type (electron-attracting) dopants	
trans-$[CH(HBr)_{0.04}]_x$ [c]	7×10^{-4}
trans-$[CHCl_{0.02}]_x$ [c]	1×10^{-4}
trans-$[CHBr_{0.23}]_x$	4×10^{-1}
cis-$[CH(ICl)_{0.14}]_x$	5×10^{1}
cis-$[CHI_{0.30}]_x$	5.5×10^{2}

(continued on next page)

TABLE I (continued)

trans-$[CHI_{0.20}]_x$	1.6×10^2
cis-$[CH(IBr)_{0.15}]_x$	4.0×10^2
trans-$[CH(AsF_5)_{0.10}]_x$	4.0×10^2
cis-$[CH(AsF_5)_{0.10}]_x^c$	1.2×10^3
cis-$[CH_{1.1}(AsF_6)_{0.10}]_x$	$\underline{ca.}\ 7 \times 10^2$
cis-$[CH(SbF_6)_{0.05}]_x$	4.0×10^2
cis-$[CH(SbCl_6)_{0.009}]_x$	1×10^{-1}
cis-$[CH(SbCl_8)_{0.0095}]_x$	1×10^1
cis-$[CH(SbCl_5)_{0.022}]_x$	2
cis-$[CH(BF_2)_{0.09}]_x^d$	1×10^2
cis-$[CH(SO_3F)_y]_x$	7×10^2
cis-$[CH(ClO_4)_{0.0645}]_x$	9.7×10^2
cis-$[CH(AsF_4)_{0.077}]_x$	2.0×10^2
cis-$[CH_{1.011}(AsF_5OH)_{0.011}]_x^e$	$\underline{ca.}\ 7 \times 10^2$
cis-$[CH_{1.058}(PF_5OH)_{0.058}]_x$	$\underline{ca.}\ 3 \times 10^1$
cis-$[CH(H_2SO_4)_{0.106}(H_2O)_{0.070}]_x$	1.2×10^3
cis-$[CH(HClO_4)_{0.127}(H_2O)_{0.297}]_x$	1.2×10^3

B. n-type (electron-donating) dopants[c]

cis-$[Li_{0.30}(CH)]_x$	2.0×10^2
cis-$[Na_{0.21}(CH)]_x$	2.5×10^1
cis-$[K_{0.16}(CH)]_x$	5.0×10^1
trans-$[Na_{0.28}(CH)]_x$	8.0×10^1

a) "cis" or "trans" refers to the principal isomeric composition before doping

b) composition by elemental analysis except where stated otherwise

c) composition by weight uptake

d) dopant used: $(SO_3F)_2$. No composition or analysis given. Anderson, L.R., Pez, G.P., and Hsu, S.L.: 1978, J.C.S. Chem. Comm., pp.1066.

e) by electrochemical doping using $[(n-C_4H_9)_4N]^+[PF_6]^-$. Nigrey, P.J., MacDiarmid, A.G., and Heeger, A.J.: 1979, unpublished observations.

pressures <1 torr are usually satisfactory. With many dopants the conductivity increases rapidly through the semiconducting regime to the metallic regime. The concentrations of the dopants given in Table I are generally the maximum or close to the maximum value readily obtainable. Doping can be terminated at any degree of lower doping level desired, with corresponding lower conductivity.

Salts containing the $(NO)^+$ or $(NO_2)^+$ ions also act as good dopants (10). For example, the SbF_6 group can be introduced readily into $(CH)_x$ simply by treating a $(CH)_x$ film (ca. 85% cis isomer) with a CH_3NO_2-CH_2Cl_2 solution of the appropriate salt. Thus, $(NO_2)^+(SbF_6)^-$ yields golden, flexible, highly conducting films of $[CH(SbF_6)_{0.05}]_x$, (Table I) with liberation of NO_2, viz.,

$$(CH)_x + 0.05x(NO_2)^+(SbF_6)^- \rightarrow [CH(SbF_6)_{0.050}]_x + 0.05xNO_2 \quad (1)$$

It has been found very recently that $(CH)_x$ films may be doped electrochemically either to the semiconducting or metallic regime (13). This is a most important development since it opens up a general, very simple, readily controllable means of doping with a wide variety of species which can not be introduced by any obvious conventional chemical means. For example, it was found that when a strip of $(CH)_x$ film (ca. 82% cis-isomer) was used as the anode in the electrolysis of aqueous 0.5M KI solution with a potential of 9 V. it was doped during ca. 0.5 hour to the metallic state, to give, by elemental analysis, $(CHI_{0.07})_x$. It is important to note that the flexible, golden-silvery films contained no oxygen (total C, H, and I content=99.8%) and hence had undergone no hydrolysis and/or oxidation during the electrolytic doping process. When the $(CH)_x$ was used as the anode in the electrolysis of 0.5M $[(n-C_4H_9)_4N]^+[ClO_4]^-$ in CH_2Cl_2 at 9 V., doping occurred during ca. 1 hour to give highly conducting (Table I), flexible films which, by elemental analysis, had the composition $[CH(ClO_4)_{0.0645}]_x$ (13). Lower doping levels obtained during shorter electrolysis times gave material having conductivities in the semiconductor region. Similar results were obtained by the electrolysis of methylene chloride solutions of $[(n-C_4H_9)_4N]^+[SO_3CF_3]^-$ and $[(n-C_3H_7)_3NH]^+[AsF_6]^-$ both of which gave highly conducting golden-silvery flexible films. The former is assumed to contain the (SO_3CF_3) and the latter, the (AsF_4) species, since elemental analysis of the film gave a composition corresponding to $[CH(AsF_4)_{0.077}]_x$. The (AsF_4) is probably formed by a reaction sequence involving proton abstraction from $[(n-C_3H_7)_3NH]^+$ by fluorine atoms from AsF_6 during the electrolysis process (13).

1.1.2. Nature of the $(CH)_x$ chains and dopant species. Raman studies show that the iodinated and brominated films should be formulated as $[(CH)^{+y}(X_3)^-_y]_x$ where X=Br or I, at least a significant portion of the halogen being present as the X_3^- ion (14). The halogen partly depopulates the pi bonding system and oxidizes the $(CH)_x$ to a polycarbonium ion chain. This conclusion is supported by carbon 1s core shifts from ESCA studies (15). The $(NO)^+$ ions are also excellent species for oxidizing the pi system of $(CH)_x$

and are capable of concomitantly introducing anions which stabilize the polycarbonium ion chains (10). For example, the
$[CH(SbF_6)_{0.050}]_x$ species given in equation 1 is more appropriately formulated as $[CH^{+0.050}(SbF_6)^-_{0.050}]_x$.

The most simple and general method for simultaneously oxidizing the $(CH)_x$ pi system and introducing stabilizing anions appears
to be that involving electrochemical doping (13). Thus, species
such as $[CH(ClO_4)_{0.0645}]_x$, $[CH(AsF_4)_{0.077}]_x$, etc. formed electrochemically as described in Section 1.1.1. are be-
lieved to contain the $(ClO_4)^-$ and $(AsF_4)^-$ ions, respectively, although the extent to which charge transfer to the anionic species
occurs may be expected to vary according to the nature of the dopant. It is interesting to note that $AgClO_4$ has also been found
to dope $(CH)_x$ films with $(ClO_4)^-$ ion, although to lower conductivity levels (16)   (ca. 3 ohm^{-1}cm^{-1}) than that obtained with
electrochemical doping. The resulting film is contaminated with
metallic silver. In this case, the Ag^+ ion acts as the oxidizing
agent, viz.,

$$(CH)_x + 0.018xAgClO_4 \rightarrow [(CH)^{+0.018}(ClO_4^-)_{0.018}]_x + 0.018Ag \quad (2)$$

Although most studies of $(CH)_x$ have been carried out on AsF_5-
or I_2-doped films, the actual chemical form in which the AsF_5
exists in the film is still not completely clear. When $(CH)_x$
film is treated with very pure AsF_5 vapor in a vacuum line pre-
treated with AsF_5, elemental analyses for C, H, As and F give an
arsenic to fluorine ratio of 1:5 (Table I) (17, 18). The sum of
the elemental analyses is 99.7% or better and hence the film contains no significant amounts of oxygen. Photoelectron spectroscopy also shows the principal arsenic species contains arsenic and
fluorine in the ratio of 1:5 (15). Since epr (19) and magnetic
susceptibility studies (20) show the paramagnetic radical anion,
$AsF_5^{\cdot -}$ is not present it seems that the AsF_5 might be in the form
of the previously unreported diamagnetic $(As_2F_{10})^{-2}$ ion. If the
$[CH(AsF_5)_y]_x$ film is treated either with AsF_5 vapor containing HF
or is immersed in 42% aqueous HF, then elemental analyses for C,
H, As and F give an arsenic to fluorine ratio of 1:6 (Table I)(17).
Again, the sum of the elemental analyses for all elements is greater than 99.7%. If, on the other hand, the $[CH(AsF_5)_y]_x$ film is
pumped for many hours in a vacuum system containing possible traces
of air, elemental analyses corresponding to $[CH_{1+y}(AsF_5OH)_y]_x$,
(Table I), are obtained. In this respect, it might be noted that
many salts containing the $[AsF_5(OH)]^-$ ion are known. The conductivity of all three types of species is essentially identical.
These experimental observations are consistent with the reactions
below:

$$(CH)_x + yAsF_5 \rightarrow [CH(AsF_5)_y]_x \quad (3)$$

$$[CH(AsF_5)_y]_x + yHF \rightarrow [CH_{1+y}^{+y}(AsF_6)_y^-] \tag{4}$$

$$[CH(AsF_5)_y]_x + yH_2O \rightarrow [CH_{1+y}^{+y}(AsF_5OH)_y^-] \tag{5}$$

The weak protonic acids, HF and HOH can be regarded as combining with the AsF_5 species to give the strong protonic acids, "$H^+(AsF_6)^-$" and "$H^+(AsF_5OH)^-$", respectively, which then dope the $(CH)_x$ portion of the material according to equations (4) and (5) (17).

Other investigators have shown on the basis of X-ray absorption and infrared data that AsF_5-doped film, of unknown elemental composition, contains the AsF_6^- ion (21). This is in no way inconsistent with the above conclusions based on elemental analyses; indeed, it supports the formulation of the $[CH_{1+y}^{+y}(AsF_6)_y^-]_x$ species given above. However, these investigators suggest that the $(AsF_6)^-$ ion arises through the reaction below which involves disproportionation of the AsF_5:

$$(CH)_x + 3yAsF_5 \rightarrow [CH^{+2y}(AsF_6)_{2y}^-]_x + yAsF_3 \tag{6}$$

Since AsF_3 is readily removed by pumping (21), the resulting material should always contain arsenic to fluorine in the ratio of 1:6. This is in conflict with the elemental analytical data for the $[CH(AsF_5)_y]_x$ material. Since $[CH(AsF_5)_y]_x$ decomposes thermally with the liberation of gaseous HF and AsF_3, it is also quite possible that $[CH(AsF_5)_y]_x$ could be converted to $[CH_{1+y}^{+y}(AsF_6)_y^-]_x$ according to equation (4) by the HF so formed under certain conditions of handling or storage of the AsF_5-doped films.

1.2. N-type doping

Electron-donating, i.e. "n-type" dopants, may also be introduced into $(CH)_x$ films (22) (Table I) simply by immersing the film in a THF solution of e.g. sodium naphthalide, viz.,

$$(CH)_x + 0.21xNa^+Npth^{\dot{-}} \rightarrow [Na_{0.21}(CH)]_x + 0.21xNpth \tag{7}$$

A very large increase in conductivity is noted but it is not as great as that observed with most p-type dopants. Alkali metals may also be introduced by, for example, allowing a liquid sodium/potassium alloy at room temperature, or molten potassium to contact a $(CH)_x$ film (23). A liquid sodium amalgam will also Na-dope the film at room temperature (23). Preliminary experiments indicate that the $(CH)_x$ pi system may also be reduced electrochemically to give n-type doping by, for example, the electrolysis of a solution of LiI in THF using a $(CH)_x$ film as the cathode, to give $[Li_y^+(CH)^{-y}]_x$ films (13).

The $(CH)_x$ chain in these materials may be considered as a poly-carbanion associated with the corresponding M^+ metal ion. They are extremely sensitive to air and moisture. This appears to be a direct result of the anionic nature of the $(CH)_x$ chain and is not directly related to the presence of the metal ion. Thus it seems likely that all n-doped $(CH)_x$ will be highly reactive regardless of the attendant metal ion, which of course is stable to air and water. Treatment of Na-doped $(CH)_x$ with D_2O results in partial hydrogenation of the carbon-carbon double bonds (17).

2. PHOTOELECTROCHEMICAL REACTIONS AT POLYACETYLENE INTERFACES

A chemical reaction involving a reduction process, e.g.,

$$S_2^{-2} + 2e^- \rightarrow 2S^{-2} \tag{8}$$

can take place with the concomitant production of an electric current when a p-type $(CH)_x$ film, immersed in a solution containing the oxidized and reduced forms of an appropriate couple, is irradiated with light of appropriate wavelength (24). In the case of the polysulfide system, the reverse (oxidation) process,

$$2S^{-2} \rightarrow S_2^{-2} + 2e^- \tag{9}$$

will take place simultaneously at the counter electrode, e.g., Pt, which is not irradiated. The ions produced at a given electrode then diffuse to the other electrode and become available for re-use at that electrode as shown in Figure 1. The process is, therefore, continuous as long as the $(CH)_x$ electrode is irradiated with light of appropriate wavelength (24,25). A definite photovoltaic effect can be observed (V_{oc} ~0.3 volts under illumination of ca. 1 sun) even with the simple set-up shown in Figure 1 if a fairly thick film of trans-$(CH)_x$ is used in order to reduce somewhat the otherwise high resistance of the $(CH)_x$ electrode. By using a different cell configuration, described in detail elsewhere (24,25), an open circuit current of ca. 40 μ amps/cm^2 may be obtained. This will undoubtedly be increased by using partially doped $(CH)_x$ and thinner films. Since $(CH)_x$ is a p-type semiconductor, photogenerated electron-hole pairs become separated at the $(CH)_x$-electrolyte interface and electrons are injected into the electrolyte as shown in Figure 2.

Preliminary experiments have been carried out (24) using aqueous solutions of the couple

$$SO_3^{-2} + 2OH^- \rightleftharpoons SO_4^{-2} + H_2O + 2e^- \tag{10}$$

with qualitatively similar results. It therefore seems highly likely that it should be possible to fabricate a variety of photo-

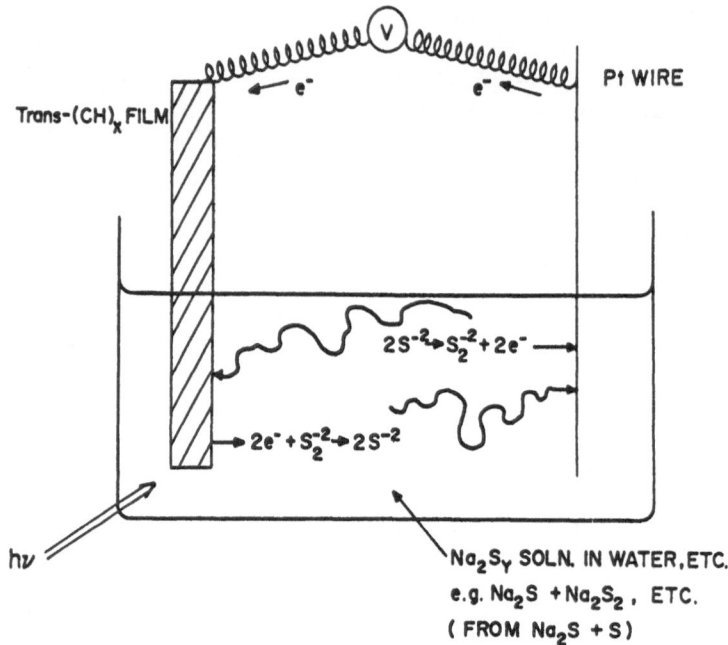

Figure 1. Simple $(CH)_x$/sodium polysulfide photoelectrochemical photovoltaic cell.

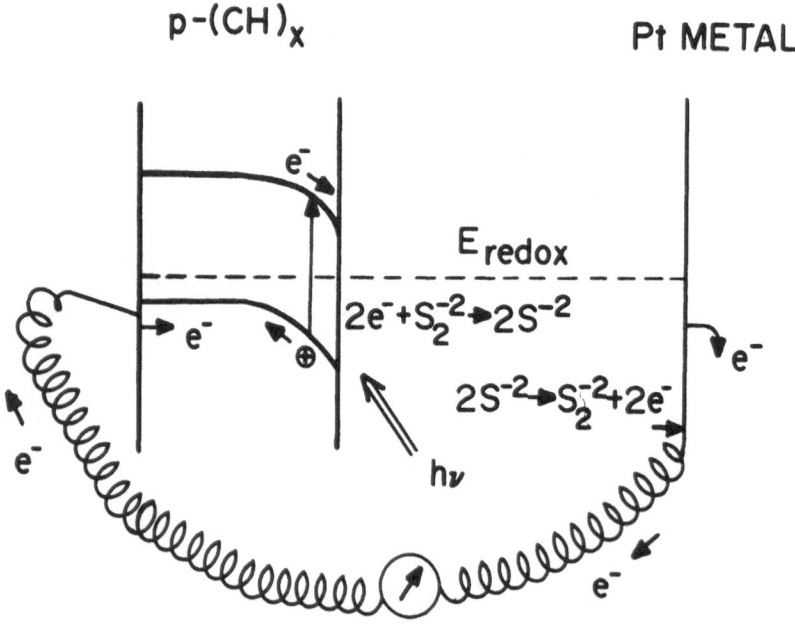

Figure 2. Electrode processes in a $(CH)_x$/sodium polysulfide photo-electrochemical photovoltaic cell.

voltaic cells using (CH)$_x$ electrodes immersed in aqueous or nonaqueous solutions of appropriate redox couples.

3. CONCLUSIONS

It can be seen clearly that (CH)$_x$ is quite remarkable in that
its conductivity can be readily modified to span an extraordinarily
large range. Considering possible polyacetylene derivatives, replacement of some or all of the hydrogen atoms in (CH)$_x$ with organic or inorganic groups, copolymerization of acetylene with
other acetylenes or olefins, and the use of different dopants
should lead to the development of a large new class of conducting
organic polymers with electrical properties that can be controlled
over the full range from insulator to semiconductor to metal.
Furthermore, there is considerable potential for the possible
application of parent or doped (CH)$_x$ to the fabrication of various
types of electronic devices, solar cells, etc.

ACKNOWLEDGMENT

This work was supported principally by the Office of Naval Research.

REFERENCES

1. Shirakawa, H. and Ikeda, S.: 1971, Polym. J. 2, pp. 231-244.
2. Shirakawa, H., Ito, T., and Ikeda, S.: 1973, Polym. J. 4, pp.
   460 462.
3. Ito, T., Shirakawa, H., and Ikeda, S.: 1974, J. Polym. Sci.
   Polym. Chem. Ed. 12, pp. 11-20.
4. Ito, T., Shirakawa, H., and Ikeda, S.: 1975, J. Polym. Sci.
   Polym. Chem. Ed. 13, pp. 1943-1950
5. Druy, M.A., Tsang, C.-H.,Brown, N., Heeger, A.J., and MacDiarmid, A.G.: 1979, J. Polym. Sci. Polym. Phys. Ed. (in press);
   Shirakawa, H. and Ikeda, S.: 1979 (to be published).
6. Park, Y.W., Druy, M.A., Chiang, C.K., MacDiarmid, A.G., Heeger,
   A.J., Shirakawa, H., and Ikeda, S.: 1979, J. Polym. Sci. Polym.
   Lett. Ed. 17, pp. 195-201.
7. Wnek, G.E., Chien, J.C.W., Karasz, F.E., Druy, M.A., Park, Y.W.,
   MacDiarmid, A.G., and Heeger, A.J.: 1979, J. Polym. Sci. Polym.
   Lett. Ed. (in press); Karasz, F.E., Chien, J.C.W., Galkiewicz,
   R., Wnek, G.E., Heeger, A.J., and MacDiarmid, A.G.: 1979, Na-
   ture, submitted.
8. Park, Y.W., Denenstein, A., Chiang, C.K., Heeger, A.J., and
   MacDiarmid, A.G.: 1979, Solid State Commun. 29, pp. 747-751.
9. Chiang, C.K., Druy, M.A., Gau, S.C., Heeger, A.J., Louis, E.J.,
   MacDiarmid, A.G., Park, Y.W., and Shirakawa, H.: 1978, J. Amer.
   Chem. Soc. 100, pp. 1013-1015.

10. Gau, S.C., Milliken, J., Pron, A., MacDiarmid, A.G., and Hee-
    ger, A.J.: 1979, Chem. Comm. (in press).
11. Chiang, C.K., Park, Y.W., Heeger, A.J., Shirakawa, H., Louis,
    E.J., and MacDiarmid, A.G.: 1978, J. Chem. Phys, 69, pp. 5098-
    5104.
12. Fincher, Jr., C.R., Ozaki, M., Tanaka, M., Peebles, D.L.,
    Lauchlan, L., Heeger, A.J., and MacDiarmid, A.G.: 1979, Phys.
    Rev. B (in press).
13. Nigrey, P.J., MacDiarmid, A.G., and Heeger, A.J.: 1979, Chem.
    Comm. (in press).
14. Hsu, S.L., Signorelli, A.J., Pez, G.P., and Baughman, R.H.:
    1978, J. Chem. Phys. 69, pp. 106-111; Lefrant, S., Lichtmann,
    L.S., Temkin, H., Fitchen, D.B., Miller, D.C., Whitwell, II,
    G.E., and Burlitch, J.M.: 1979, Solid State Commun. (in press);
    Harada, I., Tasumi, M., Shirakawa, H., and Ikeda, S.: 1978,
    Chem. Lett. 12, pp. 1411-1414.
15. Salaneck, W.R., Thomas, H.R., Duke, C.B., Paton, A., Plummer,
    E.W., Heeger, A.J., and MacDiarmid, A.G.: 1979, J. Chem. Phys.
    (in press).
16. Clarke, T.C., Geiss, R.H., Kwak, J.F., and Street, G.B.: 1978,
    Chem. Comm., pp. 489-490.
17. Pron, A., MacDiarmid, A.G., and Heeger, A.J.: 1979, unpub-
    lished observations.
18. MacDiarmid, A.G. and Heeger, A.J.: 1979, "Molecular Metals,"
    Ed., W.E. Hatfield, Plenum Press, New York, N.Y., pp. 161-186.
19. Goldberg, I.B., Crowe, H.R., Newman, P.R., Heeger, A.J. and
    MacDiarmid, A.G.: 1979, J. Chem. Phys. 70, pp. 1132-1136.
20. Weinberger, B.R., Kaufer, J., Heeger, A.J., Pron, A. and Mac-
    Diarmid, A.G.: 1979, Phys. Rev. B 20, pp. 223-230.
21. Clarke, T.C., Geiss, R.H., Gill, W.D., Grant, P.M., Macklin,
    J.W., Morawitz, H., Rabolt, J.F., Sayers, D.E., and Street,
    G.B.: 1979, Chem. Comm., pp. 332-333.
22. Chiang, C.K., Gau, S.C., Fincher, Jr., C.R., Park, Y.W., Mac-
    Diarmid, A.G., and Heeger, A.J.: 1978, Appl. Phys. Lett. 33,
    pp. 18-20.
23. Gau, S.C., MacDiarmid, A.G., and Heeger, A.J.: 1979, unpub-
    lished observations.
24. Chen, S.N., Heeger, A.J., Kiss, Z., MacDiarmid, A.G., Gau,
    S.C., and Peebles, D.L.: 1979, Appl. Phys. Lett. (in press).
25. Heeger, A.J. and MacDiarmid, A.G.: 1979, Proceedings NATO ASI
    on Low Dimensional Solids, Tomar, Portugal, Aug. 1979, pp. 353.

APPLICATIONS OF CHARGE-TRANSFER COMPLEXES AND RELATED COMPOUNDS

Serge Flandrois[*] and Luis Alcácer

[*]CRPP, CNRS, 33405 Talence, France
  Lab. Fis. Eng. Nucl., Sacavém, Portugal

I - INTRODUCTION

Fundamental research in a given field is always more or less followed by applications. The field of organic conductors is no exception to this rule, although there are not as yet spectacular applications of these materials.

In the fifties, research on the so-called organic semiconductors aimed to surpass the inorganic semiconductors. One dreamed of cheap materials with variable macroscopic shapes (fibers, thin films , tissues ...), with good mechanic resistance and interesting electronic properties : a new state of matter in the service of modern electronics.

In France, in the early sixties, a group of solid state physicists and chemists of macromolecules was formed for this purpose under the direction of Prof. Champetier who was the first to synthesize polyacetylene in 1929 (1). We know what happened. The group did not last long.

This did not affect the development of fundamental research. TCNQ had just appeared : a new family of synthetic materials with exciting electronic properties. Scientists were more or less encouraged (and may be they still are !) by Little's ideas on high temperature superconductivity (2). In any case, there was not much need to justify fundamental research in this field along these years particularly favorable.

In the last few years however, due to economic difficulties things have changed. Funds for research are not so easy to get if there is no application in view. The question is : do charge trans-

403

L. Alcácer (ed.), The Physics and Chemistry of Low Dimensional Solids, 403–412.

fer complexes by their actual and potential applications justify
the enormous effort that is being made at present ?

Our purpose is to examine among the numerous published papers
those which have some relation to applications. We attribute to
the word "applications" a broader sense than usually given. In general, people think first of technological applications. But there
is another kind of applications no less important : the use of the
particular properties of those compounds on the study of other substances, i.e. applications to fundamental research.

II - APPLICATIONS TO FUNDAMENTAL RESEARCH
1 . Study of surfaces

The ability of TCNQ to accept an electron can be used for investigating some surface properties. In particular a number of
metal oxides are used in catalysis. It is well known that electron-
transfer sites exist on the surfaces of these oxides. TCNQ adsorption may thus be used for characterizing the electron-donor property of metal oxides.

When TCNQ is adsorbed from its acetonitrile solution, the
surface of the metal oxide acquires a coloration characteristic
of each oxide (3). The coloration is caused by the formation of
TCNQ anion-radicals, confirmed by EPR and electronic spectra

The table 1 shows the results obtained for silica, alumina
and titania systems (4). It is clearly seen that the electron -
donor property of the oxides may be estimated and the oxides classified following this property.

TABLE 1 . TCNQ adsorption by metal oxides

Sample	Surface area (m^2 / g)	Color of the sur-face after TCNQ adsorption	Relative ra-dical con -centration/ m^2
Silica	600	Yellow	2.8
Alumina	263	Blue green	1000
Titania	64	Violet	45
$36\%SiO_2-64\%Al_2O_3$	279	Brown	7.2
$19\%SiO_2-81\%Al_2O_3$	367	Grey	70
$79\%SiO_2-21\%TiO_2$	348	Pink	0.5
$45\%SiO_2-55\%TiO_2$	343	Colorless	0.1
$72\%Al_2O_3-28\%TiO_2$	331	Orange	8.3
$38\%Al_2O_3-62\%TiO_2$	337	Red	1.4

Other acceptors than TCNQ may be used. However their electron
affinity must be high as shown in table 2 (5).

TABLE 2 . Electron acceptor adsorption on alumina

Acceptor	E_A (eV)	Adsorbed amount $(\text{mole/m}^2) \times 10^{-7}$	Radical concentration (spins/m^2)
TCNQ	2.8	9.9	1.5×10^{18}
2.5 dichloro-p-benzoquinone	2.3	6.2	6.3×10^{15}
p-dinitrobenzene	1.8	0.25	2.0×10^{13}
m-dinitrobenzene	1.3	negligible	negligible

2 . Study of micelles

The critical micelle concentration (CMC) of surfactants is generally determined by measuring the change in some physical property of the solution such as conductance or surface tension.

When TCNQ is solubilized in surfactant solutions a coloration appears above the CMC, while the solution is colorless below the CMC (6). The appearance of color results from the formation of TCNQ-micelles charge transfer complexes. Optical spectra show new absorption bands in the range 400-900 nm and the intensity measurements of these bands allow a precise determination of the CMC : a break in the curve absorbance vs.log (concentration of surfactant) occurs at the CMC.

The CMC values estimated from this break point are in good agreement with those obtained from other methods. Tables 3 and 4 give some examples of CMC determination for nonionic and ionic surfactants resp.

TABLE 3 . CMC of nonionic surfactants : polyethylene glycol n-Dodecyl ethers ($C_{12}E_n$) (7)

Surfactant	CMC (mole/l) $\times 10^5$ from Surface tension	TCNQ solubilization
$C_{12}E_5$	6.5	6.6
$C_{12}E_6$	6.8	6.7
$C_{12}E_7$	6.9	6.9
$C_{12}E_8$	7.2	7.0

TABLE 4 . CMC of ionic surfactants : dodecyl sulfate (8)

Surfactants	CMC (mole/l) $\times 10^{-3}$ from :		
	Conductance	Surface tension	TCNQ solubilization
Li DS	8.4	7.0	7.6
Na DS	8.1	6.8	7.0
K DS	7.4	6.6	5.8

However, it has been shown recently that the TCNQ method is not suitable for some surfactants such as alkylammonium propionates, which are capable of reacting with TCNQ (9).

3 . Applications to analytical chemistry

Among the applications of C.T. salts to analytical chemistry we emphasize their use in ion-selective electrodes and in thin layer and paper chromatography as the most serious.

TCNQ salts have been considered good ion-selective electrodes materials for the determination of Ag^+, Pb^{2+}, Cd^{2+}, K^+, Na^+, UO_2^{2+}, $(Et_4N)^+$, etc., showing good Nernstian behavior, but limited by the solubility of the TCNQ salts (10)(11).

Other C.T. salts have been proposed as ion selective electrode materials, such as the Mnt complex salts. For example:

$Ag_2[Ag_2S_4C_4(CN)_4]$ was considered a good Ag^+ electrode, showing no interference for other atomic species (11).

$(SN)_x$ has been studied and proposed as an electrode material for electroanalytical chemistry in aqueous solution, more precisely in cyclic voltammetry and cyclic chronoamperometry (12) (13).

TCNQ has been used in thin layer and paper chromatography and in colorimetry in air-pollution studies, for the analysis of free radical precursors, mercaptans, prolines , carbazoles, acridines, thiosemicarbazones, polynuclear aromatic carbons, amino acids, etc. (14).

4 . Catalysis of polymerization

C.T.C. are known to induce radical polymerization. A good example is given by the system N,N-dimethylaniline N-oxide(A) and TCNQ(B), which induces radical polymerization of vinyl monomers, especially methylmethacrylate and ethylacrylate (table 5) (15)

TABLE 5 . Polymerization of vinyl monomers with the system A-B in acetonitrile at 60°C (concentration $[A] = [B] = 5$ mmol. l^{-1}).

Monomer	Reaction time (h)	Yield (%)
Methylmethacry- late	4	11.2
Methylmethacry- late	8	21.1
Ethylacrylate	8	17.4

It was demonstrated that catalysis takes place via a radical mechanism through the formation of the C.T.C. $[A\ B]$.

Perylene cation radical is also an initiator of cationic polymerization (16) for styrene (slow), isobutylvinylether (very
rapid), N-vinyl carbazole (instantaneous), isobutene, THF, etc...

On the other hand some C.T.C. act as inhibitors on polymeri-
zations. That is the case of the C.T.C. of N,N-dimethylacriline
or triethylamine with chloranil on the polymerization of methylmethacrylate (17).

5 . Uses in organic synthesis
Due to its chemical properties TCNQ can be used in organic
synthesis. An example is given by the dehydrogenation of benzyl-
type alcohols and hydroaromatic compounds (table 6)(18). In these
dehydrogenations TCNQ is reduced to p-benzenedimalononitrile.

TABLE 6 . Dehydrogenation by TCNQ at 140°C in dioxane

H donor	Reaction time (h)	Yield of dehy- drogenation product (%)	Recovery of H donor (%)
9,10-dihydroan- thracene	6	91	5
1,4-dihydrona- phtalene	6	86	5
1,2-dihydrona- phtalene	6	26	43
1-phenylpropanol	44	75	18

6 . Ion-radicals as spin labels
The most widely used spin labels are certainly nitroxyde radicals. These probes have revealed very useful in the elucidation
of the structure of membranes. However only the hydrophobic core
of the membranes can be investigated.

Conversely ion-radicals are good candidates for the investigation of polar heads. Thus TCNQ was used for the study of polar
heads transition in some phospholipids (19). For example, EPR
spectra of aqueous solutions of TCNQ and dimyristoylphosphatidyl-
choline (DMPC) exhibit a hyperfine structure with 9 lines at high
temperatures, which disappears with decreasing temperature. Parametrization of.the spectra shows that the linewidth of individual
lines varies sharply at about 19°C, which is characteristic of a
phase transition. As TCNQ must interact with the ammonium groups
of polar heads, the transition may be assumed to be an order-di-
sorder transition of polar heads. This result is in agreement
with recent NMR experiments.

Similarly TCNE has been proposed recently (20) as a good anionic spin probe for the study of lyotropic liquid crystals.

7 . Nucleation kinetics
Nucleation in solids is very complex and differs from that in
liquids because of the presence of structural defects and aniso -
tropic effects. Furthermore, it is generally not possible to observe directly the formation of a nucleus and indirect methods are
used, such as measurements of the change in a physical property
during the nucleation process. A more direct method consists in
counting the transformed domains when they have reached an observable size. In this way transformations in alloys and glasses have been investigated. However, measurements were made on sections
through bulk specimens, so that calculations of the number of nuclei per volume lead to additional errors. For these various reasons experimental studies of nucleation in solids are scarce.

A number of TCNQ salts exhibit first order phase transitions
which are very suitable for studying nucleation kinetics. One of
them, diethylcyclohexylammonium (TCNQ)$_2$, was investigated recently (21). The main advantage of this system is that the transition
velocity is determined only by the nucleation frequency, i.e. the
growth velocity is high compared with the nucleation frequency so
that a single nucleus is responsible for the complete transformation of a single crystal. Thus by observing a collection of single crystals and counting the crystals transformed in dependence
of time, one directly obtains the nucleation frequency. Moreover
the salt was produced in the form of a powder consisting of small
flat crystals. Since in polarized light the crystals change their
color the transformation can easily be observed with a light mi -
croscope. From the data obtained with collections of about 1000
crystals it could be concluded that there is a distribution of
stationary nucleation frequencies. Comparison with other phase
transitions in metals and alloys showed that this result does not
seem an exceptional case but could be general in heterogeneous
kinetics.

III - TECHNOLOGICAL APPLICATIONS
We wanted to emphasize the applications of C.T.C. to basic
research because they are probably less known. However, we cannot pass in silence over technological applications.In the following we will describe some of them which seem to us the most
interesting.

1 . Optical printing of colored and conductive characters
When a donor compound such as TTF is dissolved in a halocarbon solvent such as CCl_4 , a photooxydation reaction of the donor
can be produced by irradiation with UV light. The reaction product
is a colored and conductive mixed-valence halide such as $(TTF)Cl_{0.7}$.

The coloration depends on the nature of the halocarbon and the donor (TTF derivatives (22) or TTT and TSeT (23)).

Thus by applying to a substrate (paper, glass ...) a solution of neutral donor in a halocarbon solvent, covering with a mask and then exposing to UV radiation, a colored and conductive image is obtained.

These conductive images can be used in turn in the fabrication of printed circuits through electrolytic deposition of a metal (24). During metal deposition a simultaneous dissolution of the organic film occurs ; this results in high adherence of the metal film to the substrate.

2 . Batteries
Shortly after the discovery of TCNQ salts a proposal of electrolytic nonaqueous cell was made (25). This cell based on TEA - $TCNQ_2$ dissolved in acetonitrile has a voltage of 0,11 volts at room temperature. Its interest is mainly academic.

Other cells were proposed with a mixture of TCNQ and graphite (or a conducting TCNQ salt) at the cathode, the anode being a metal such as Mg, Cu, Ag (26)(27). With a Mg anode the cell voltage is 1.5 volt.

Of greater interest are the cells with a cathode based on a C.T.C. of iodine (28), such as Phenazine-I_3 , which acts as an iodine reserve for the cell reaction. The anode is a metal, such as Ag, and the electrolyte the metal iodide which is also the reaction product. Owing to the high potential of the reaction $Li \rightleftharpoons Li^+ + e^-$ it is obvious that the use of Li as anode is advantageous. Indeed a Li cell has been developped with C.T.C. of poly(vinylpyridine) and iodine at the cathode. This cell is currently finding use in cardiac pacemakers (29).

3 . Solar cells
Recent proposals have been made with $(SN)_x$ (30) and $(CH)_x$ (31) as Schottky barriers and p-n junctions. They are discussed in other papers of this volume.

4 . Electrolytic capacitors
The replacement of MnO_2 by conducting TCNQ salts leads to improvment in dissipation factor or impedance at low temperatures and at high frequencies (32).

5 . Miscellaneous
- conductive polymers (for antistatic coatings) (33)
- thermal stabilizers (in rubber) (34)
- thermistors (35)
- electron multiplier (36)
- fertilizers (!) (37)

IV - CONCLUSION

The purpose of this paper was to give an overview of application of C.T.C. We hope we have shown that these compounds have interesting possibilities both in basic research and in technological applications. The field is still open ; no doubt that advances will be made in the near future. .

AKNOWLEDGEMENT: This work was partially supported through a NATO Research Grant

REFERENCES

1 . A. Job and G. Champetier, C.R. Acad. Sci. Paris, 189,1089 (1929) ; Bull. Soc. Chim. Fr., 47, 279 (1930)

2 . W.A. Little, Phys. Rev. A 134, 1416 (1964)

3 . H. Hosaka, T. Fujiwara and K. Meguro, Bull. Chem. Soc. Jap., 44, 2616 (1971)

4 . H. Hosaka, N. Kawashima and K. Meguro, Bull. Chem. Soc. Jap., 45, 3371 (1972)

5 . K. Meguro and K. Esumi, J. Coll. Int. Sci., 59, 93 (1977)

6 . S. Muto, K. Deguchi, Y. Shimazaki, Y. Aono and K. Meguro, Bull. Chem. Soc. Japan, 44, 2087 (1971)

7 . K. Deguchi and K. Meguro, J. Coll. Int. Sci., 38, 596 (1972)

8 . S. Muto, Y. Aono and K. Meguro, Bull. Chem. Soc. Jap., 46, 2872 (1973)

9 . O.A. El Seoud, M.J. Da Silva and M.I. El Seoud, J. Coll. Int. Sci., 62, 119, (1977)

10 . M. Sharp and G. Johansson, Anal. Chim. Acta, 54, 13 (1971) M. Sharp, Anal. Chim. Acta, 85, 17 (1976)

11 . L. Alcacer, M.R. Barbosa, R.A. Almeida and M.F. Marzagao, Rev. Port. Quim., 15, 192 (1973)

12 . R.J. Nowak, W. Kutner, H.B. Mark and A.G. Mac Diarmid, J. Electrochem. Soc., 125, 232 (1978)

13 . C. Bernard and G. Robert, Bull. Soc. Chim. Fr., 395 (1978)

14 . E. Sawicki, C.R. Engel and W.C. Elbert, Talanta, 14, 1169
     (1967)
     M. Guyer and E. Sawicki, Anal. Chim. Acta, 49, 182 (1970)

15 . T. Sato, M. Yoshioka and T. Otsu, Makromol. Chem., 177,
     2009 (1976)

16 . E. Oberrauch, T. Salvatori and S. Cesca, J. Polymer Sci.,
     Pol. Lett., 16, 345 (1978)

17 . A.A. Yassin and N.A. Rizk, J. Polymer Sci., Polymer Chem.,
     16, 1475 (1978)

18 - A. Ohki, T. Nishiguchi and K. Fukuzumi, J. Org. Chem., 44
     766 (1979)

19 . P. Delhaes, C. Lussan, M.O. Valiron and J. Amiell, FEBS
     Letters, 69, 252 (1976)

20 . G.F. Pedulli and C. Zannoni, Mol. Cryst. Liq. Cryst. Lett.,
     41, 275 (1978)

21 . S. Flandrois and G. Sauthoff, J. Phys. Chem. Solids, 34,
     1779 (1973)

22 . E.M. Engler, F.B. Kaufman and B.A. Scott, U.S. Patent (1977)

23 . M. Masson, Thesis, Bordeaux (1979)

24 . J.C. Mc Groddy and B.A. Scott, U.S. Patent (1977)

25 . W.R. Wolfe, U.S. Patent (1963)

26 . F. Gutmann, A.M. Hermann and A. Rembaum, J. Electrochem. Soc.
     114, 323 (1967)

27 . P. Weidenthaler and E. Pelinka, Coll. Cz. Chem. Com., 34;
     1482 (1969)

28 . M. Pampallona, A. Ricci, B. Scrosati and C.A. Vincent, J.
     Appl. Electrochem., 6, 269 (1976)

29 . A.M. Hermann and E. Luksha, J. Card. Pulm. Tech., 6,15
     (1978)
     A.A. Schneider, W. Greatbach and R. Mead, Power Sources, 651
     (1974)

30 . R.A. Scranton, J.B. Mooney, J.O. Mc Caldin, T.C. Mc Gill and
     C.A. Mead, Appl. Phys. Lett., 29, 47 (1976)
     M.J. Cohen, and J.S. Harris, Appl. Phys. Lett., 33, 812
     (1978)

31 . C.K. Chiang, S.C. Gau, C.R. Fincher, Y.W. Park, A.G. Mac
     Darmid and A.J. Heeger, Appl. Phys. Lett., 33, 18 (1978)

32 . Y. Itoh and S. Yoshimura, J. Electrochem. Soc., 124, 1128
     (1977)
     S. Yoshimura, Molecular metals, Ed. W.E. Hatfield, P. 471
     (1979)

33 . See for example : S. Ikeno, M. Yokoyama and H. Mikawa, J.
     Polym. Sci., Polym. Phys., 16, 717 (1978)

34 . Numerous patents. See for example : Russian Patent, 304, 278
     (1971) ; Japan. Kokai, 7617, 935 (1976)

35 . See for example ; Y. Kishimoto and F. Oda, Japan. Kokai (1976)

36 . T. Hayashi, M. Hashimoto, K. Yamamoto, W. Shimotsuma, H. Mo-
     riga and T. Shimizu, Ger. Offen. (1970)

37 . South African Patent, 6905, 048 (1970)

ORGANIC CONDUCTORS AS ELECTRON BEAM RESIST MATERIALS

Y. Tomkiewicz, E. M. Engler, J. D. Kuptsis,
R. G. Schad and V. V. Patel

IBM T. J. Watson Research Center
Yorktown Heights,
New York, 10598, U. S. A.

Conducting organic charge-transfer salts made up of organic π donors as TTF and TTT and halogen acceptors are shown to be electron beam resist materials having a combination of unique features with no parallel among conventional resist materials.

The microelectronics industry in general, and the computer industry in particular, require miniaturization of electronic circuits. The small size is essential both for packaging and speed. The current requirement of the technology is micron or even submicron resolution. The only way to achieve this resolution, is to use materials which develop a differential mass loss between areas which are exposed to some sort of ionizing radiation and the unexposed areas. Thus, a fine pattern can be "written" which then can be further treated by sophisticated technological steps to achieve the desired final product. The first essential step, though, is the high resolution lithography. A material which meets the requirement of the differential mass loss is called a resist - a positive resist if the mass loss occurs in the exposed areas, and a negative resist if it occurs in the unexposed areas. Different resists are sensitive to different kinds of radiation and accordingly they are grouped into three different groups: photoresists, electron-beam resists and x-ray resists. The commonly used materials in lithography are polymers that change their solubility upon exposure to radiation. The chemical reaction taking place in the exposed area is either chain cross-linking or chain scission. The sensitivity of the resist depends on the yield of the chemical reaction as well as on the differential solubility of the starting material in comparison to the final product. All the presently used E-beam resists are polymeric insulators. In their high sensitivity mode, they give rise to a positive pattern. The insulating nature of resist gives rise to a build-up of static charge which in turn causes a deterioration of resolution.

L. Alcácer (ed.), The Physics and Chemistry of Low Dimensional Solids, 413–418.

This paper describes a new class of E-beam resist materials comprised of conducting organic charge transfer salts with very unique features. The materials, which fit into this category, are salts made up of such donors as TTF and TTT and their derivatives, and halogen acceptors such as bromine and chlorine.

The lithographic process is based on the following mechanism: Exposure of the charge transfer salt to an electron beam having sufficient current density causes the loss of the halogen from the film; a reaction occurs described by equation (1). We will refer to this reaction as a reverse charge transfer since it is the reverse of the reaction forming the conducting salt.

$$(Donor)_{1-x}(Donor^+)_x X_x^- \; \frac{E-beam}{irradiation} \; Donor + X_2 \qquad (1)$$

$$X = halogen \qquad 0 < x < 1$$

The effect demonstrated in Figure 1 shows the characteristic bromine x-ray emission of a TTF-Br$_{0.76}$ film before and after exposure to an electron beam. Figure 1a shows Br L$_\alpha$ emission before exposure while Figure 1b shows the emission from the same sample after a fraction of its area was exposed to the electron beam. It is clearly seen that practically no bromine x-ray emission does exist in the exposed area. No change in the sulfur x-ray emission was observed using this particular beam density.

The reverse charge transfer reaction introduces differential solubility between the exposed and unexposed areas, since the ionic charge transfer salts are soluble in polar solvents (alcohol for example) while neutral donors are soluble in nonpolar solvents (toluene, halocarbons, etc.). Hence, nonpolar organic solvents will preferrentially dissolve the exposed areas. Alternately, the exposed areas will preferrentially dissolve in polar solvents.

An alternative way to develop a resist pattern is to take advantage of the differential vapor pressure between the neutral donor and the charge transfer salt. In the investigated case of TTF-Br$_{0.76}$, the low sublimation temperature of TTF, 80°C, enables one to develop a pattern in situ by utilizing the energy of the electron beam. In this mode, which requires high current density for subliming TTF, the resulting resist is positive.

The differential solubility between the exposed and unexposed areas is enhanced by cross-linking of the neutral donor molecules occurring after the bromine evaporation. Actually the sublimation of the neutral donor and the cross-linking are competing processes, the latter requiring at least an order of magnitude smaller current density. The cross-linking mode yields a negative resist, an example of which is shown in Figure 2. The resolution of TTF-Br$_{0.76}$ in the negative mode is about 0.5μ and its sensitivity is about 10^{-5} Coulomb/cm^2. Thus, a TTF-Br$_{0.76}$ resist can be used as a positive and also as a negative resist.

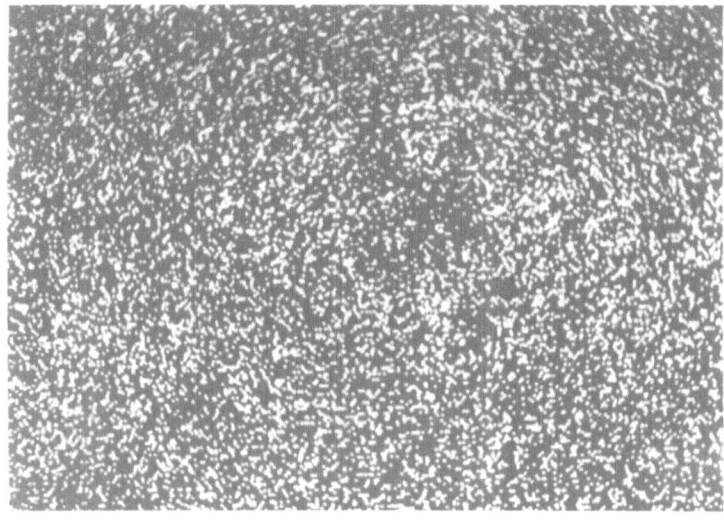

Fig. 1a   Br L_α scan taken on a TTF-$Br_{0.76}$ film before it was exposed to an electron beam.

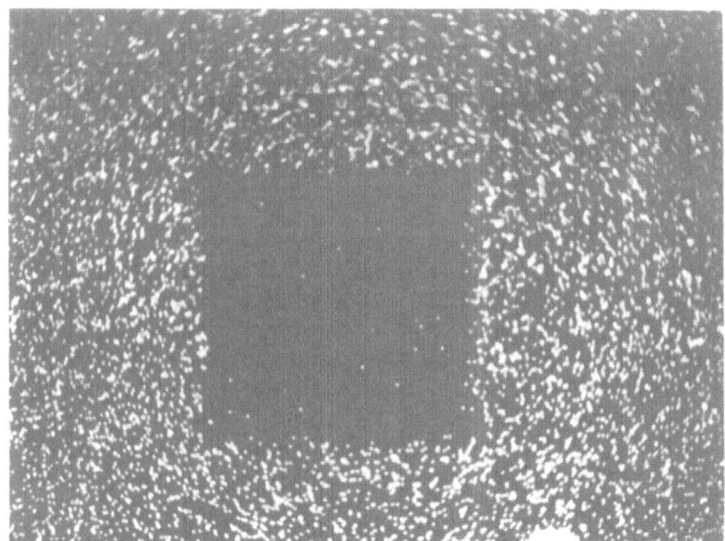

Fig. 1b   Br L_α scan taken on a TTF-$Br_{0.76}$ film after a small area of it was exposed to an electron beam.

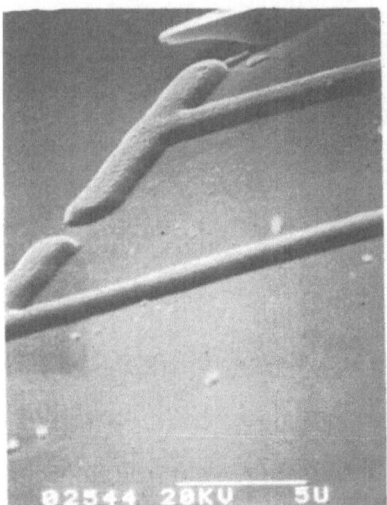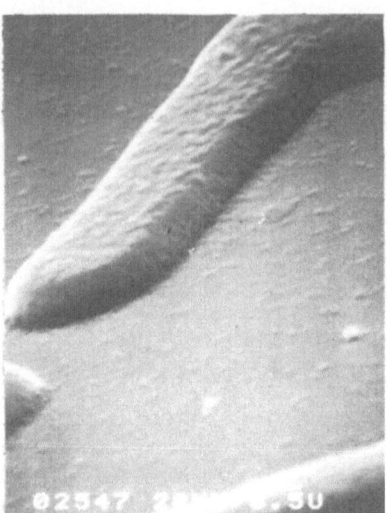

Fig. 2.   High resolution scanning electron micrographs of a developed film of
TTF-Br$_{0.76}$.

A TTF-Br$_{0.76}$ resist is unique in comparison to the conventionally used resists
in its method of preparation.   While the other resists are spin-coated on the wafers,
TTF-Br$_{0.76}$ is sublimed.   If there is an existing structure on the wafer, which is quite
common, spin-coating yields a resist with varying thickness, while sublimation yields
a uniform thickness which follows the contours.   Since both resolution and sensitivity are thickness dependent, it is quite important to have a resist of uniform thickness independent of the texture of the surface.

Another unique feature of the TTF halide resist is that it is conducting with
conductivities   of the order of 10-20 $\Omega^{-1}cm^{-1}$.   The importance of conductivity in
lithography is demonstrated in Fig. 3.   Fig. 3a shows scanning electron micrographs
of a film in which five dots were exposed to electron beam.   Since a reverse charge
transfer does occur in the exposed areas, the dots are no longer conducting.   Actually, the difference in the conductivity of the exposed and unexposed areas is
$10^9\Omega^{-1}cm^{-1}$.   In Fig. 3b, the beam explores simultaneously the exposed and
unexposed areas.   The charging effect in the unexposed part, causing a blurred
image, is clearly seen.   The unexposed part of the resist is conducting and, therefore, no build-up of charge does occur.   The build-up of charge causes a loss of
resolution.

The work presented in this paper concentrates on TTF-Br$_{0.76}$.   However, other
charge transfer salts with different donor molecules such as TTT and other halides

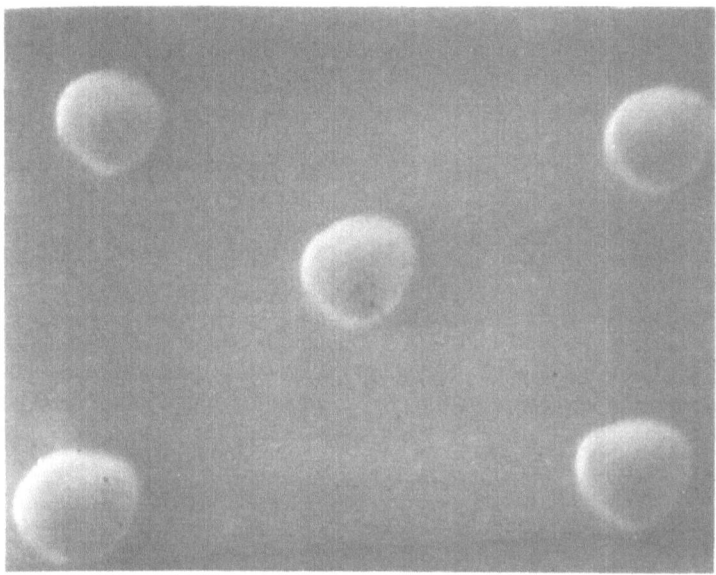

Fig. 3a   Scanning electron micrograph of a TTF-$Cl_{0.8}$ film.   The dots are the areas of the film which were exposed to the beam.

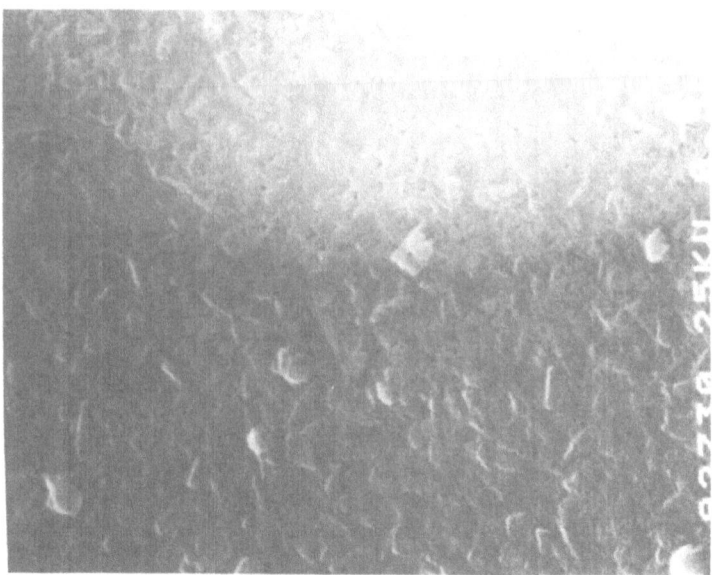

Fig. 3b   High resolution scanning electron micrograph of an exposed area and its boundary.

e.g. chlorine and iodine, also show the reverse charge transfer reaction upon exposure to an electron beam. Thus, the effect seems to be quite general. Therefore, the different characteristics of the different available salts can be utilized to fine tune the resist to a particular application.

In conclusion, we have shown that organic charge transfer salts can offer a unique combination of features for the resist technology with no parallel among existing materials. The most outstanding features of these resists are 1. Negative images are formed with 0.5μ resolution. 2. Positive images can be developed in situ. 3. The resists are highly conducting.

ACKNOWLEDGMENTS

Sensitivity measurements by Dr. Hatzakis are greatly appreciated. Special thanks to Ms. Marino for her help in meeting the deadline for this paper.

ELECTRICAL SWITCHING AND MEMORY PHENOMENA IN SEMICONDUCTING ORGANIC CHARGE-TRANSFER COMPLEXES

R. S. Potember, T. O. Poehler, D. O. Cowan* and
A. N. Bloch*

Applied Physics Laboratory, Johns Hopkins University
Laurel, Maryland 20810, U.S.A.

*Department of Chemistry, Johns Hopkins University
Baltimore, Maryland 21218, U.S.A.

INTRODUCTION

The observation of electrical switching and memory phenomena is known to exist in a wide variety of inorganic semiconducting thin films. These inorganic materials include the metal oxides of nickel, silicon, aluminum, titanium, zirconium, and tantalum, all of which exhibit a voltage controlled negative resistance when arranged in a metal-oxide-metal sandwich structure.[1,2] The mechanism of switching is believed to be due to a forming process. The electrical behavior of these materials is not stable, reproducible, or independent of polarity. The amorphous alloys, including chalcogenide glasses, are inorganic semiconductors which also show switching behavior. These glasses contain up to four elements often including arsenic and/or tellurium and exhibit a current-controlled negative resistance. A typical example of an amorphous alloy which exhibits switching behavior is $Te_{40}As_{35}Ge_6Si_{18}$.[3] These materials have been extensively investigated and the mechanism of conduction and switching is believed to be due to filaments which are described by both thermal and electronic models.

There are also several reports in the literature of electrical switching in organic polymers and polycrystalline dielectrics. One of the earliest reports of switching pertains to insulating films of mylar.[4] The behavior in these films is believed to be associated with an electrical breakdown due to weak spots in the material. Other materials reported in the literature which exhibit a switching or a breakdown phenomena include anthracene,[5]

L. Alcácer (ed.), The Physics and Chemistry of Low Dimensional Solids, 419–428.

lead phthalocyanine,[6] tetracene,[7] and polystyrene.[8] In all of these cases, the electrical characteristics are either erratic in nature or not very reproducible. To date, none of these organic devices has been shown to be comparable to the inorganic amorphous glasses.

SWITCHING AND MEMORY DEVICE: MATERIALS AND FABRICATION

This paper is a report on stable and reproducible current-controlled bistable electrical switching and memory phenomena observed in polycrystalline metal-organic semiconducting films. The effects are observed in films of either copper or silver complexed with the electron acceptors tetracyanonaphthoquinodimethane (TNAP), tetracyanoquinodimethane (TCNQ),[9] or other TCNQ derivatives shown below. The character of the switching in going from

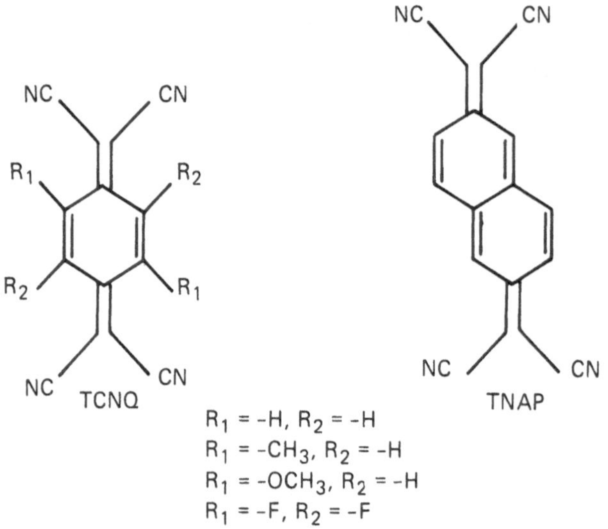

$$R_1 = -H, R_2 = -H$$
$$R_1 = -CH_3, R_2 = -H$$
$$R_1 = -OCH_3, R_2 = -H$$
$$R_1 = -F, R_2 = -F$$

a high to a low impedance state in these organic charge transfer complexes is believed to be comparable in many respects to existing inorganic materials. The basic configuration of the device, shown in Figure 1, consists of a 5-10 μm thick polycrystalline aggregate of a copper or a silver charge-transfer complex sandwiched between two metal electrodes. Electrical connection is made to the two metal electrodes through silver conducting paste or through liquid metals of mercury, gallium or gallium-indium utectic. Fabrication of the device consists of first mechanically removing any oxide layers and organic contaminants from either a piece of copper or silver metal foil. The cleaned metal foil is

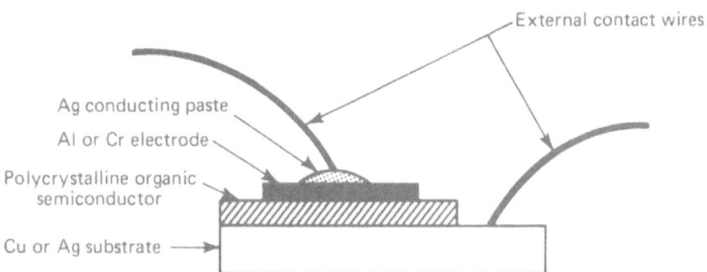

Figure 1.   Schematic diagram of an organic switching device.

then placed in a solution of dry and degassed acetonitrile which
has been saturated with a neutral acceptor molecule, for example,
TCNQ°. The neutral acceptors used in all of these experiments
are recrystallized twice from acetonitrile and then sublimed under
a high vacuum prior to their use.[10] When the solution saturated
with the neutral acceptor is brought in contact with a metal
substrate of either copper or silver, a rapid oxidation-reduction
reaction occurs in which the corresponding metal salt of the ion-
radical acceptor molecule is formed. The basic reaction is shown
in Equation 1 for copper and TCNQ°.

$$Cu^0 + \quad\underset{\text{TCNQ}}{\begin{array}{c}NC\\NC\end{array}\diagup\hspace{-0.3em}\diagdown\hspace{-0.3em}\underset{CN}{\overset{CN}{\diagdown\diagup}}} \quad \xleftarrow{} \quad Cu^+ \left[\begin{array}{c}NC\\NC\end{array}\diagup\hspace{-0.3em}\diagdown\hspace{-0.3em}\overset{CN}{\underset{CN}{\diagdown\diagup}}\right]^{\cdot-} \qquad (1)$$

      This technique of forming semiconducting films by direct
oxidation-reduction is used to grow highly microcrystalline films
directly on the copper or silver substrates. These films show a
metallic sheen and can be grown to a thickness of 10 μm in a
matter of minutes. Once the polycrystalline film has been grown
to the desired thickness, the growth process can be terminated by
simply removing the metal substrate containing the organic layer
from the acetonitrile solution; this terminates the redox reaction.
The two component structure is gently washed with additional
acetonitrile to remove any excess neutral acceptor molecules and
dried under a vacuum to remove any traces of solvent. Elemental
analysis performed on polycrystalline films of Cu–TCNQ and Cu–TNAP
reveals that the metal/acceptor ratio is 1:1 in both complexes.[11]
Finally, the three component structure is complete when a top

metal electrode of either aluminum or chromium is evaporated or
sputtered directly on the organic film.

ELECTRICAL BEHAVIOR

        Threshold switches and memory switches are two terminal
devices that can exist in either a high or low impedance state.
The transition from the high to the low impedance mode occurs
when an applied field in excess of a threshold value is surpassed.
The transition must be reversible and reproducible. The difference between threshold and memory switching is determined by the
response of the material, once in the highly conductive state, to
an applied bias. A memory switch is a device that will remain in
the low impedance state when the initial applied field is removed;
the original high impedance state is restored by the application
of a high current pulse through the device. On the other hand,
a device that immediately returns to the high impedance state
when the applied field drops below a minimum holding value is said
to display threshold switching.

        Threshold and memory behavior is observed in these materials
by examining current as a function of voltage across the two
terminal structure. Figure 2 shows a typical dc current-voltage

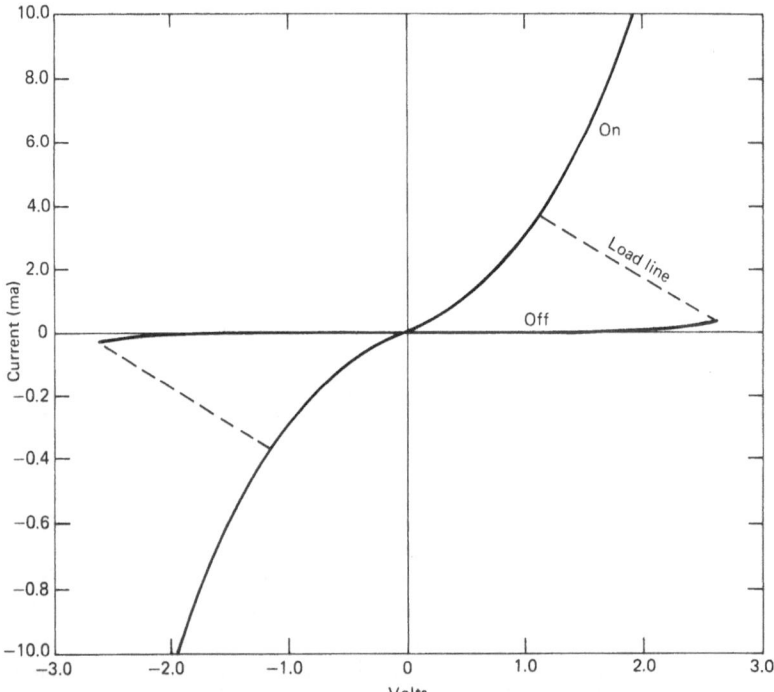

Figure 2.   Typical dc current-voltage characteristic showing high-
            and low-impedance states for a 3.75 μm Cu-TNAP sample.

curve for a 3.75 μm thick Cu/Cu-TNAP/Al system. The trace in
Figure 2, as well as all other I-V measurements presented in this
paper, are made with a 10^2-Ω load resistor in series with the
device. Figure 2 shows that there are two stable non-ohmic resistive states in the material. These two states, labeled "OFF"
state and "ON" state, are essentially insensitive to moisture,
light, and the polarity of the applied voltage. A rapid switching
is observed from the "OFF" to the "ON" state along the load line
when an applied field across the sample surpasses a threshold
value (V_{th}) of 2.7 V. This corresponds to a field strength of
approximately 8.1×10^3 V/cm. At this field strength the initial
high impedance of the device, 1.25×10^4 ohms, drops to a low impedance value of 190 ohms. This rise in current to 4 ma and
concurrent decrease in the voltage to approximately 1.2 V along
the load line is observed in the Cu-TNAP system. It is representative of the switching effects observed in all of the metal
charge-transfer salts examined and is characteristic of all two
terminal S-shaped or current-controlled negative-resistance
switches.[12]

In addition, it has been observed in all of the materials
investigated that once the film is in the "ON" state it will
remain in that state as long as an external field is applied. In
every case studied, the film eventually returned to its initial
high impedance state after the applied field was removed. It
was also found that the time required to switch back to the initial state appeared to be directly proportional to the film thickness, duration of the applied field, and the amount of power dissipated in the sample while in this state.

Three general trends are noted in the "ON" state character
of the copper and silver complexes as related to the different
acceptor molecules. The first is that the copper salts consis-
tently exhibited greater stability and reproducibility over the
corresponding silver salts of the same acceptor. Second, it is
possible to correlate the preferred switching behavior of the
different complexes to the reduction potential of the various
acceptors. This plot is shown in Figure 3 using copper as a donor
in each case. It appears that for devices made from weak electron
acceptors, the switching behavior is usually of the threshold type,
i.e., when the applied voltage is removed from a device in the
"ON" state, the device will immediately return to the "OFF" state.
On the other hand, for strong electron acceptors a memory effect
is observed. This memory state remains intact from a few minutes
up to several days and can often be removed by the application of
a short pulse of current in either direction. For intermediate
strength acceptors, it is possible to operate the device as either
a memory switch or a threshold switch by varying the strength or
the duration of the applied field in the low impedance state.
Third, it also recognized that the field strength of the switching

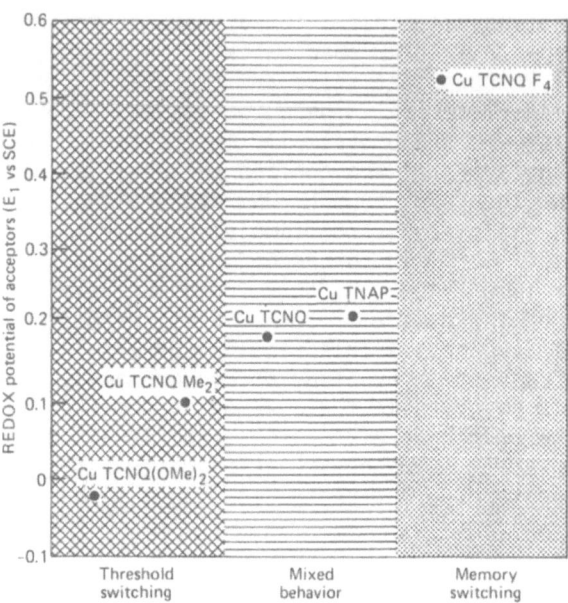

Figure 3. Type of switching behavior plotted versus the reduc-
           tion potential of the acceptor.

threshold tends to parallel the strength of the acceptor. For
instance, the copper salt of TCNQ-(OMe)$_2$ switches at a field
strength of approximately 2×10^3 V/cm, while the copper salt of
TCNQF$_4$ is found to switch at a field strength of about 2×10^4 V/cm.
It is clear that these three trends are related to the reduction
potential of the acceptor calculated from solution redox potentials.[13] However, as these values do not always parallel the
values found in the solid phase, a more quantitative description
relating to the switching behavior to the acceptor can not be made
unless the various contributions to the binding energy of the
different ion-radical salts are considered.

      The dependence of the threshold voltage required for switch-
ing by a single pulse is shown for Cu-TNAP in Figure 4. For long
pulses, the threshold voltage is identical to that shown in the
dc characteristic of Figure 2. As the pulse length is decreased,
the threshold voltage increases sharply for pulses of 1-5 μsec
duration. For pulses of nanosecond duration, switching is still
observed, however, the value of the required threshold voltage is
increased slightly and is somewhat erratic.

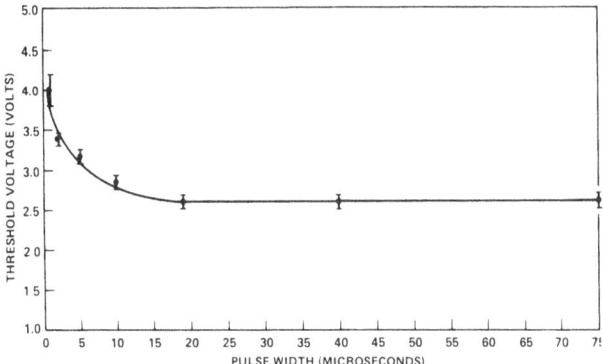

Figure 4. Threshold voltage dependence versus plane length in
           Cu-TNAP.

    The response to a very short pulse is perhaps better exem-
plified in the next figure. Figure 5 is an oscilloscope trace
showing both the leading edge of a voltage pulse and current pulse

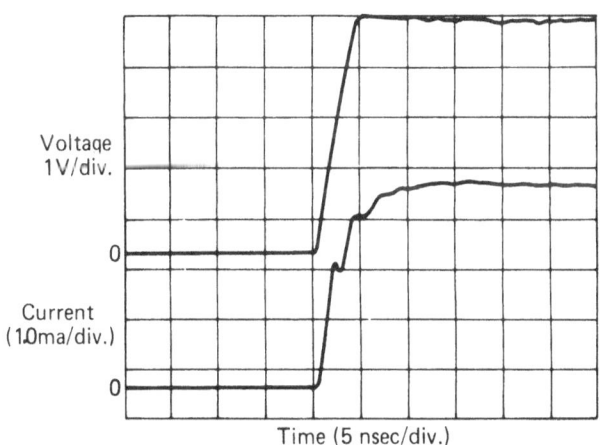

Figure 5. Transient response to a 4 nsec rise time rectangular
           pulse.

versus time for a Cu-TNAP sample in response to a rectangular
voltage pulse with a 4 nsec rise time. This voltage pulse switched
the sample from the high to the low impedance state and contained
a 1.0 V overvoltage to eliminate any current oscillations between

the "OFF" and "ON" states. Current oscillations arise when the applied voltage is set very close to V_{th}. It is not possible from this experiment to determine values for the conventional delay times and rise times, because the combined delay and rise time appears to be less than 4 nsec (the limiting rise time of the pulse generator). This experiment suggests that the mechanism of the switching phenomena is not due to thermal effects,[14] which are used to describe switching and memory phenomena in many other systems. From Figure 5, it appears that the delay time is shorter than reported values for inorganic semiconductors under the same experimental conditions. A recent example of delay times in an inorganic amorphous material is given for the composition $Te_{10}As_{35}$-$Ge_7Si_{17}P_1$,[14] approximately 1 μm thick, sandwiched between two molybdenum electrodes. A typical delay time reported for this device in response to a single 12 V pulse is about 2 μsec. To reduce the delay time to a value of 10 nsec, a 30 V pulse (18 V overvoltage) was required.

An experiment was designed to determine if the device generates an open-circuit voltage or electromotive force (emf) when returning from the low to the high impedance mode. The appearance of a spontaneous emf[16] would indicate that an electrochemical reaction was responsible for switching phenomena. In this experiment:

1)   an applied voltage in excess of the threshold voltage was used to place a Cu-TNAP sample into a low impedance state where it would remain for a short time after the applied voltage was removed, i.e., memory state;

2)   the sample was then externally short circuited to eliminate any capacitive effects, and finally;

3)   a high input impedance storage oscilloscope was used to measure open-circuit discharge voltage when the sample spontaneously returned to its original high impedance state. The oscilloscope was set to trigger whenever a voltage exceeding a few millivolts appeared across the sample.

The results are shown in Figure 6 where the spontaneous open circuit voltage measured by the oscilloscope is reproduced and is seen to have a maximum voltage discharge of approximately 0.3 volts.

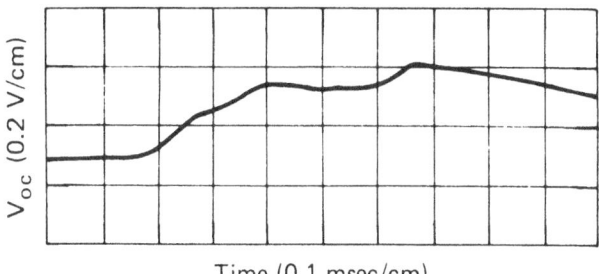

Figure 6. Spontaneous open circuit potential generated in a
Cu-Cu-TNAP/Al sample at room temperature.

The open circuit voltage of 0.3 volts observed in this experiment does show that the mechanism by which the switching
occurs is consistent with a field induced solid-state reversible
electrochemical reaction associated with the metal charge-transfer
salts.

CONCLUSIONS

It is postulated that mixed-valence species or complex
salts[17] formed as a result of this field induced redox reaction
control the semiconducting behavior of these films and these
complex salts exist in a solid-state equilibrium with the simple
1:1 salt. Since non-integral oxidation states are common in
solids, it is difficult to predict exact stoichiometry in the
equilibrium equation, but a likely equation for switching in Cu-
TCNQ, for example, may involve

$$[Cu^+(TCNQ^{\overline{\cdot}})]_n \rightleftharpoons Cu^\circ_x + [Cu^+(TCNQ^{\overline{\cdot}})]_{n-x} + (TCNQ^\circ)_x \; .$$

In addition, an ionic or a molecular displacement associated with
this equilibrium would explain the observed memory phenomena and
the fact that all the devices show only two stable resistive
states.

ACKNOWLEDGMENTS

We gratefully acknowledge support by the National Science
Foundation (DMR 76-84238) and the Department of the Navy (N00024-
78-C-5384).

REFERENCES

[1] T. W. Hickmott, J. App. Phys. 35, 2118 (1964).

[2] J. F. Gibbons and W. E. Beadle, Solid-State Electron. 7, 785 (1964).

[3] S. R. Ovshinsky, Phys. Rev. Lett. 21, 1450 (1968).

[4] Y. Invishi and D. A. Powers, J. App. Phys. 28, 1017 (1957).

[5] A. R. Elsharkawi and K. C. Kao, J. Phys. Chem. Solids 38, 95 (1977).

[6] C. Hamann et al, Phys. Stat. Sol. (a) 50, K189 (1978).

[7] A. Szymanski, D. C. Larson, and M. M. Labes, App. Phys. Lett. 14, 88 (1969).

[8] H. Carchano, R. Lacoste, and Y. Sigui, App. Phys. Lett. 10, 414 (1971).

[9] R. S. Potember, T. O. Poehler, and D. O. Cowan, App. Phys. Lett. 34, 405 (1979).

[10] R. V. Gemmer, D. O. Cowan, T. O. Poehler, A. N. Bloch, R. E. Pyle, and R. H. Banks, J. Org. Chem. 40, 3544 (1975).

[11] Elemental analysis was performed by Galbraith Laboratories, Inc., Knoxville, Tenn. 37291.

[12] A. E. Owen and J. M. Robertson, IEEE Trans. Electron. Devices 20, 105 (1973).

[13] Values for the reduction potential of acceptor were taken from R. C. Wheland and J. L. Gillson, J. Am. Chem. Soc. 98, 3916 (1976).

[14] W. D. Buckley and S. H. Holmberg, Solid-State Electron. 18, 127 (1975).

[15] D. K. Reinhard, App. Phys. Lett. 31, 527 (1977).

[16] A spontaneous electrochemical reaction is reported in magnesium-TCNQ salts. See F. Gutmann, A. M. Herman, and A. Rembaum, J. Electrochem. Soc. 114, 323 (1967).

[17] For a discussion of complex TCNQ salts see O. H. LeBlanc, Jr., J. Chem. Phys. 42, 4307 (1965).